Lecture Notes in Geoinformation and Cartography

For further volumes:
http://www.springer.com/series/7418

Jacynthe Pouliot · Sylvie Daniel
Frédéric Hubert · Alborz Zamyadi
Editors

Progress and New Trends in 3D Geoinformation Sciences

Springer

Editors
Jacynthe Pouliot
Geomatics
Université Laval
Quebec, QC
Canada

Frédéric Hubert
Geomatics
Université Laval
Quebec, QC
Canada

Sylvie Daniel
Geomatics
Université Laval
Quebec, QC
Canada

Alborz Zamyadi
Geomatics
Université Laval
Quebec, QC
Canada

ISSN 1863-2246 ISSN 1863-2351 (electronic)
ISBN 978-3-642-29792-2 ISBN 978-3-642-29793-9 (eBook)
DOI 10.1007/978-3-642-29793-9
Springer Heidelberg New York Dordrecht London

Library of Congress Control Number: 2012945737

Printed on acid-free paper

Springer is part of Springer Science+Business Media (www.springer.com)

Contents

Modelling 3D Topographic Space Against Indoor Navigation Requirements

Gavin Brown, Claus Nagel, Sisi Zlatanova and Thomas H. Kolbe

Abstract Indoor navigation is growing rapidly with widespread developments in the collection and processing of sensor information for localisation and in routing algorithms calculating optimal indoor routes. However, there is a general lack of understanding about the requirements for topographic space information to be used in indoor navigation applications and thus the suitability of existing information sources. This work presents a structured process for the identification of topographic space information starting with use cases that support the complete capture of requirements, thus allowing existing models to be evaluated against these requirements and conceptual semantic and constraint models developed. A proposal is put forward for the implementation of topographic space semantic and constraints models as a CityGML Application Domain Extension (ADE) that will be integrated into the Multilayered Space-Event Model (MLSEM), a flexible framework supporting all indoor navigation tasks.

Keywords Indoor navigation · Topographic space · Building modelling · Indoor routing · 3D

G. Brown (✉) · C. Nagel · T. H. Kolbe
Institute for Geodesy and Geoinformation Science, Technische Universität Berlin, Berlin, Germany
e-mail: gbrown1@mailbox.tu-berlin.de

C. Nagel
e-mail: claus.nagel@tu-berlin.de

T. H. Kolbe
e-mail: thomas.kolbe@tu-berlin.de

S. Zlatanova
OTB, Research Institute for the Built Environment, Delft University of Technology, Delft, The Netherlands
e-mail: s.zlatanova@tudelft.nl

J. Pouliot et al. (eds), *Progress and New Trends in 3D Geoinformation Sciences*, Lecture Notes in Geoinformation and Cartography, DOI: 10.1007/978-3-642-29793-9_1,

1 Introduction

The field of indoor navigation is now a major research topic with research taking place on the development of localisation sensors and techniques, routing algorithms and display and dissemination of navigation information to a user. Topographic Space is a fundamental part of indoor navigation, representing the interior environment of buildings and its semantic decomposition into building elements (e.g. rooms and storeys) for route planning and use in combination with additional sensor information. Indoor environments are increasingly being modelled in 3D using Industry Foundation Classes (IFC) and CityGML and therefore components of these indoor environments are inherently represented in 3D. When considering an Unmanned Aerial Vehicle (UAV) being routed through a large indoor airport terminal, we have a real-world navigation object (represented as a 3D geographic feature) interacting with real-world topographic space features (e.g. a door opening) that therefore must be described by geographic features with a 3D representation in Euclidean space (Nagel et al. 2010). A number of developing indoor navigation techniques are reliant upon a constant, rich 3D information model for building interiors, considered within the wider context of indoor. Currently topographic space information is frequently being provided by building models captured for the purposes of urban/building modelling. These current sources of information create a number of potential problems including incomplete/inconsistent topographic space features and incompatibility of all information sources required for tackling the complete set of indoor navigation tasks (localisation, route planning and route homing). As an example when considering the use case of routing a person from a start point to an end point within a single building during an emergency evacuation scenario, a number of requirements are created including the need to define that elevators are commonly out of use during this scenario. In existing building models (e.g. CityGML), all semantic features are not always captured and there is a general lack of support for defining complex navigation constraints (e.g. that an 'elevator' = 'inaccessible' if 'scenario' = 'emergency').

The lack of suitable information models is complicated by the lack of understanding of the use cases for topographic space information and the corresponding requirements. Therefore there is a need to improve the understanding of the semantics and constraints required for topographic space. Standardised building models are increasingly being used to provide the topographic space information, even though these models have not been developed considering this specific application. The evaluation of existing building models will provide us with a detailed understanding of the comparable suitability of building models and the developments required to fully meet these requirements. The problem is also complicated by topographic space information only being a sub-part of the information required for full indoor navigation. Therefore the integration of the extended building model within a flexible indoor navigation framework is required to ensure that the information provided works in combination with other

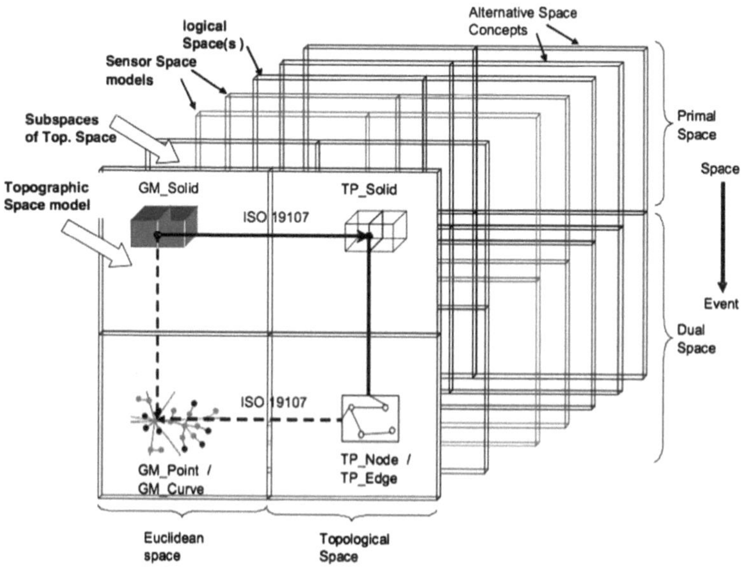

Fig. 1 Multilayered space-event model combining differing space layers (topographic, sensor, logical space etc.) (Nagel et al. 2010)

information sources to fully support the requirements for all indoor navigation tasks. A Multilayered Space-Event Model (MLSEM) framework has been proposed in order to fully support all of the navigation tasks (as detailed in Nagel et al. 2010). Crucial aspects of this framework are the flexibility in integrating multiple space layers (topographic space, sensor space, and logical space), the clear separation of these space layers and the integration of user context information (e.g. modes of locomotion and user groups). This framework allows a range of different information sources to be used for space layers (see Fig. 1), including both CityGML and IFC for a topographic space layer. This model allows arbitrary space cells to be captured but however, lacks the semantic information required for differing topographic space features. The MLSEM does not aim to provide this level of semantics, instead preferring that a suitable existing building model provide the required semantic information. Therefore future work will look at extending existing building models that can be integrated into the MLSEM for full indoor navigation support.

In Sect. 2 the use cases and corresponding requirements for modelling 3D topographic spaces are discussed in detail, as a prerequisite for the assessment of the suitability of related models. Existing topographic space semantic and constraint models are introduced and evaluated in Sect. 3. In Sect. 4 our conceptual approach for modelling semantic topographic space objects and constraints, with respect to the identified requirements is presented. Linked hierarchical conceptual models have been developed for semantic topographic space objects, including all relevant spaces and objects relevant for indoor routing and topographic space constraints, including all factors that can be used to define the level of navigability through/around indoor spaces and obstacles. In Sect. 5 we draw conclusions and give an outlook to future work including the implementation of the conceptual models for an existing building model.

2 Topographic Space Requirements for Indoor Navigation

In order to develop a customised building model suitable for use in indoor navigation, a structured set of use cases for indoor navigation is required. Those papers proposing models for indoor navigation space do not include use case analysis for the developed semantic models (Tsetsos et al. 2006; Meijers et al. 2005; Goetz and Zipf 2011; Yang and Worboys 2011). Therefore a process needs to be developed to identify the uses of topographic space information and the resulting requirements. Only use cases within the scope of indoor routing are considered, with those use cases considering navigation guidance, visualisation of information etc. viewed as being outside the scope of this work. Requirements can then be drawn out from the identified use cases and test cases developed to ensure that a customized/extended building model is fit for use as the topographic space information model.

2.1 Indoor Navigation Topographic Space Use Cases

Planning a route to single/multiple destinations is one of the fundamental tasks of indoor routing. For this task, a user wants to calculate an optimal route to a single/ multiple known destinations considering parameters including the mode of locomotion, current scenario (e.g. emergency evacuation), time of day and access permissions of the user. Therefore this task can be broken down into use cases (see Table 1) to abstract the detailed requirements for topographic space information model. In Table 1, five core use cases (use cases 1–5) are introduced with use cases 6–8 relating to this core set.

2.2 Indoor Navigation Topographic Space Requirements

From the use cases, defined in Sect. 2.1, an extensive list of requirements for topographic space information has been identified. The completeness of this list will be subject to further investigation along with the determination of dependencies between requirements. The indoor environment requirements identified are as follows:

Table 1 Use cases for indoor navigation topographic space information with accompanying navigation constraints

No.	Use case title	Example scenario	Navigation constraints
1	Route a user within a single room in a building	Route Person A from check-in desk 14 to gate A2 in London Heathrow airport, considering that the space can contain fixed (e.g. pillars), movable (e.g. furniture) and dynamic obstacles (e.g. crowds of people)	Spatial extent of space and obstacles Surface material of floor Supporting weight of floor/ furniture
2	Route a user between separate rooms in a single storey	Route Person A between office 0.02 (Ground floor) and office 0.12 (Ground floor) of a multi-storey office building	Temporal, user specific, access type, directional, current state and spatial extent restrictions on door and window (to a lesser extent) spaces
3	Route a user between different storeys within a building	Route Person A from the main entrance (ground level) to Platform 12 (2 storeys below ground level) of Berlin main train station using ramps, stairs, escalators and lifts where appropriate	Space types (e.g. power assisted escalator, stairs etc.)
4	Route a user from outside a building to inside a building and from inside to outside a building	Route Person A from Office 5.12 in the main building of the TU Berlin to fire evacuation point 4 (outside the building) during an emergency evacuation scenario	Type of door/window spaces (e.g. interior/exterior)
5	Route a user between separate buildings (e.g. from a start point in Building A to a destination in Building B)	Route a user from a parking space in a car park to office 6.13 in a neighbouring office block, requiring a user to walk outside	
6	Route a user with specific requirements (e.g. human on foot, human in a wheelchair, UAV, emergency services worker)	Route Person A travelling in a wheelchair from departures entrance 1 of Berlin Tegel airport to check in desk B2	User groups/mode of locomotion
7	Route a user considering a specific scenario (e.g. emergency evacuation, rush hour journey etc.)	Route Person A from supermarket 1 to men's clothing shop 2 within a large shopping centre, at a peak shopping time	Scenario type for routing Persistency and current state of obstacles (e.g. walls, furniture)
8	Route a specific user within a building where access to certain parts is controlled	Route Person A with limited security clearances, considering restricted access for specific person/directional access/temporal access constraints, between room 1 and room 10	User specific access permissions

- **Requirement 1**: An indoor environment model shall capture the general semantic information for a specific building and be represented by all spaces belonging to this indoor environment (relates to use cases 3, 4 and 5).
  Example Scenario: When route planning all space objects belonging to a specific building (e.g. a Hospital) will need to be able to be identified for use with routing algorithms.
- **Requirement 2**: All spaces belonging to an indoor environment shall be represented both semantically and geometrically, defining spatial properties of physical spaces (relates to use cases 1 and 6).
  Example Scenario: Indoor navigable spaces (e.g. rooms) must be semantically classified and have a geometry so that navigable space can be identified for different modes of locomotion (e.g. user in a wheelchair).
- **Requirement 3:** Spaces belonging to an indoor environment shall be categorised according to specific pre-defined space types (relates to use cases 2 and 3).
  Example Scenario: All space types will need to be broken down into pre-defined space types for the definition of common constraints (e.g. power-assisted movable doors).
- **Requirement 4**: All spaces belonging to an indoor environment shall be able to be decomposed into smaller space parts (relates to use cases 1 and 7).
  Example Scenario: A large indoor navigable space (e.g. an airport) will need to be subdivided into smaller space parts for the definition of start and end points for a route.
- **Requirement 5**: All spaces belonging to an indoor environment shall be able to be extended with additional semantic attributes (does not relate directly to any single use case).
  Example Scenario: Future requirements from routing algorithm developers could require that additional semantic attributes be represented (e.g. speed penalty traversing for dynamic obstacles).
- **Requirement 6**: Storeys within an indoor environment should be represented and associated to all spaces belonging to a specific storey within an indoor environment (relates to use cases 2 and 3).
  Example Scenario: When routing between Room A and Room B on different storeys, the storeys these rooms are located on are required to support the analysis of whether elevators can be used to travel between these 2 storeys.
- **Requirement 7**: An indoor environment model should be able to be seamlessly used with outdoor spatial information providing transport networks, navigable areas etc. (relates to use cases 4 and 5).
  Example Scenario: When routing a user from a space in Building A to a space in a separate Building B, outdoor information must be used together with the indoor information to define outdoor navigable routes between the entrance/exits of these buildings.

The indoor space requirements identified are as follows:

- **Requirement 8**: Storage of semantic information for the function, usage and occupants of an indoor space (relates to use case 2).

Example Scenario: When planning a route it is important that the usage of a room is known, so that a user is not navigated through meeting rooms when unoccupied rooms are available instead.

- **Requirement 9**: Specialised types of indoor space shall be used to differentiate levels of connectivity of indoor spaces (relates to use cases 2 and 3). This information could be derived but is required for the categorisation of connected spaces.
  Example Scenario: When planning a route between two rooms, only the spaces that connect together multiple spaces must be considered when creating a route between a start and an end position.
- **Requirement 10**: Specialised types of connecting space with specific semantics shall be used for vertical (e.g. staircase) and horizontal (e.g. corridor) and fixed (e.g. ramp), assisted (e.g. escalator) and transfer (e.g. elevator) connecting spaces (relates to use cases 3 and 5).
  Example Scenario: A vertical staircase space requires different specialist attributes for the spatial properties of the stairs, number of flights of stairs and the staircase types, to determine if this space is navigable for a wheelchair user in an emergency scenario.

The transfer space requirements identified are as follows:

- **Requirement 11**: Transfer spaces (e.g. a door opening space between two rooms) shall be separated into both physical (e.g. door or window opening spaces) and virtual opening spaces (e.g. airport security gate) for which specialist attributes can be defined (relates to use cases 2 and 4).
  Example Scenario: Virtual opening spaces are required when no physical boundaries exist between two indoor spaces or indoor and outdoor spaces. A virtual opening could define the potential access points into an indoor environment.

The indoor obstacle space requirements identified are as follows:

- **Requirement 12**: Indoor obstacle spaces should be semantically categorised as fixed (e.g. pillar), movable (e.g. small table) and dynamic (e.g. fire) obstacle spaces, with physical attributes representing the spatial extent, supporting weight, persistency, current state and scenario type (relates to use cases 1, 6 and 7).
  Example Scenario: Persistency of obstacle spaces is required, as certain types of wall (a fixed obstacle space) could be removed in an emergency evacuation scenario, if required.
- **Requirement 13**: Fixed position obstacle spaces will have the surface material and specialist semantics defined for interior and external walls, floors, ceilings, stairs, ramps and general fittings (e.g. light fittings) (relates to use case 1), allowing constraints to be defined for these features.
  Example Scenario: The surface material of a floor surface is required to determine the suitability of a floor surface for use by a wheelchair user.

- **Requirement 14**: Movable obstacle spaces will have semantics including physical weight and specialist semantics defined for windows, doors, furniture, construction work etc. (relates to use cases 1, 6 and 7) allowing constraints to be defined for moving this obstacle space.
  Example Scenario: A movable furniture obstacle requires physical weight and other attributes to determine the movability of this obstacle by different user groups.
- **Requirement 15**: Door and window (movable obstacle spaces) should have specialist semantics allowing constraints to be defined according to the type, opening mechanism, sub-parts, directionality of opening, current state, accessibility (users with access, times of access, access type and direction) and usability in scenarios (relates to use cases 2, 6 and 7).
  Example Scenario: A movable door obstacle must be able to capture the users that have access permissions for opening a door in a specific direction.

3 Related Models

Indoor navigation requires a detailed topographic space model including both semantics and constraints, to meet the requirements specified in Sect. 2.2. In the following section, we will examine the existing building models, semantic topographic space models and constraint models against these requirements.

3.1 Semantic 3D Building Models

We will focus on the international standards CityGML and IFC only. These semantic building models have the potential to provide part or all of the topographic space information required for indoor navigation through semantic enrichment. Only semantic models are considered as the requirements have defined that a detailed set of semantics are required for topographic space. 3D graphics formats will only be considered in comparison to these semantic building models.

3.1.1 CityGML

CityGML is an Open Geospatial Consortium (OGC) standard based on GML3. This multi-purpose information model is used for describing geometric, topologic and semantic aspects of city models in a 3-dimensional way (Kolbe et al. 2005). The building model is the most detailed thematic concept of CityGML and

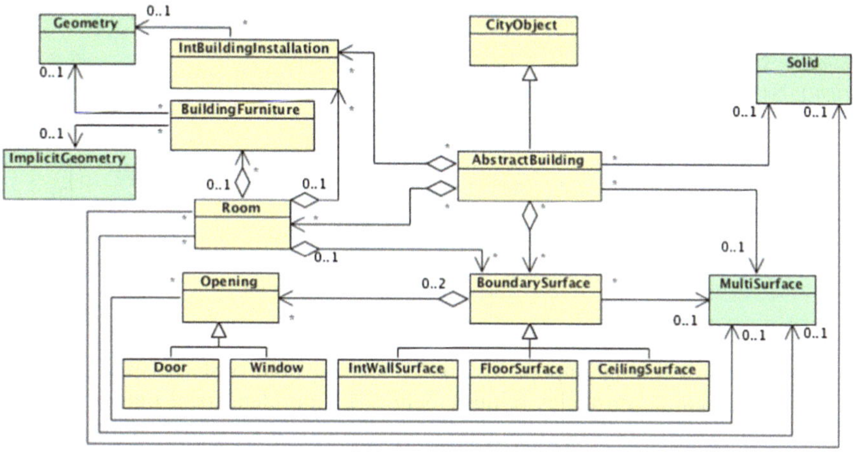

Fig. 2 Simplified CityGML LOD4 building model

supports 4 levels of detail (LOD). A LOD4 building model provides the semantics and geometry for the interior of a building (see Fig. 2).

The semantic and constraint features of a CityGML building model include an *AbstractBuilding*, with each building being able to be composed of Rooms and *IntBuildingInstallations* (requirement 1). Indoor spaces (*Rooms*) and transition spaces (*Openings*) can be fully semantically and geometrically represented in CityGML. All required obstacles can only be partially represented by *Building-Furniture*, *IntBuildingInstallation* and indirectly from *BoundarySurfaces* (requirement 2). CityGML does not provide a specific concept for the representation of storeys, as is implemented for IFC. A storey can though be represented as an explicit aggregation of all building features on a certain height level using CityGMLs notion of *CityObjectGroups* (Gröger et al. 2008). This *CityObject-Groups* may also have a defined geometry. However, if building features are associated to a specific storey, this may require the vertical fragmentation of these features, one part per storey (Gröger et al. 2008). CityGML also supports the use of a world coordinate system, allowing outdoor and indoor spatial information to be used seamlessly together to route a user outdoors between buildings (requirement 7).

Room features contain attributes allowing the function and usage of these indoor spaces to be defined (requirement 8). CityGML does not include predefined connected room types, however, through the aggregation associations between a *Room* and *BoundarySurface* and *BoundarySurface* and *Opening*, some information on the connectivity of indoor spaces can be derived (requirements 9 and 10).

For transition spaces, CityGML defines *Openings* (windows and doors) in the *BoundarySurfaces* and can create virtual openings through the use of *Closure-Surfaces* (requirement 11).

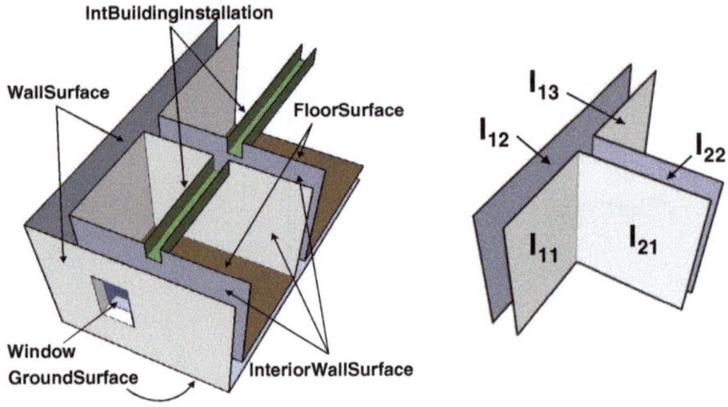

Fig. 3 CityGML model of a building storey (*left*) and *BoundarySurfaces* for rooms (*right*) (Nagel et al. 2009)

CityGML has limited support for fixed and movable obstacle spaces and no support for dynamic obstacles (requirement 12). Complete fixed indoor obstacle spaces (e.g. walls) are indirectly partly represented in CityGML by *Boundary-Surfaces*, which capture only the visible surfaces of a room, see Fig. 3. Therefore indoor obstacle space can be derived as being the Space of a building minus the indoor spaces (e.g. rooms). As a result of this wall, ceiling and floor spaces have no semantics and are unable to be decomposed into sub-parts (requirement 4). The movable components of a window or door are not modelled separately to the opening in CityGML and lack detailed semantic information. The limited semantic information for doors and windows and the lack of support for constraints prevents navigation constraints on topographic features (e.g. wheelchair only being able to traverse through a power assisted door) from being defined, as needed for requirements 12, 13, 14 and 15.

3.1.2 IFC

The term Building Information Modeling (BIM) describes the process of generating and managing building data (Ashcraft 2007), using 3D modeling approaches. A commonly used format for BIM is the IFC, describing a neutral and open specification, registered as ISO 16739 (IAI 2008). IFC defines an entity-relationship model providing an abstract and conceptual representation of data, consisting of around 900 entity classes organized into an object-oriented hierarchy (Goetz and Zipf 2011). IFC provides detailed semantics for constructive building elements, including beams and walls (Fig. 4).

The semantic features of an IFC building model include an *IfcBuilding* that should have one or more *IfcBuildingStorey* (requirement 6), with each *IfcBuildingStorey* having zero or more *IfcSpaces* related to it (requirement 1). All indoor spaces (*IFCSpaces*), obstacles (*IfcBuildingElement* and *IfcFurnishingElement)* and transition spaces (*IfcOpeningElement*) are represented in IFC semantically and allowing multiple geometric representations (requirement 2 and 3). IFC supports complex space groups, spaces and partial spaces (requirement 4). IFC models are not normally used to model complete urban environments, but workarounds exist

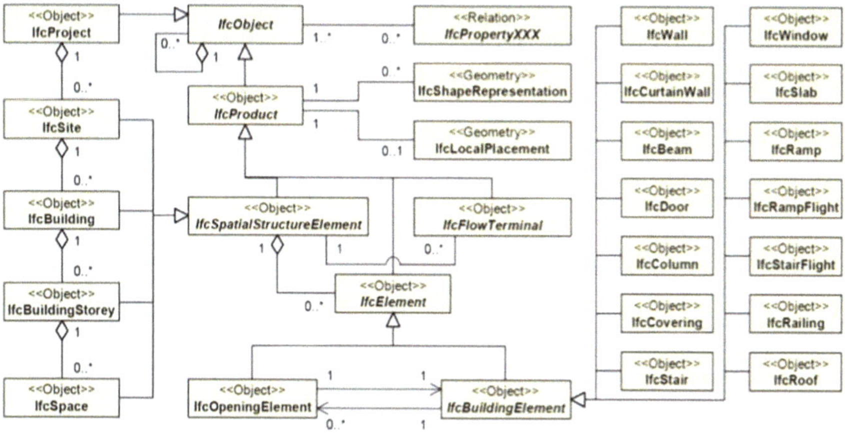

Fig. 4 Subset of IFC classes relevant topographic space information (Benner et al. 2005)

to support the modelling of sets of buildings within a real world coordinate system (requirement 7).

IfcSpaceType allows the function of specific spaces to be defined (requirement 8). IFC building models have limited support for connected indoor spaces, with IFC entities and relation classes (*IfcRelConnects*) defining general applicable object types for the connectivity relationship (IAI 2008) that can express some information on the connectivity between spaces (requirements 9 and 10).

For transition spaces IFC supports the capture and representation of openings (window and door) and can indirectly create virtual openings through the utilization of *IfcVirtualElement* (requirement 11).

IFC has no support for dynamic obstacles (requirement 12). Detailed semantics are provided for fixed obstacles (*IfcBuildingElements*) and furniture (movable) obstacles (as specified in *IfcFurnitureType* for furniture elements) however, other movable obstacles (e.g. construction work and indoor vehicles) are not supported (requirements 13 and 14). *IfcDoor* and *IfcWindow* provide detailed semantics including the opening direction, operation type (e.g. double swing) and operation type, hinge location and construction material. IFC does lack support for complex topographic space constraints (requirements 12, 13, 14 and 15).

3.1.3 Summary

CityGML, IFC and 3D graphics formats (e.g. kml, collada, X3D etc.) were quantitatively evaluated against the requirements specified in Sect. 2.2 (see Table 2), based upon the knowledge of and experience gained from working with these models/formats.

CityGML and IFC are both versatile data model that aim at spatio-semantic coherent models but also allow the representation of 3D models at various degrees

Table 2 Evaluation of CityGML, IFC and 3D graphics formats against topographic space requirements (++ requirement fully met, + requirement partially met, and o requirement not met)

Requirement no.	CityGML	IFC	3D graphics formats
Indoor environment:			
1	++	++	o
2	+	++	o
3	+	+	o
4	+	++	o
5	++	++	o
6	++	++	o
7	++	++	o
Indoor space:			
8	++	++	o
9	o	o	o
10	o	o	o
Transition space:			
11	++	+	o
Indoor obstacle space:			
12	+	+	o
13	+	++	o
14	+	+	o
15	o	+	o

of geometric and semantic complexity (Stadler and Kolbe 2007). In this evaluation full spatio-semantic coherent IFC and CityGML building models are used. Table 2 clearly shows that an IFC building model fulfils slightly more of the overall requirements than CityGML. The minimal support for fixed obstacles (e.g. walls) in CityGML can be summarised as being a highly significant differences between these building models. 3D graphics formats are included in this evaluation to show that visualisation models are not sufficient as they lack semantics.

3.2 Semantic Indoor Navigation Topographic Space Models

Semantic models and ontologies are increasingly being developed for indoor navigation topographic space. An Indoor Navigation Ontology (INO) is included in the OntoNav framework, to describe the basic spatial and structural concepts of indoor environments (Tsetsos et al. 2006). INO introduces concepts that are relevant for indoor navigation (excluding guidance) including: Space (e.g. Building, Room, Floor etc.); Path_Element (e.g. Corridor_Segment, Escalator, and Door); and Obstacle (e.g. table and closed elevator). The Path_Element concept models the physical or conceptual elements of a navigation path. Passage, a type of Path_Element, is any spatial element that is part of a path and has specific accessibility properties (requirement 9). These are separated into: Horizontal (e.g. connecting corridors); Vertical (e.g. ramp); and Motor passages (requirement

10). This semantic model does not discuss the requirement for the decomposition of spaces into smaller parts (requirement 4) or the semantics and constraints for indoor obstacles including doors and windows (requirements 12, 13, 14 and 15).

Meijers et al. (2005) present a semantic model of interior spaces for facilitating the calculation of evacuation routes. This semantic model has a building composed of an aggregation of complexes of sections (e.g. a storey) or of sections (requirements 1 and 6). Three types of sections exist: end (with only one entrance/exit); connector (with more than one entrance/exit) and non-accessible (no entrance/exit) sections (requirements 9 and 10). These sections are geometrically defined by 3D polygons normally representing walls and are classified according to persistence (potential for temporary removal), existence (real and virtual walls), access granting (non-granting, limited and granting access) and types of passing (uni and bi-directional), partly fulfilling requirements 13 and 15.

In Goetz and Zipf (2011) a 3D Building Ontology (3DBO) is introduced for indoor environments, intended for use with OpenStreetMap (OSM). In this ontology the 3D Building representation is defined as having a distinct number of levels (requirement 6), with each level having BuildingParts for spatial elements belonging to a distinct level (requirement 2 and 3). These BuildingParts are categorised as rooms, halls, corridors, vertical passages and horizontal passages (requirement 8, 9 and 10). BuildingParts can also be fixed and movable obstacles and have both windows and doors, partly fulfilling requirements 11 and 12. This model does not consider how building parts can be decomposed into sub-parts and additional constraints added to fixed and movable obstacles (requirements 13, 14 and 15).

Yang and Worboys (2011) have started work on developing ontologies for a navigation model in a unified indoor and outdoor space. Four levels of ontologies are developed: upper (general event, object, state, setting concepts); domain (structure of spaces); navigation task (concepts for navigation guidance); and application (e.g. for indoor navigation of pedestrian). The domain ontologies include a structure ontology for indoor spaces. In this ontology the highest-level features modelled are: Surface (e.g. floor); Portal (e.g. window or entrance); ControlDevice (e.g. key or lock); Container (e.g. elevator or room); Obstacle (e.g. wall or internal door). This ontology captures indoor spaces as rooms and passages, transition spaces as doorways and window spaces and obstacles as fixed and movable barriers. In this model there is no support for complex constraints (e.g. persistency of wall obstacles in an emergency scenario).

To summarise, the existing semantic models for indoor navigation topographic space align much more closely with the requirements than the building models evaluated as they were developed for these specific tasks. The modelling and method for integrating navigation constraints in a semantic model was generally lacking in the semantic models evaluated.

3.3 Constraint Models

The analysis of the requirements for topographic space information showed that there is a need to be able to add both simple and combined constraints to topographic space entities (requirements 12, 13, 14 and 15). A simple constraint expresses a single condition/restriction for a single topographic space element, whilst a combined constraint expresses multiple constraints on a single feature or single/multiple constraints on a series of features. An example of a simple constraint is the users with access permissions for a specific door.

The ISO Geographic Data File (GDF) standard (ISO 14825:2011) uses constraints when modelling features relevant for outdoor routing. This method and structure used for defining constraints in GDF can be evaluated and considered for use in modelling constraints of topographic space information. A combined navigation constraint for Prohibited Manoeuvres has been implemented as a GDF Relationship, defining a manoeuvre that is physically possible but prohibited legally. This GDF Relationship is specified for at least 2 road elements, a junction and a traffic sign feature. A similar principal may be able to be applied for uni-directional access through a door space (requirement 15).

In the OntoNav system, user context is modelled using a developed User Navigation Ontology (UNO) (Tsetsos et al. 2006). This ontology contains user classes and elements of user context. Only physical navigation rules, applied to discard any paths that are not physically accessible to a user, are within the scope of this work. This ontology links to the OntoNav INO model, fulfilling requirements 12, 13, 14 and 15. An example of a possible rule is for excluding Stairways for wheelchair users (requirement 2).

Stoffel et al. (2007) developed a semantic spatial model for pedestrian indoor navigation and introduced the concept of annotating nodes and region graphs with further attributes (e.g. list of key-value-pairs) to provide further context information. Modelling of Boolean constraints is introduced for doors and windows, which can be locked or require access authorisation, have temporal opening times, and used only in an emergency scenario (all constraints fitting to requirement 15).

To summarise, limited work has been undertaken on understanding the constraints needed for topographic space, hierarchically modelling these conceptual constraints and implementing a method for all navigation constraints to be defined for semantic entities (IAI 2008; Gröger et al. 2008; Yang and Worboys 2011; Meijers et al. 2005).

4 Indoor Navigation Topographic Space Model

Existing models have been shown to be lacking semantic and constraint entities (Sect. 3). Initial work has started on the conceptual modelling of the topographic space features and constraints needed to fully meet the requirements, defined in

Fig. 5 Visualisation of CityGML LOD4 building model for Berlin main train station

Sect. 2.2. These conceptual models would support the customisation/extension of existing building models. In order to introduce the semantic and constraint concepts an example environment, Berlin main train station, is used (see Fig. 5).

4.1 Conceptual Semantic Indoor Navigation Topographic Space Model

We define an indoor environment as an abstraction of the collection of all real-world topographic objects being relevant indoor environment components. The conceptual modelling of an indoor environment and its components complies with ISO 19109 and the General Feature Model (GFM) concept, with featureType, Constraint and geometry stereotypes used.

The basic unit for modelling topographic space (*IndoorNavigationTopographicSpaceObject*) is an abstract concept mapping topographic space features to the GFM feature types. All *IndoorNavigationTopographicSpaceObjects* are aggregated together as a collection of units (*IndoorNavigationTopographicSpaceModel*), see Fig. 6. All abstracted topographic space components (e.g. train station platform spaces, staircases and doors) can be aggregated to an *IndoorEnvironment* (requirement 1). This central feature is an extension of the Building feature used in other semantic models (Meijers et al. 2005; Tsetsos et al. 2006; Yang and Worboys 2011; Goetz and Zipf 2011; IAI 2008; Gröger et al. 2008), to include additional environments (e.g. underground transport systems).

An *IndoorEnvironment* is composed of multiple *SpaceUnits*, with these hierarchically categorised and able to be decomposed into sub-space parts (requirement 2 and 4).

Existing building and semantic models use various approaches for modelling a *Storey* feature, including the aggregation of an indoor environment into storeys (with a geometry) and storeys aggregated into spaces (IAI 2008; Meijers et al. 2005; Goetz and Zipf 2011) and the approach discussed for CityGML (see

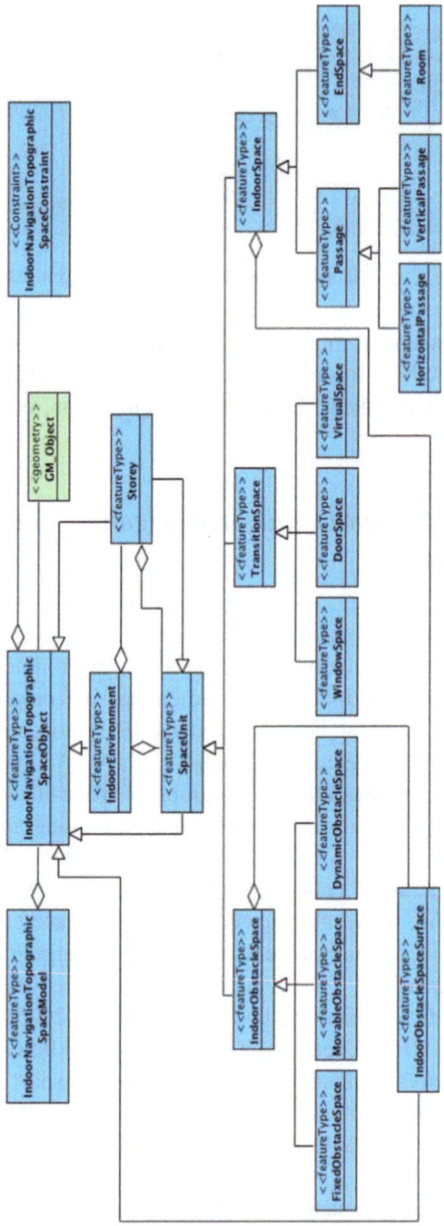

Fig. 6 Conceptual semantic indoor navigation topographic space model

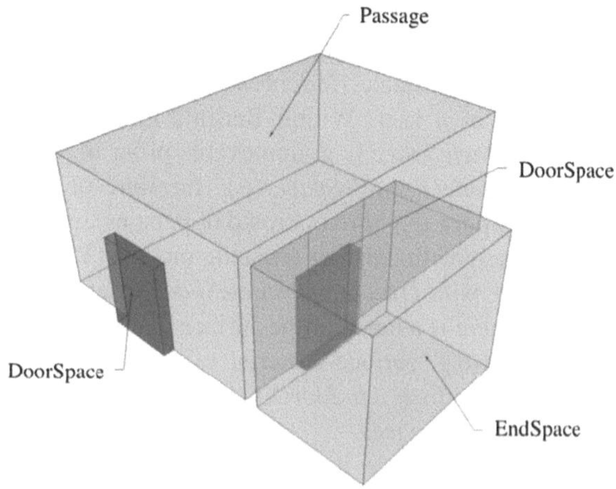

Fig. 7 Specialised types of *IndoorSpace*: EndSpace with one *DoorSpace*; and *Passage* with multiple *DoorSpaces*

Sect. 3.1.1) whereby a CityObjectsGroup could be used to associate all CityObjects belonging to a storey (with or without an individual geometry). The defining of the geometry of a *Storey* requires that all indoor spaces are aggregated to a single storey. This requirement results in this approach not being considered at this point in time, with a *Storey* currently represented as a collection of indoor *SpaceUnits* (requirement 6).

One of the types of *SpaceUnit* belonging to an indoor environment is *IndoorSpace* and is defined as a volume of space that has the potential to be navigated through by a user. *IndoorSpace* and sub-space parts are able to have geocoded addresses, function and usage attributes stored, supporting the search for spaces and routing considering the semantics of *IndoorSpaces* (requirements 7 and 8). *IndoorSpace* is categorised into both *EndSpaces* and *Passages*. This categorisation uses the concept of end and connector space as introduced by Meijers et al. (2008) (requirement 9), see Fig. 7. An *EndSpace* is a unit of bounded indoor space that only has a single entrance/exit. A connector space (*Passage*) has multiple entrance/exits and thus is connecting together multiple indoor spaces (e.g. corridor). Similar categorisations are used in existing semantic models (Tsetsos et al. 2006; Yang and Worboys 2011; Goetz and Zipf 2011), with spaces and passages being separated. The classification of a corridor space varies within semantic models, with a corridor either considered a space (Tsetsos et al. 2006; Goetz and Zipf 2011) or a Passage (Meijers et al. 2005; Yang and Worboys 2011). The use of the connector space concept in this semantic model results in a corridor or connected room (to 2 or more indoor spaces) being defined as a passage. A *Passage* is categorized by the direction of passage, with both *HorizontalPassage* (e.g. corridor, moving sidewalk) and *VerticalPassage* (e.g. staircase and elevator) being defined (requirement 10). This same categorisation is adopted in existing semantic models (Tsetsos et al. 2006; Goetz and Zipf 2011). For both *HorizontalPassges* and *VerticalPassages* further specialist space objects can be categorised and defined according to whether these passages are *Fixed* (e.g. staircase), *Assisted* (e.g. escalator) or *TransferSpace* (e.g. elevator), requirement 10.

An indoor obstacle (*IndoorObstacleSpace*) is any object that can restrict the movement of a user. Within Berlin's main train station obstacles will include *FixedObstcacleSpace* (e.g. unmovable pillar in the centre of a room or an interior wall), *MovableObstacleSpace* (e.g. furniture or construction work) and *DynamicObstacleSpace* (e.g. fire or crowd of persons), requirement 12. This categorisation fits in closely with existing semantic models (Yang and Worboys 2011; Goetz and Zipf 2011), with *DynamicObstacleSpace* extending the categories defined in these models. There is a need for the surface materials of some *FixedObstcacleSpaces* to be defined (e.g. surface material of a floor object is important for wheelchair users). *Therefore IndoorObstacleSpaceSurfaces* may be aggregated to *IndoorObstacleSpaces*.

A *TransitionSpace* is an opening space providing passage between *IndoorSpaces*. Similar concepts are termed Portal (Yang and Worboys 2011) and Polygon (Meijers et al. 2005) but represented significantly differently in other existing semantic models (Tsetsos et al. 2006; and Goetz and Zipf 2011). *TransitionSpace* has 3 subclasses: *WindowSpace*; *DoorSpace*; and *VirtualSpace* (requirement 11). Window and door transition spaces can be related to a window or door movable obstacle, which represents the actual door or window entity.

4.2 Conceptual Constraints Model

Initial work on the modelling of topographic space constraints follows on from the modelling of topographic space semantics. Topographic space constraints are linked to the topographic space semantic model through the relation of *IndoorNavigationTopographicSpaceConstraints* to *IndoorNavigation-TopographicSpaceObject*, shown in Fig. 6. When considering topographic space constraints, the basic unit is *IndoorNavigationTopographicSpaceConstraint*, an abstract concept mapping topographic space constraints to the GFM feature types. Single *IndoorNavigationTopographicSpaceConstraints* can be aggregated together as a collection of units for a more complex constraint (*CombinedIndoorNavigation-TopographicSpaceConstraint*), see Fig. 8.

From the requirements for topographic space information for indoor navigation, different categories of constraints can be identified: *AccessConstraint*; *PhysicalConstraint*; *ScenarioConstraint*; and *SpaceConstraint*, shown in Fig. 8.

AccessConstraints allow the access properties for Door and Window *TransitionSpace* objects to be defined, requirement 15. A sub-type of *AccessConstraints* is *Users,* supporting the provision of access permissions for individual/groups of users. The second sub-type is *Temporal*, allowing the definition of one-off or repetitive times where access through a door is permitted. *AccessType* supports the definition of whether access is possible, partially (in one direction) or not possible.

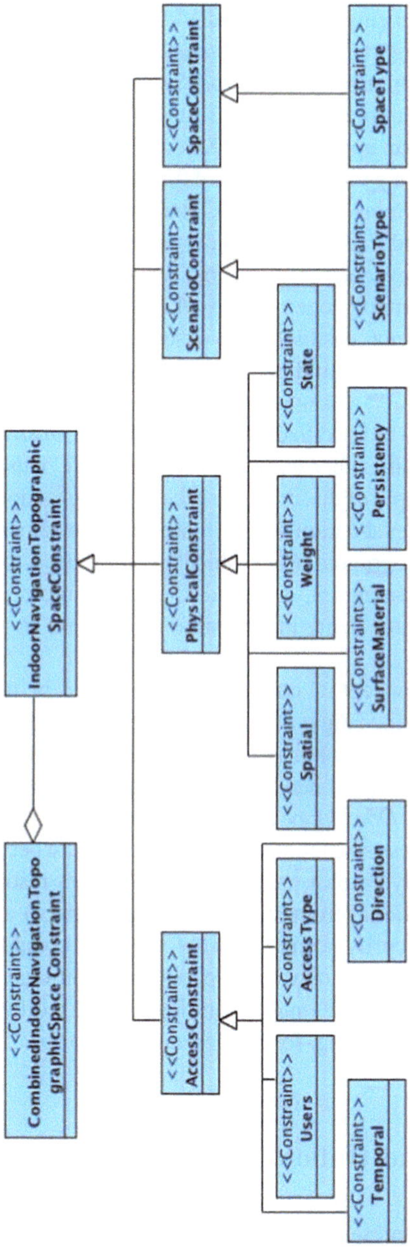

Fig. 8 Conceptual model of indoor navigation topographic space constraints

Direction of access can be combined with other constraints to allow a specific user group to be allowed to have access to pass in one direction through a door.

PhysicalConstraints support the determination of the physical usability of spaces and obstacles, requirements 2, 12, 13 and 14. The sub-types of *PhysicalConstraints* include: *Spatial*; *Weight*; *State*; *Persistency*; and *SurfaceMaterial*. *Spatial* constraints allow the minimum required *IndoorSpace* to be defined for a particular mode of locomotion. For example a wheelchair can only pass through a door space where the width is greater than approximately 83 cm. *Weight* constraints include both the physical weight (e.g. of a movable piece of furniture) and the supporting weight of a fixed *IndoorObstacleSpace* (e.g. floors). *Persistency* reflects the possibility of an obstacle being removed if required, including the removal of a thin glass interior wall during an emergency evacuation scenario. *State* defines whether a fixed or movable indoor obstacle is currently open or closed (e.g. a window being open). Specific *SurfaceMaterials* cannot easily be traversed by certain modes of locomotion, including wheelchairs and blind persons, and therefore the surface material needs specifying.

A scenario type can be specified using the *ScenarioConstraint* and is designed for use in combination with other constraints to form a *CombinedTopographicSpaceConstraint*. An example of this could be access through a door only being available for all users when an evacuation scenario occurs, requirements 12 and 15.

SpaceTypes of topographic space objects can be used to differentiate between single objects, requirements 3, 8, 9 and 15. An example is an electric powered elevator type of elevator being inaccessible to all users in an emergency evacuation scenario.

The hierarchical conceptual constraint model could be used to create single or combined constraints for single or a series of semantic topographic space entities. An example could be a combined constraint limiting the passage through a doo space. This constraint would need to be created for an ordered series of semantic entities (*IndoorSpace*, *DoorSpace* and *IndoorSpace*). The complex constraint could include *Temporal* access restrictions (e.g. only accessible between 08.00 and 16.00), *Users* access (e.g. only employees), *Directional* access restriction (e.g. navigation constraint is unidirectional from start space to end space) and *ScenarioType* (e.g. only in heightened security scenario). Only when all components of this complex constraint are fulfilled, is this door space passable. Additional constraints can also be created for the same series of semantic entities but with a difference in the properties of the constraint elements (e.g. for a emergency scenario).

5 Conclusions and Outlook

In this paper we have presented a structured process for defining topographic space requirements, the evaluation of existing topographic space models and conceptual semantic topographic space and constraints models. Further work is though required to analyse the completeness and dependencies between the defined requirements. The approach explained for deriving conceptual features from a requirement capture driven by use cases, does not only allow the derivation of the required conceptual data model entities and their relations, but also supports the identification of requirements and thus use cases for specific entities in the conceptual models. This enables us to answer questions like if we do not have this type of information in our dataset/building model, which requirements and ultimately which use cases cannot be realised.

In Sect. 2, a detailed set of semantic and constraint requirements were identified from use cases for the task of planning a route to single/multiple destinations, supporting the review of existing models. The evaluation of existing semantic and constraint models for topographic space (Sect. 3) showed that none of the models are sufficient to meet all the requirements identified, with a general lack of support for the modelling of constraints. Whilst topographic space semantic models have been implemented (Sect. 3.2), building data is normally acquired in accordance with a standardised building information model (Sect. 3.1). Therefore the development of a customised/extended building model from a developed conceptual semantic and constraint model is required in order to avoid duplicating existing standards for the capture of interior environments. The IFC data model allows additional elements to be created as standard generic IFC elements. However, the creation of specialised IFC elements requires a change to the IFC core classes, as there is not an extension mechanism in place for IFC. An extension mechanism is available for CityGML through the use of Application Domain Extensions (ADEs). Therefore only CityGML is considered as being suitable for extension in order to provide all the required topographic space information.

This work has produced conceptual models for both semantic topographic space and constraints. These models will be implemented as a CityGML ADE and will be used by routing algorithms for the calculation of shortest/fastest/optimal routes that are possible between a start and an end location. The focus of this work is not on defining user preferences (soft constraints) but instead focuses on defining the physically passable spaces and traversable/movable obstacles (hard constraints). A routing algorithm can then determine the shortest/fastest/optimal/preferred route, considering additional user preferences created independently of the topographic space information.

Indoor route planning is though only one of the three main navigation tasks: determination of position (localisation); determination of best route (route planning); and route tracking (homing) (Becker et al. 2009). For localisation additional information is required to represent sensor spaces (e.g. for WiFi and Bluetooth sensors). Therefore this information must be integrated with topographic space information to support the complete set of indoor navigation tasks. The MLSEM will provide the framework for integrating all available indoor navigation information sources together. The work presented in the paper has identified the use cases, requirements and conceptual models independent to the MLSEM. Future work will need to look at understanding and defining the relationships between an extended CityGML model (implementing the conceptual semantic topographic space and constraint models) and the MLSEM.

Acknowledgments The presented work was started during an EU COST Action funded Short Term Scientific Mission (STSM) to TU Delft in March 2011. During this STSM cooperation and support from Sisi Zlatanova and Liu Liu have helped to develop this work.

References

Ashcraft H (2007) Building information modelling: a framework for collaboration. Constr Lawyer 28(3):1–14

Benner J, Geiger A, Leinemann K (2005) Flexible generation of semantic 3D building models. In: Gröger G et al (eds) Proceedings of the 1st international workshop on next generation 3D city models, Bonn

Becker T, Nagel C, Kolbe TH (2009) A multilayered space-event model for navigation in indoor spaces. In: Kolbe TH et al (eds) Advances in 3D geo-information sciences. Springer, Berlin

Goetz M, Zipf A (2011) Extending openStreetMap to indoor environments: bringing volunteered geographic information to the next level. In: Rumor M, Zlatanova S, ledoux H (eds) Urban and regional data management, Udms Annual 2011, Delft, The Netherlands

Gröger G, Kolbe TH, Czerwinski A, Nagel C (2008) OpenGIS city geography markup language (CityGML) encoding standard. Version 1.0.0, OGC doc no. 08-007r1

IAI (2008) Industry foundation classes (IFC) model documentation. http://buildingsmart-tech.org/ifc/IFC2x3/TC1/html/index.htm. Accessed 1 Oct 2011

ISO 14825:2011 Intelligent transport systems—geographic data files (GDF). http://www.iso.org/iso/catalogue_detail.htm?csnumber=54610. Accessed 10 Jan 2012

Kolbe TH, Gröger G, Plümer L (2005) CityGML—interoperable access to 3D city models. In: Oosterom et al (eds) International symposium on geo-information for disaster management, Delft, The Netherlands

Meijers M, Zlatanova S, Preifer N (2005) 3D geoinformation indoors: structuring for evacuation. In: Proceedings of next generation 3D city models. Bonn, Germany

Nagel C, Becker T, Kaden R, Li K, Lee J, Kolbe TH (2010) Requirements and space-event modelling for indoor navigation. OpenGIS discussion paper, doc ref OGC 10-191r1

Nagel C, Stadler A, Kolbe TH (2009) Conceptual requirements for the automatic reconstruction of building information models from uninterpreted 3D models. In: Kolbe TH, Zhang H, Zlatanova S (eds) Academic track of Geoweb 2009—cityscapes, international archives of the photogrammetry, remote sensing and spatial information sciences, vol XXXVIII 3-4/C3

Stadler A, Kolbe TH (2007) Spatio-semantic coherence in the integration of 3D city models. http://www.isprs.org/proceedings/XXXVI/2-C43/Session1/paper_Stadler.pdf. Accessed 14 Dec 2011

Stoffel EP, Lorenz B, Ohlbach HJ (2007) Towards a semantic spatial model for pedestrian indoor navigation. In: Hainaut JL et al (eds) Advances in conceptual modeling—foundations and applications, vol 4802. Springer, Berlin

Tsetsos T, Anagnostopoulos C, Kikiras P, Hadjiefthymiades S (2006) Semantically enriched navigation for indoor environments. Int J Web Grid Serv 2(4):453–478

Yang L, Worboys MF (2011) A navigation ontology for outdoor–indoor space. http://www.worboys.org/publications/ISA2011Yang.pdf. Accessed 11 Jan 2012

Enhancing the Visibility of Labels in 3D Navigation Maps

Mikael Vaaraniemi, Martin Freidank and Rüdiger Westermann

Abstract The visibility of relevant labels in automotive navigation systems is critical for orientation in unknown environments However, labels can quickly become occluded, e.g. road names might be hidden by 3D-buildings, and consequently, the visual association between a label and its referencing feature is lost. In this paper we introduce five concepts which guarantee the visibility of occluded labels in 3D navigation maps. Based on the findings of a pre-study, we have determined and implemented the two most promising approaches. The first approach uses a transparent aura to let the label shine through occluding objects. The second method lets the feature, e.g. the roads, glow through the 3D environment, thus re-establishing the visual association. Both methods leave the 3D world intact, preserve visual association, retain the label's readability, and run at interactive rates. A concluding user study validates our approaches for automotive navigation. Compared to our baseline—simply drawing labels over occluding objects—both approaches perform significantly better.

Keywords Labeling · Label · Occlusion · Visibility · 3D cities · Navigation · Maps

M. Vaaraniemi (✉)
BMW Research and Technology GmbH, Munich, Germany
e-mail: mikael.vaaraniemi@bmw.de

M. Freidank
University of Koblenz-Landau, Koblenz, Landau, Germany
e-mail: mfreidank@uni-koblenz.de

R. Westermann
Technische Universität München, Munich, Germany
e-mail: westermann@tum.de

J. Pouliot et al. (eds), *Progress and New Trends in 3D Geoinformation Sciences*,
Lecture Notes in Geoinformation and Cartography, DOI: 10.1007/978-3-642-29793-9_2,

1 Introduction

Automotive navigation devices started appearing in the mid-1980s. The first commercially available device, the Etak Navigator introduced in 1986, guided drivers with an annotated 2D map and guidance arrows to their destination Thielmann (2006). Since then, textual annotations in maps have been helping the driver navigate through unknown environments. They are essential for the exploration of navigation maps. The visualization has improved gradually and nowadays, 3D navigation maps have become omnipresent. Several competing companies, like Sygic or Navigon, include terrain and 3D city models in their latest navigation devices. In these systems, labels are usually rendered over occluding 3D elements (e.g. road names over buildings). This approach makes them easily readable, but the visual association to their corresponding feature is lost. As labels appear in front of occluding objects, depth perception is hindered and spatial orientation becomes difficult. In this paper, our primary goal is to preserve the visibility of labels in 3D navigation maps. Hence, deduced from cartographic rules by Imhof (1975) and our expert study from Sect. 4, we define the following rules for labeling 3D navigation maps:

- All labels should be readable, even occluded labels.
- The visual association between the label and its feature should be guaranteed.
- Labels should not occlude other labels or important features.
- Depth cues of the 3D world should be preserved.
- Labels should support spatial orientation.

Our main contribution are two approaches fulfilling these rules and, consequently, enhancing the visibility of occluded labels in 3D navigation maps. The first approach creates a transparency aura around every label and lets labels shine through occluding objects (see Fig. 1a). The second method lets the referenced features, e.g. the roads, glow through the 3D environment, thus creating a visual association (see Fig. 1b). Both methods leave the 3D world intact, preserve visual association and retain the labels' readability. Also, they are able to run at interactive framerates. The enhancements of these approaches are validated in a user study.

2 Labeling Techniques

2.1 World-Space and Screen-Space Labels

Annotations can be placed in World-Space (WS) or in Screen-Space (SS) into the 3D world. SS labels (or 2D labels) are placed parallel to the screen (see Fig. 2a, b). They can be thought as being part of a Head-Up-Display (HUD), overlaid over the 3D scene. WS labels (or 3D labels) are part of the 3D world (see Fig. 2c, d). As such, they are transformed by the perspective projection. Chen et al. (2004) compare both types of labels. They show that SS labels are better for naive search

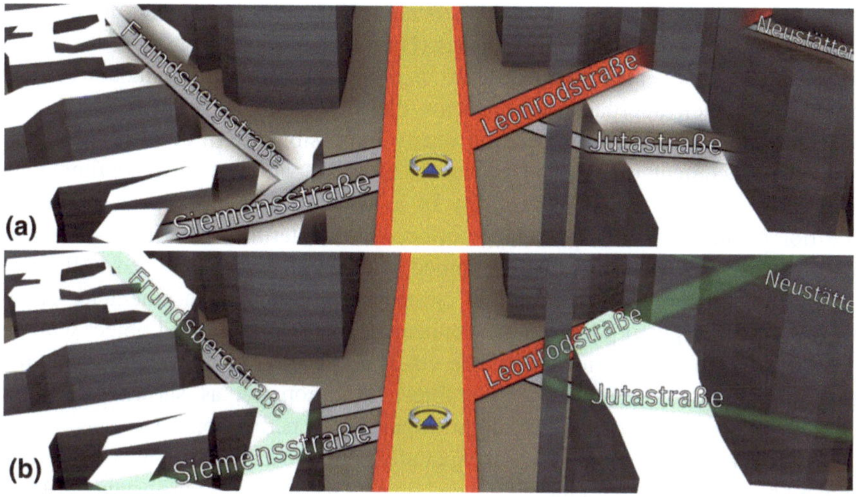

Fig. 1 The two selected approaches to preserve the visibility of textual labels in a 3D world. **a** Transparency label aura: the labels blend out occluding 3D objects. **b** Glowing roads: the roads shine through occluding 3D objects

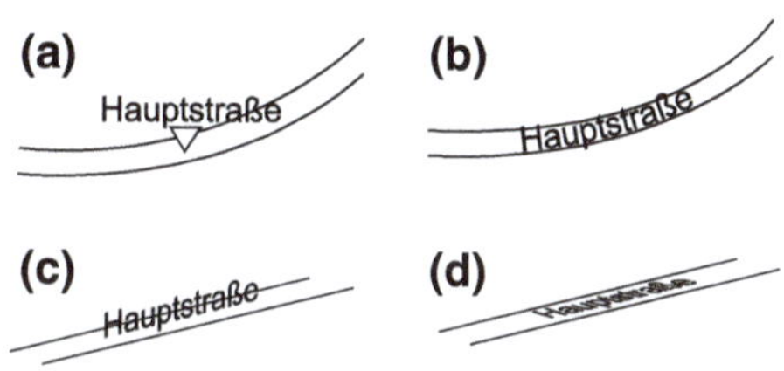

Fig. 2 World-space (WS) and screen-space (SS) labeling techniques used in our approaches. **a** External SS label with a *triangle* anchor. **b** Internal SS label following the road. **c** Internal WS label placed upright. **d** Internal WS label laid onto the road

tasks in densely packed scenes. Also, they are easy to read because they are always facing the viewer. In contrast, as WS labels are part of the 3D scene, they exhibit occlusion problems and can be very difficult to read (e.g. when they follow the object's curvature). However, because they provide strong association cues, they improve the visual association to the referenced feature (Goldstein 2009). Polys et al. (2005) evaluate both techniques and state, that even tough WS provides tight coupling, SS performs better across all tested tasks.

2.2 External and Internal Labels in 3D Worlds

External Labels. Fekete and Plaisant (1999) introduce external labels to annotate dense sets of points. Connected with an anchor (e.g. a line or a triangle), they are displayed beside (or outside) the referenced objects (see Fig. 2). Hence, they do not hide the referenced object. Because they are primarily displayed as SS labels they are also easy to read. External labels are mainly used for annotation of single 3D objects, e.g. in scientific illustrations (Hartmann et al. 2004; Ali et al. 2005). However, Maass and Döllner (2006b) use external labels to annotate virtual landscapes. Their approach creates dense clusters of labels and long connecting lines which makes visual association nearly impossible. Stein et al. (2008) compute the placement of external SS labels in a 3D world with an optimization algorithm. To determine the visibility of a label, a sphere is placed at the 3D position of the anchor. Its percentage of occlusion determines the transparency of the label. If the sphere is fully occluded, the feature is not labeled. All these approaches use greedy algorithms to compute an optimum placement for annotations. The computed positions are connected with the referenced object with an anchor line. This connection makes the visual association more difficult compared to a placement directly beside the object. Additionally, as shown by Maass et al. (2007), using anchor lines might impair depth perception.

Internal Labels. Internal labels are spatially bound to an object. This allows for a direct visual association to the referenced object (see Fig. 2b). For instance, Maass and Doellner (2006a) annotate 3D buildings intuitively with billboards in WS. They introduce an approach to annotate line features in WS (2007). They determine the placement of labels on the fly using sample points. But, changing the view results in different label placements and thus in a temporally incoherent layout. They present an approach to integrate labels directly onto the hulls of 3D buildings by taking their shape into account Maass and Döllner (2008). This creates internal WS labels which are part of the world. In general, internal labels depict the visual extent of an object. Ropinski et al. (2007) and Cipriano and Gleicher (2008) introduce internal WS labels to annotate e.g. medical illustrations. However, these labels hide parts of the referenced object and their readability depends on distortion and the viewing angle.

Hybrids. Bell et al. (2001) and Götzelmann et al. (2005, 2006) present similar hybrid approaches, which use internal and external labels. Bell et al. annotate virtual 3D cities while Götzelmann et al. annotate scientific illustrations. External labels with anchor lines are used when the viewer is far away. When the viewer gets closer and the objects' dimensions allow it, they use internal labels. In contrast, Google Earth Google Inc (2012) uses SS external labels for cities and WS internal labels for streets. This makes street names difficult to read at low viewing angles.

2.3 Summary

None of the presented approaches satisfy our stated goals in Sect. 1. In particular, the goal to preserve readability of labels which are being occluded in a 3D world. The computations of most SS layouting algorithms are done solely in screen space. They do not take into account the occlusion between labels and a 3D scene. SS approaches to annotate scientific illustrations place external labels around single objects, hence, are not affected by occlusion problems (Hartmann et al. 2004; Ali et al. 2005; Götzelmann et al. 2005, 2006). Most SS approaches for labeling 3D worlds ignore occlusion problem by rendering labels over the scene (similar to a HUD) (Maass and Döllner 2006b; Google Inc 2012). Only newer SS algorithms take the visibility of the anchor into account (Stein and Décoret 2008). On the other hand, internal WS approaches try to find visible positions for labels at runtime (Maass and Döllner 2006a, 2007, 2008). However, if unsuccessful, the object remains unlabeled.

3 Concepts

In this section we introduce several concepts which assure the visibility and thus preserve the readability of labels occluded by objects of the 3D world.

3.1 Baseline

The first concept we introduce represents our baseline. It consists of drawing the labels over the 3D world (see Fig. 3). Hence, all occlusion created by objects from the 3D world is ignored. We chose it as a baseline, because it is a straightforward solution for resolving occlusion problems. Also, it is used in almost all existing navigation systems, e.g. Sygic GPS Navigation Sygic (2012) and Google Earth Google Inc (2012).

3.2 Cutaways

Our second concept is cutaways (see Fig. 4). This method is inspired by 2D magic lenses which were first introduced by Bier et al. (1993). These lenses highlight focus regions by modifying their representation. One such approach Bier et al. depicts, is the wireframe representation inside the focus region. Viega et al. (1996) extend these to 3D environments with flat and volumetric lenses. Coffin and Höllerer (2006) introduce perspective cutaways for 3D scenes.

The resulting holes are rendered with the correct perspective as if they were cut in the occluding object. Our approach is very similar to the perspective cutaways. Every label creates a focus region which cuts away all occluding objects in a perspectively correct manner.

3.3 Transparency Label Aura

The next concept creates a smoothly blended transparency aura around the labels. It is similar to Krüger et al. (2006) interactive focus + context method called Clear-View. Their approach is directly inspired by magic lenses. They create a semi-transparent area around the focus region while the remaining parts stay opaque to

Fig. 3 Baseline: drawing labels over the 3D world in bird's eye with SS (*left*) and snail view with WS labeling (*right*)

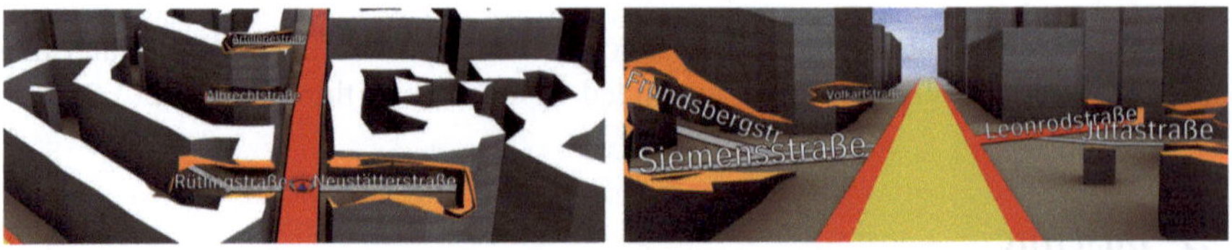

Fig. 4 Cutaways: labels create perspective cut aways in occluding objects of the 3D world in bird's eye with WS (*left*) and snail's view with SS labeling (*right*)

Fig. 5 Transparency label aura: labels create a transparent region in the occluding objects in bird's eye with SS (*left*) and snail's view with WS labeling (*right*)

Fig. 6 Glowing labels: labels are glowing through the 3D world with a distinct color in bird's eye with SS (*left*) and snail's view with WS labeling (*right*)

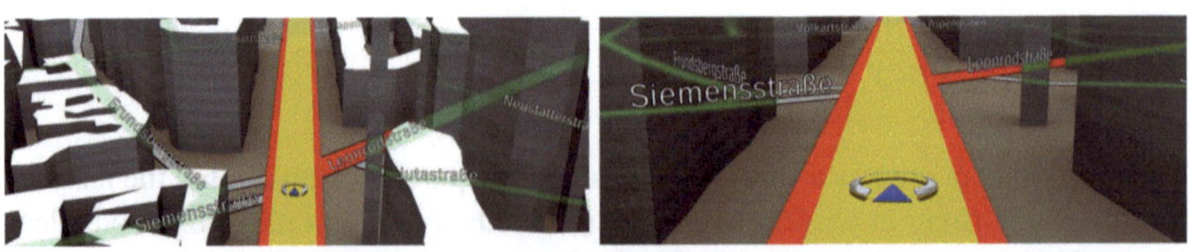

Fig. 7 Glowing roads: roads are glowing through the 3D world in bird's eye with SS (*left*) and snail's view with WS labeling (*right*)

preserve context information. Elmqvist et al. (2007) evaluate such X-ray vision and state that it leads to faster and better object discovery. Analogously, we define in our concept a transparency region around the label (similar to a focus area). All objects of the 3D world lying in front of this region become transparent. This x-ray vision lets the user read every label. Because we define the region to be larger than the label, the referenced feature (e.g. the road) can be seen partially.

This preserves the context of the focus region. Hence, the visual association to the referenced feature is retained.

3.4 Glowing Labels

In our third concept we let labels glow through occluding objects (see Fig. 6). This method is inspired by augmented reality (AR) applications. Kalkofen et al. (2007, 2009) present an approach to augment real objects with context + focus information. This helps recreate the spatial relationship between reality and virtual information. We note that this approach is used in almost all isometric strategy PC games (e.g. Command and Conquer, Age of Empires).

Units being hidden by structures (e.g. buildings) are usually tinted with a different color. Similarly, we tint the occluded parts of labels with a color distinct from the surrounding world.

3.5 Glowing Roads

The baseline concept makes the labels visible but thereby loses the visual association to its referenced feature, e.g. the road.

Our fourth concept tries to solve this problem by adding glowing roads to the baseline. Again, in a similar fashion to the approaches by Kalkofen et al., we let the occluded parts of the roads shine through the 3D world (see Fig. 7). This method recreates the missing context of the labels.

4 Expert Study

We conducted an initial expert study. Our goal was to determine which of the introduced concepts fulfills our rules for labeling a 3D navigation map (stated in Sect. 1). Also, we wanted to form an opinion about the usability and aesthetics of each method from our domain experts. Besides, the preferred labeling space (SS or WS) was surveyed. Two engineers working for over five years on automotive navigation were chosen as experts. Also, as further subjects, we selected three research engineers working on human machine interaction systems.

4.1 Study Design

We presented the four concepts introduced in Sect. 3: cutaways (see Fig. 4), transparent label aura (see Fig. 5), glowing labels (see Fig. 6) and glowing streets (see Fig. 7). Each concept was compared to our baseline: rendering labels over the 3D scene (see Fig. 3).

Movies. Movement is an important aspect which greatly affects the way a 3D concept is perceived. Animation can cause occlusion and creates an important depth cue: the motion parallax. Hence, to improve the value of our study, we chose to create animated sequences lasting 20 to 30 seconds. Each movie was shown with SS- and WS-labeling. We presented each movie with the same flight path in two perspectives: a snail view closer to the ground and a bird's eye view. All these combinations culminated to sixteen different animated sequences. To each subject we showed these concepts in a fixed order as they are introduced in Sect. 3. In an ensuing discussion, we queried all statements and asked for a ranking of the presented concepts (see Fig. 8).

Conceptual Details. We selected a light violet color for the glowing labels (see Fig. 6). Usually, such a color is not present in a 3D navigation visualization, yet it still remains an aesthetically pleasing color. The hidden parts of the glowing road concept are drawn slightly blurred in a light green color, similar to HUD designs (see Fig. 7). Still images from the presented movies can be seen from in Figs. 4, 5, 6, 7.

4.2 Discussion

In both views, glowing streets was ranked highest. 4 of 6 experts chose this as the best approach in both perspectives (bird and snail). Two experts stated that this concept improves orientation. Another expert liked how the glowing roads improve readability by creating an enhanced contrast to the background. One expert criticized the chosen color and suggested to continue the road in its original color. Finally, the last expert described this approach as being too colorful.

The second place is shared between the concept transparency label aura and our baseline. The former performs well in the snail's view, where labels are frequently hidden by 3D buildings. Our baseline sufficed in bird's eye view where occlusion plays a minor role and the spatial relationship is not needed.

Generally, the concept glowing labels was not approved and always ranked last. Three experts stated that the label seemed lost in the world and the coloring makes the visual association even more difficult. Two different experts did not approve that occluded parts should be marked with a different color. Finally, two experts criticized the color as being too vivid and distracting.

Our last concept, cutaways, was quickly dismissed by all experts, because it introduces too much animation. Every movement leads to new cut outs in the 3D buildings, thus removing parts of the world. When a lot of labels are present, the 3D world falls more and more apart.

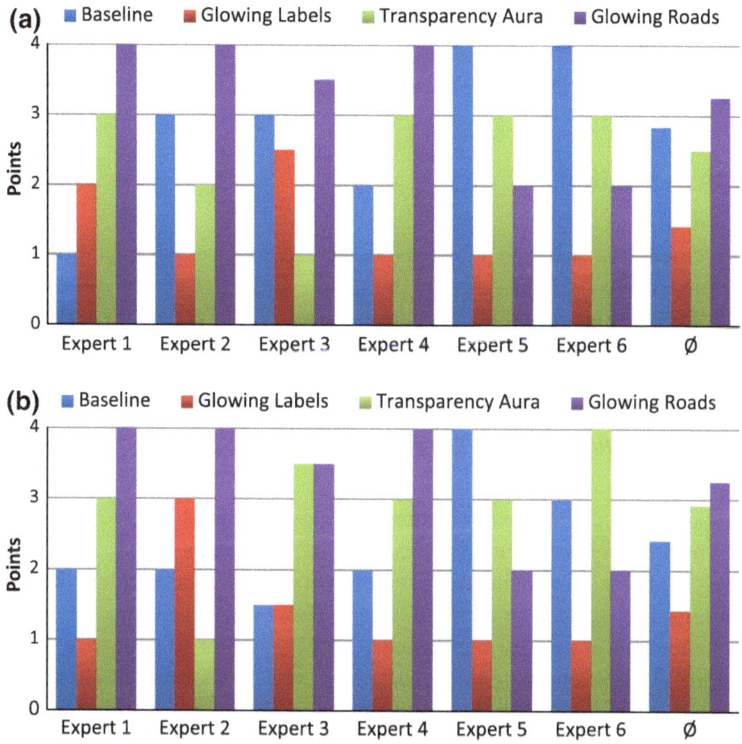

Fig. 8 Ranking of our concepts according to our six experts. Each concept was presented as a short movie. The concept glowing roads ranks first in both viewing perspectives. **a** Bird's eye. **b** Snail's view

When deciding which labeling space was best, 5 of 6 experts voted SS in bird's eye and 5 of 6 experts voted WS in snail's view. All but one expert agreed that in snail's view WS labeling was better despite the restricted readability.

4.3 Results

Concepts. As a first consequence, we dismiss two approaches: glowing labels and cutaways. In the experts' opinion, the disadvantages of the glowing labels concept (e.g. unaesthetical, bad visual association) outweigh the readability improvements. Cutaways introduce too much movement and destroy huge parts of the 3D world.

Visual association. Displaying the referenced feature besides the label is an important requirement for our implementation. One expert liked the transparency aura mainly because he was seeing the referenced road. The glowing labels ranked last because the association to the road becomes lost. In contrast, the concept glowing road recreates this reference.

Labeling technique. The last conclusion we draw, is the need to combine both SS and WS labeling in a 3D navigation. We choose SS in bird's eye and WS in snail's view. In snail's view the WS labels fits into the world's 3D space. In the bird's eye we hover at higher altitudes in which the world flattens. Therein, the better readability of 2D SS labels outweigh the deteriorated spatial relationship.

5 Implementation

We implement the selected concepts in an existing research platform for the visualization of navigation data. In this framework, the central processing unit (CPU) helps loading and preparing data for rendering. To ease the CPU load, our approaches run on the graphics processing unit (GPU) using shader programs.

5.1 Transparency Label Aura

In this concept, occluding parts of the buildings are faded out.

Overview. Our implementation consists of four steps. First, every building occluding a label is drawn into an offscreen buffer. In the second step, the entire set of buildings are again rendered offscreen. However, this time, we discard all fragments located in front of the occluded label – similar to an inverse depth test. In the third step, we combine these buffers to create a transparent aura around the label. Finally, we composite the result into the existing 3D world.

Implementation. The first rendering pass is trivial: we create an offscreen buffer and render all occluding 3D buildings into it. The second pass performs our inverse depth test in a fragment shader on the GPU. For this step, we need a texture (buffer) containing the depth information of all labels. We approximate each label with an object-oriented bounding-box (OOBB). And, because our experts stated in Sect. 4.3 that the referenced objects should be seen, we slightly enlarge the bounding-box of each label. Then, we render all OOBBs of every visible label into a depth-only offscreen buffer. Finally, all buildings are drawn. In the fragment shader we compare the incoming depth value (of our buildings) $z_{building}$ with the depth value of our OOBBs (our labels) z_{label}. If $z_{label} > z_{building}$ the building occludes the label and we can discard this fragment. For the third step, we create a smooth blending in the transparency aura by rendering the OOBBs with a gradient texture. Finally, using this fullscreen alpha mask, we composite the results of the prior steps and render it over the current scene (Fig. 9).

5.2 Glowing Streets

In this concept, all occluded parts of the roads are glowing over the 3D world.

Overview. The implementation consists of two steps. First, we detect which parts of the roads are being occluded. These parts are drawn with a selected color (e.g. light green). Then, optionally, a blurring filter is applied. Finally, the result is composited over the existing 3D world and all labels are rendered.

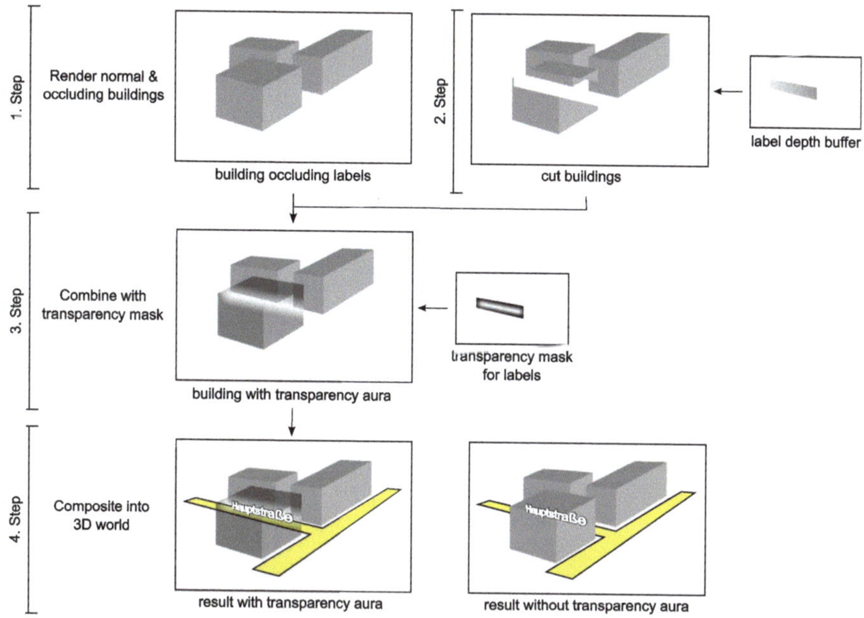

Fig. 9 GPU implementation of the transparency label aura approach

Implementation. Initially, we need the depth values of all rendered 3D buildings $z_{building}$ and roads z_{road}. Then, a fragment shader compares both depth values: If $z_{building} < z_{road}$, then the road is occluded and has to be drawn as a glowing road. If the glowing road is drawn with a single color, we simply output a constant color to an offscreen buffer. If we render the roads in their original color we first have to fetch this color. The resulting buffer can be smoothed with a blur shader and finally, composited with the existing 3D world. After these steps, all labels are drawn on top with a disabled depth test (Fig. 10).

6 Results

6.1 Benchmark

We benchmarked the approaches transparency aura and glowing roads. Our goal was to evaluate the performance scalability and suitability for real-world scenarios.

Configuration. The evaluation was done on an Intel Core 2 Duo E8400 3 Ghz CPU with 4GB RAM and Windows XP SP3. The GPU was a NVIDIA Quadro FX 580 (driver v275.89). To reduce the impact of data loading we preloaded all the needed data. Our performance measurement were done with a flight over a 3D city with roads, 3D buildings and labels. Figure 12 shows the resulting performance graph during a flight of 20 s. We compare the baseline with the

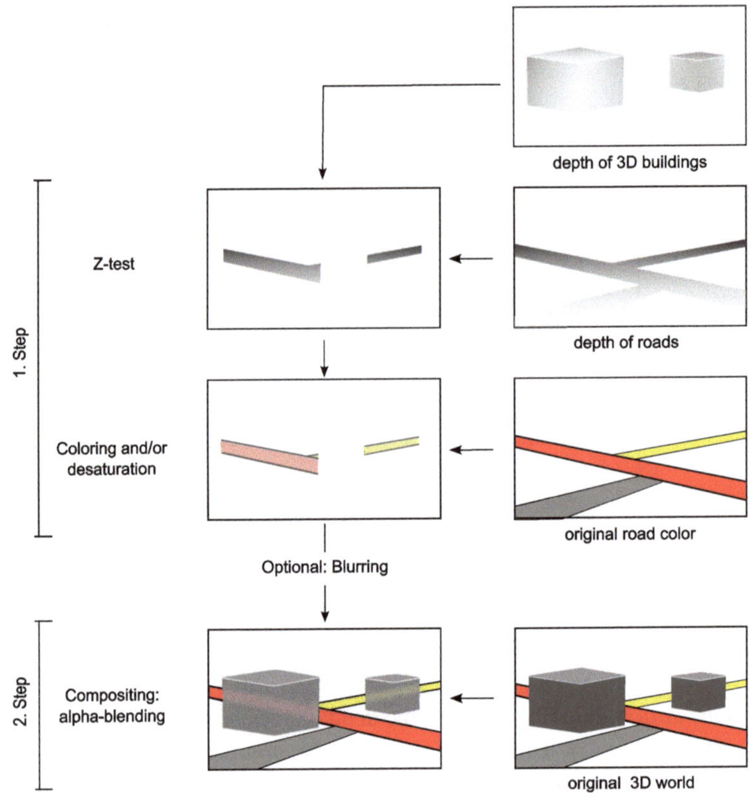

Fig. 10 GPU implementation of the glowing roads approach. Each step represents a shader pass

transparency aura and two variants of the glowing roads: using a single color and using the original road color. We measured the framerate for low $1,024 \times 768$ (Fig. 12, top) and high resolution 1680×1050 (Fig. 12, bottom). During this run we tracked the number of buildings, road meshes and labels (see Fig. 12, middle).

Results. At low resolution ($1,024 \times 768$) our new approaches behave similar to the baseline. Compared to our baseline, they incur a performance drop between 10 and 30 %. The average performance decrease for every approach and for two resolutions can be seen in Table 1. Our approaches are fillrate bound. At approximately twice the fragments (0.8–1.8 MP) we have a 50 % performance decrease for every approach. Also, the increased number of 3D buildings, roads and labels do not impact the framerate as much as the increase in resolution (see Fig. 12, middle).

6.2 User Study

Our goal was to evaluate the usability, attractiveness and novelty of our approaches.

Table 1 Average performance of the implemented concepts and framerate decrease (drop) compared to the baseline

Approach	Framerate				
	$1,024 \times 768$ (fps)	Diff (%)	$1,680 \times 1,050$ (fps)	Diff (%)	Resolution impact (%)
Baseline	110	–	59	–	−46
Transparency label aura	82	−25	43	−27	−47
Glowing roads (single color)	90	−18	46	−22	−49
Glowing roads (road's color)	84	−24	42	−29	−50

Also, we list the performance impact when changing the resolution from $1,024 \times 768$ to $1,680 \times 1,050$. We determine that both approaches are fillrate bound

Participants. We conducted an user study lasting 20 min with 24 persons aged between 17 and 45 consisting of 20 men and four women. About one third worked in the GIS domain. There were 9 students, 12 engineers, two programmers and one manager. Everyone had experience with commercial 3D navigation systems (Fig. 11).

Study Design. These candidates tested the fully working prototypes of our baseline and the two implemented concepts: transparency label aura and glowing roads. In the first part of our evaluation, every subject flew three times the same 30 s lasting route through a 3D city. First, the baseline approach was active. Then, both new methods were shown in a changing order. After every flight the candidates had to fill out an AttrakDiff questionnaire (see Fig. 13). In the second part of the study, we let the subjects choose manually between all three concepts during a flight of two minutes. Finally, they completed a second informal questionnaire (see Fig. 14).

AttrakDiff. After experiencing the prototype, every candidate completed the AttrakDiff questionnaire from Hassenzahl et al. (2003), Hassenzahl (2007). They had to choose repeatedly between two different statements (e.g. attractive vs dull). These pairs were given by the AttrakDiff questionnaire to measure the perceived hedonic quality (HQ) and pragmatic quality (PQ). PQ is an indicator of the perceived usability of our concepts. HQ is divided into identity (HQ-I) and stimulation (HQ-S): HQ-I describes the user's identification, HQ-S defines the novelty of the tested concept. Finally, the questionnaire measures the overall attractiveness (ATT).

6.2.1 Results

Figure 13a presents the averaged results of the AttrakDiff questionnaire. Compared to our baseline (orange), both approaches increase significantly every quality aspect and the overall attractiveness. The boxes in Fig. 13b indicate the overall classification in HQ and PQ. Therein, a placement in the top-right quadrant defines a very desired product. The size of the light boxes indicate the variability of the answers. In our case, the small box size of the baseline (orange) and glowing roads (blue) indicates a consistent opinion. In contrast, answers about the transparency aura (red) display more variation. In both figures, glowing roads (blue) achieve the best usability impact (PQ) and attractiveness (ATT). Overall, this validates the ranking of our experts from our pre-study.

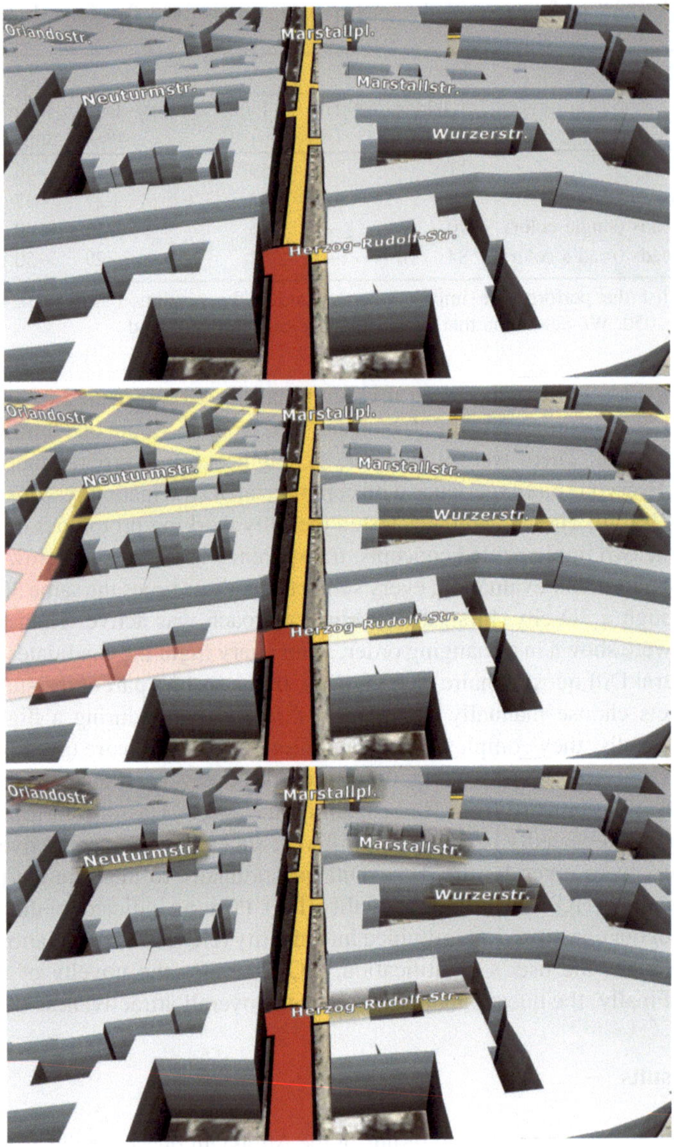

Fig. 11 Comparison of the implemented approaches in bird's eye with World-Space labeling: baseline (*top*), glowing roads (*middle*) and transparency label auras (*bottom*). As concluded from a conducted user study, the last two methods increase attractiveness and usability compared to the baseline

Informal Questionnaire. Figure 14 depicts the results of our second questionnaire. The majority state that the application of both approaches create an advantage compared to our baseline, create a better orientation and are aesthetically pleasing. The glowing roads display a higher distraction and are less calm

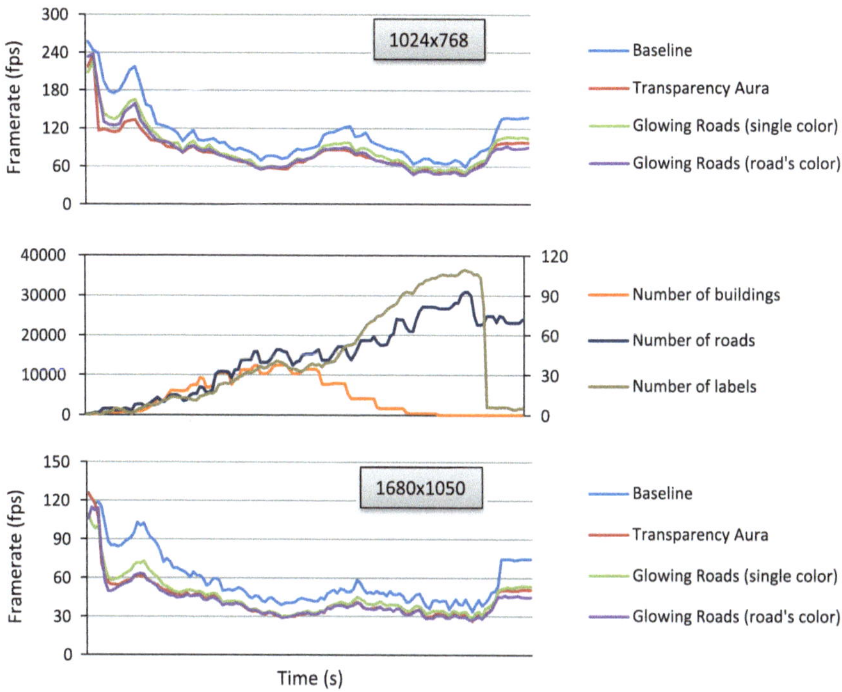

Fig. 12 Benchmark results of the GPU implementation: both approaches are fillrate-bound

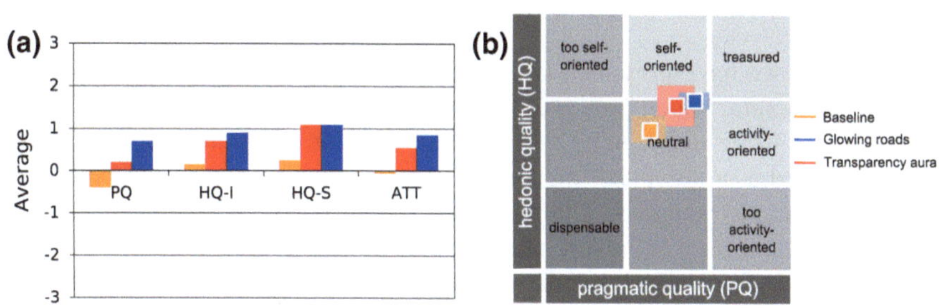

Fig. 13 Resulting AttrakDiff questionnaire from our conducted user study. PQ describes the perceived pragmatic quality (≈ usability), HQ-I the hedonic quality based on identity (≈ user's identification), HQ-S the hedonic quality provided through stimulation (≈ innovative) and ATT describes the concepts overall attractiveness. Compared to our baseline, both our presented approaches improve significantly the HQ, PQ and attractiveness. **a** Averaged values of the perceived qualities of the presented concepts. **b** Quality classification of the concepts and variability of the given answers

than the transparency label aura. Our subjects would more likely use these approaches in a GIS than in a car. Overall, the proposed methods are perceived as a significant improvement compared to the baseline: 86 % see transparency label aura and 77 % glowing roads as enhancement (Fig. 15).

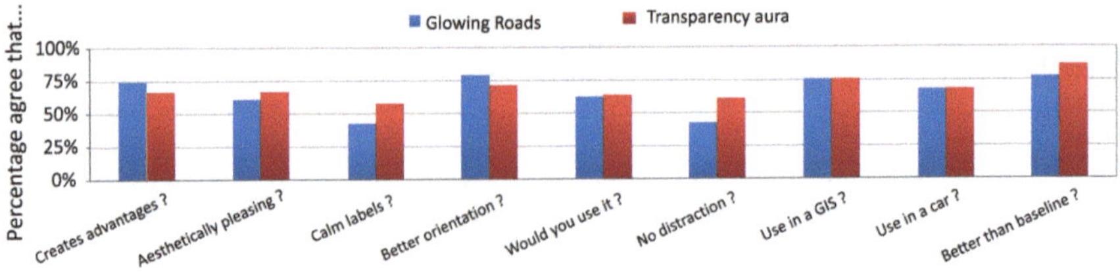

Fig. 14 Informal questionnaire answered by our 24 test candidates

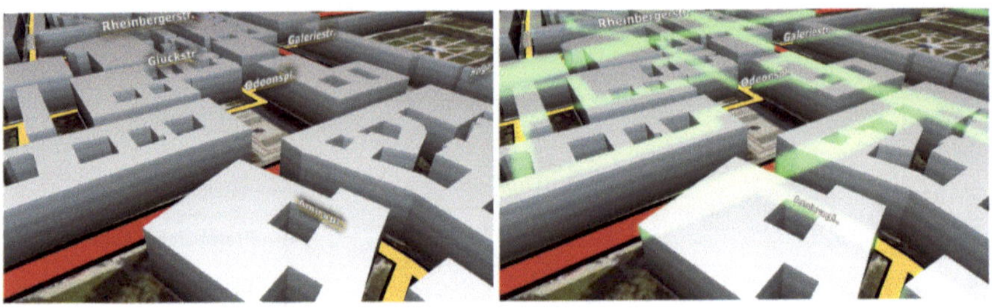

Fig. 15 Comparison of transparency label aura (*left*) and single colored glowing roads (*right*). Both figures are in bird's eye viewing space with WS labeling

6 Conclusion

In this paper we have presented two new approaches, glowing roads and transparency label aura, which preserve the readability of occluded labels in 3D navigation maps while maintaining the reference to their corresponding object. We have described a prototypical implementation of both methods on the GPU running at interactive framerates. Our profiling has shown that these implementations are fillrate-bound. In a following user study including 24 subjects we compared them to our baseline: simply rendering all labels over the world, as done e.g. by Google Earth and almost every commercial navigation system. We have revealed that both our methods innovate and improve significantly the usability and overall attractiveness. Over 86 % deem the approach glowing road better than our baseline. In further research, we plan to evaluate these approaches in real-world scenario, e.g. while driving through a city. Furthermore, a combination of both concepts could create new approaches, e.g. transparent road auras.

Acknowledgments We would like to thank Marco Matt for preparing and conducting the expert and user studies at our research labs. Also, we are grateful for the continuous feedback from Christopher Roelle while designing the labeling techniques. Finally, we thank Philipp Promesberger for developing the basis of the map rendering framework.

References

Ali K, Hartmann K, Strothotte T (2005) Label layout for interactive 3d illustrations. J WSCG 13(1):1–8

Bier EA, Stone MC, Pier K, Buxton W, Derose TD (1993) Toolglass and magic lenses: the see-through interface. In: Proceedings of the 20th annual conference on computer graphics and interactive technique, ACM Press, pp 73–80

Bell B, Feiner S, Höllerer T (2001) View management for virtual and augmented reality. In: Proceedings of the 14th annual ACM symposium on User interface software and technology, ACM, pp 101–110

Chen J, Pyla P, Bowman D (2004) Testbed evaluation of navigation and text display techniques in an information-rich virtual environment. In: IEEE proceedings on virtual reality, pp 181–289

Cipriano G, Gleicher M (2008) Text scaffolds for effective surface labeling. IEEE Trans Visual Comput Graphics 14(6):1675–1682

Coffin C, Höllerer T (2006) Interactive perspective cut-away views for general 3d scenes. In: Proceedings of IEEE symposium 3D user interfaces (3DUI 06), IEEE CS, Press, pp 25–28

Elmqvist N, Assarsson U, Tsigas P (2007) Employing dynamic transparency for 3D occlusion management: design issues and evaluation. In: Baranauskas C, Palanque P, Abascal J, Barbosa SDJ (eds) Proceedings of INTERACT, series LNCS, vol 4662. Springer, pp 532–545

Fekete J, Plaisant C (1999) Excentric labeling: dynamic neighborhood labeling for data visualization. In: Proceedings of the SIGCHI conference on human factors in computing systems: the CHI is the limit, ACM, pp 512–519

Goldstein E (2009) Sensation and perception. Wadsworth Pub Co

Google Inc (2012) Google Earth v6.1.0.5001. (Software). http://earth.google.com/

Götzelmann T, Ali K, Hartmann K, Strothotte T (2005) Adaptive labeling for illustrations. In: Proceedings of Pacific Graphics 2005:64–66

Götzelmann T, Hartmann K, Strothotte T (2006) Agent-based annotation of interactive 3D visualizations.In: Smart Graphics, Springer, pp 24–35

Hartmann K, Ali K, Strothotte T (2004) Floating labels: applying dynamic potential fields for label layout. In: Smart Graphics, Springer, pp 101–113

Hassenzahl M (2007) Attrakdiff (tm). URL http://wwwattrakdiffde/en/AttrakDiff/

Hassenzahl M, Burmester M, Koller F (2003) Attrakdiff: Ein fragebogen zur messung wahrgenommener hedonischer und pragmatischer qualität. Mensch and Computer, pp 187–196

Imhof E (1975) Positioning names on maps. Cartography Geogr Inform Sci 2(2):128–144

Kalkofen D, Mendez E, Schmalstieg D (2007) Interactive focus and context visualization for augmented reality. In: Proceedings of the 2007 6th IEEE and ACM international symposium on mixed and augmented reality, IEEE Computer Society, pp 1–10

Kalkofen D, Mendez E, Schmalstieg D (2009) Comprehensible visualization for augmented reality. IEEE Trans Vis Comput Graph 15(2):193–204

Krüger J, Schneider J, Westermann R (2006) Clearview: an interactive context preserving hotspot visualization technique. IEEE Trans on Vis Comput Graph 12(5):941–948

Maass S, Döllner J (2006a) Dynamic annotation of interactive environments using object-integrated billboards. In: 14th international conference in central Europe on computer graphics, visualization and computer vision, WSCG, pp 327–334

Maass S, Döllner J (2006b) Efficient view management for dynamic annotation placement in virtual landscapes. In: Smart Graphics, Springer, pp 1–12

Maass S, Döllner J (2007) Embedded labels for line features in interactive 3d virtual environments. In: Proceedings of the 5th international conference on Computer graphics, virtual reality, visualisation and interaction in Africa, ACM, pp 53–59

Maass S, Döllner J (2008) Seamless integration of labels into interactive virtual 3d environments using parameterized hulls. In: 4th International symposium on computational Aesthetics in graphics, Lisbon, pp 33–40

Maass S, Jobst M, Döllner J (2007) Depth cue of occlusion information as criterionfor the quality of annotation placement in perspective views. The European information society pp 473–486

Polys N, Kim S, Bowman D (2005) Effects of information layout, screen size, andfield of view on user performance in information-rich virtual environments. In:Proceedings of the ACM symposium on virtual reality software and technology, November, ACM, pp 07–09

Ropinski T, Praßni J, Roters J, Hinrichs K (2007) Internal labels as shape cues for medical illustration. In: Proceedings of the 12th international fall workshop on vision, modeling, and visualization (VMV07), pp 203–212

Stein T, Décoret X (2008) Dynamic label placement for improved interactive exploration. In: Proceedings of the 6th international symposium on non-photorealistic animation and rendering, ACM, pp 15–21

Sygic (2012) Sygic GPS navigation. (Software). http://www.sygic.com/

Thielmann T (2006) you have reached your destination! position, positioning and superpositioning of space through car navigation systems. Navigation 2:27–62

Viega J, Conway MJ, Williams G, Pausch R (1996) 3d magic lenses. ACM Symposium on user interface software and technology, pp 51–58

Semantic 3D Modeling of Multi-Utility Networks in Cities for Analysis and 3D Visualization

Thomas Becker, Claus Nagel and Thomas H. Kolbe

Abstract Precise and comprehensive knowledge about 3D urban space, critical infrastructures, and belowground features is required for simulation and analysis in the fields of urban and environmental planning, city administration, and disaster management. In order to facilitate these applications, geoinformation about functional, semantic, and topographic aspects of urban features, their mutual dependencies and their interrelations are needed. Substantial work has been done in the modeling and representation of aboveground features in the context of 3D city and building models. However, standardized models such as CityGML and IFC lack a rich information model for multiple and different underground structures. In contrast, existing utility network models are commonly tailored to a specific type of commodity, dedicated to serve as as-built documentation and thus are not suitable for the integrated representation of multiple and different utility infrastructures. Moreover, the mutual relations between networks as well as embedding into 3D urban space are not supported. The Utility Network ADE of CityGML as proposed in 2011 provides the required concepts and classes for the integration of multi-utility networks into the 3D urban environment. While the core model covers only the topological and topographic representation of network entities, the functional and semantic classification of network objects is now introduced in this paper. This paper will show how concepts and classes can be defined to fulfill the requirements of complex analyses and simulation, and how

T. Becker (✉) · C. Nagel · T. H. Kolbe
Institute for Geodesy and Geoinformation Science, Technische Universität Berlin,
Straße des 17. Juni 135, 10623 Berlin, Germany
e-mail: thomas.becker@tu-berlin.de

C. Nagel
e-mail: claus.nagel@tu-berlin.de

T. H. Kolbe
e-mail: thomas.kolbe@tu-berlin.de

J. Pouliot et al. (eds), *Progress and New Trends in 3D Geoinformation Sciences*,
Lecture Notes in Geoinformation and Cartography, DOI: 10.1007/978-3-642-29793-9_3,

properties of specific networks can be defined with respect to 3D topography but also network connectivity and functional aspects.

Keywords 3D Data models · 3D City models · 3D Utility networks · Multi-utility networks · Critical infrastructures · Disaster management · Emergency response · CityGML

1 Introduction and Motivation

The range of applications of city models reaches far beyond pure visualization today. Applications such as energy consumption analysis, carbon balancing, risk and disaster management as well as future applications like city life cycle management require an extensive "inventory list" of urban space. Nowadays city models are used to give the administration, disaster managers, and companies access to the city's inventory. The inventory comprises buildings, streets, vegetation objects, plants, classified land uses, and elevation models. Typically, real world objects above ground can be modeled using existing standards for 3D city modeling such as CityGML or the forthcoming INSPIRE data specifications in the future (Gröger et al. 2008). Assuming that a city can be understood as a system in terms of an organized structure regarded as a whole and consisting of interrelated and interdependent elements (components, entities, factors, members, parts, etc.), thus, a city model can be seen as an abstract representation of such a real existing system (ISO19109 2005). CityGML and IFC represent such an abstraction of a system. Whereas CityGML can represent many elements of the city system (respectively a part of a system), IFC represents a system within a system. Nevertheless, the overall concept of having objects representing physical (building) and conceptual entities (city), giving them contextual information by setting them into relationships and assigning characteristic properties is valid for both models. The IFC model includes object connectivity, processes, etc. that is still an ongoing task within the city wide model—CityGML. The Utility Network ADE of CityGML represents a first approach to extend the abstract model of a city by integrating utility infrastructures into the urban space and to make their network topology and topography explicit.

The core model (Becker et al. 2011a, b) of this application domain extension establishes the relation, or—to be more precise—the connection between aboveground and belowground urban inventory with respect to utilities. The core of this ADE defines the modeling environment by making relevant features and their mutual relations explicit and allowing the 3D topographical modeling of entire networks, sub-networks and network features as well as their graph representations. The consequent treatment of network features as abstraction of real world objects (topographic point of view) as well as a graph object, represented by its own network graph, makes the model more flexible as the models realized in existing GIS utility systems. The module *Networkcomponent* of the Utility

Network ADE will extend the core concept by classes that will describe the entities of any utility network in a semantical-functional way.

Already existing utility network models represent utilities in a semantically rich way, but their components do not interact with or have explicit relations with urban features. Those networks are very detailed and rich of semantics; some of them dedicated for daily use in utility companies, some used as data exchange models and others just represent utility networks within parts of urban space (buildings). Each of them is an abstraction and reduction of a system (model) in itself but they do not provide links to the higher context and thus are not feasible for analysis or simulation purposes in terms of urban energy consumption analysis, carbon balancing, risk- and disaster management, and city life cycle management. A model feasible for those purposes has to meet the following requirements and should represent an eligible generalization and subset of reality:

- The elements of such a model must have functional as well as structural relationships between each other.
- The model must represent independent but interrelated elements in order to enable simulation and complex analysis.
- The model must be valid for different, heterogeneous types of utility networks.
- The model must reduce the complexity on the one hand but preserve the required information for usage in simulations, analysis, calculations, and cartographic visualization in disaster case.

Some popular data models for representing, exchanging, and storing utility networks are introduced and discussed briefly in Sect. 2. The ArcGIS utility models stands proxy for popular GIS-based utility solutions, the IFC model as proxy for building wide supply system, and the INSPIRE network model as proxy for a city or country wide supply system. In Sect. 3, a short overview about the core model of the Utility Network ADE is given, laying out the basis for the introduction of the specific extensions of the Utility Network ADE—the *NetworkComponents* (Sect. 4) and the *NetworkProperties* (Sect. 5). Section 6 sketches the implementation of the model in ArcGIS. Finally, we draw conclusions and point to future work (Sect. 7).

2 Analysis of Existing Geospatial Utility Network Models

2.1 INSPIRE Network Model

The "Network" package of the INSPIRE data specifications (INSPIRE 2010a) defines the basic application schema for networks which is extended by additional, domain specific spatial data schemas (INSPIRE 2010b, c). The central class is *NetworkElement*, which may be any entity that is relevant for a network. The network package consists of further classes that are required for modeling networks, such as *Network, Link*, and *Node*.

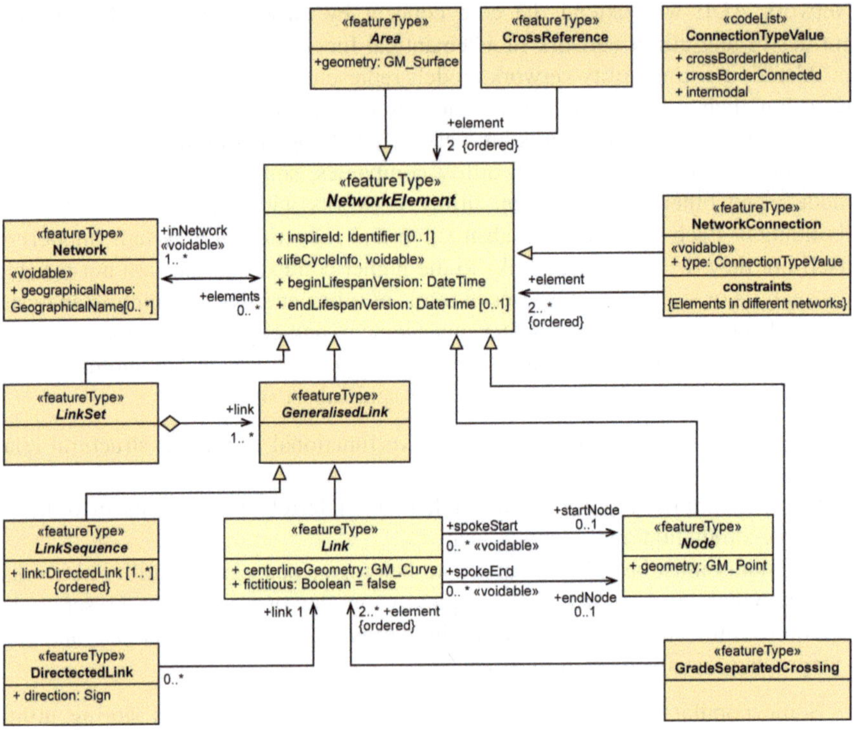

Fig. 1 Network application schema of INSPIRE (adapted from (INSPIRE 2010a))

A *Network* is a collection of *NetworkElements* that is the superclass for elements like *Area*, *Node*, and some special classes such as *GeneralisedLink*, *LinkSet* and *GradeSeparatedCrossing* (see Fig. 1).

Thus, a simple network may only consist of *Nodes* and *Links*, where a *Link* must be bounded by exactly two nodes. The clear distinction of *NetworkElements* into point-like (*Node*) and line-like (*Link*) objects and the lack of a feature aggregation schema does not allow for hierarchical decompositions of network components within the core model. For example, a point-like object cannot consist of other point-like or line-like objects. The hierarchical modeling of a line-like object is supported by the class *LinkSet*. A hierarchical modeling of a network and the modeling of interdependencies is possible by using the class *NetworkConnection*. *NetworkConnections* are also *NetworkElements* relating two or more arbitrary *NetworkElements* facilitating the modeling of hierarchical networks (see (INSPIRE 2010a, p. 93).

The INSPIRE application schema "Utility Networks" including the sub schemas for Electricity, Oil and Gas, Sewer, Telecommunications, and Water Networks extends the Generic Network Model (GNM, see Fig. 2) besides transportation networks now also by utility networks (INSPIRE 2010a, b, c, 2011). The utility networks application schema extends the classes provided by the GNM by utility

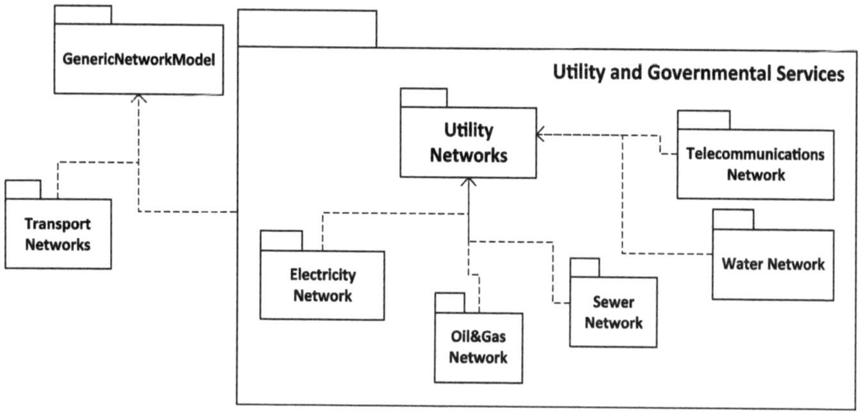

Fig. 2 Principle structure of modeling networks using the INSPIRE specification. The dependencies are established by creating subclasses from referred packages

specific abstract classes such as *UtilityLinkSet*, *UtilityLinkSequence*, *UtilityLink*, *UtilityNode*, *UtilityNodeContainer* and *UtilityNetworkElement*. For further information, reference is made at this point to INSPIRE (2011) consolidated UML Model.

In addition to this, the model provides common features (called *CommonTypes*) to the subsequent application schemas for electricity, water, and so on. Those *CommonTypes* are types such as pipe, duct, manhole, pole, and diverse enumerations that describe material, exterior shape, and type of features that occur in other utility networks as well. They serve as container features for entities of other utility networks. The further semantic specialization of needed utility entities is then done within the respective application domain schema. Since the core model only provides a 2D representation of network elements, the theme "Utility and Governmental Services" allows for the modeling of an *ElevationLine* or *ElevationPoint* to make the relative height of a network component with respect to the terrain explicit. However, the model lacks of an explicit 3D topographic representation of network objects useful for collision detection, simulation of impacts of blast, and 3D visualization.

Basically, only a specialization of the distribution entities such as pipes and cables, devices (called appurtenance) and of the respective network is done. There is no further domain specific classification into other pipes, devices, or other domain specific entities as shown in Fig. 3. Type attributes being available for both distribution elements and appurtenance entities take over the further specialization of those main elements of networks. However, the functionality of those network entities cannot be derived directly or is not obvious. The named use cases for the representation of utility networks in INSPIRE are mapping and documentation, e.g. to facilitate information portals which make information about cables and lines available for contractors that are planning excavation works. Thus, the focus of the model is on describing the topography of distribution elements and appurtenances and not on modeling their functionality and interdependencies.

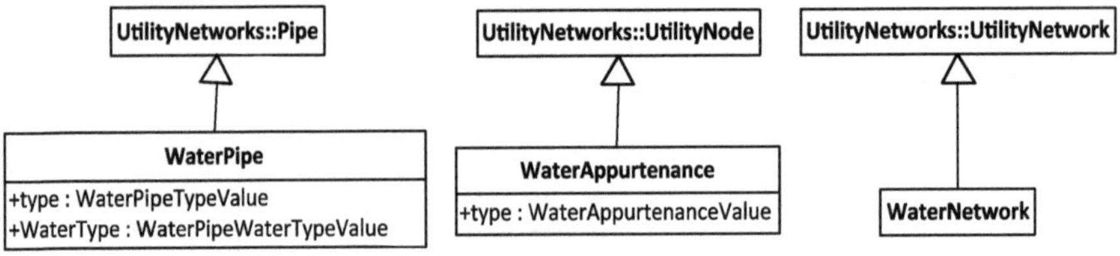

Fig. 3 Specialization within the domain specific extensions of the INSPIRE UtilityNetwork model (excerpt)

In summary, the structure of the INSPIRE modeling approach for utility networks can be partitioned into three parts. The first part is the Generic Network Model dedicated to the modeling of topological relationships between any network entities. The second part defines utility network specific entities such as *UtilityLinks* and *UtilityNodes* and provides more entity specific attributes as well as common types for use in domain specific application schemas. Those schemas form the third part of utility network modeling and specify the domain specific entities and attributes for the respective domain (cf. Fig. 3).

2.2 IFC Utility Model

The most important standard for data exchange of buildings in the field of architecture and civil engineering are the Industry Foundation Classes (Liebich 2009). The IFC represent logical building structures, accompanying properties (attributes) with 2D and 3D geometry as well as utilities. As Liebich et al. (2007), Liebich (2009), Becker et al. (2011a, b), and Hijazi et al. (2011) point out, IFC offers two different ways of connectivity in order to build up a network that may be a physical or logical connection between building service elements. In general, a logical connection realizes the linkage of two components via so-called Ports, whereas a physical connection is established by a realizing element such as *IfcFlowFitting*. The connectivity concept of IFC comprises both the physical connection between elements (*IfcRelConnectsElements*) and the logical connection of building service items on the level of their ports (*IfcRelConnectsPorts*).

Besides the connectivity concept that realizes the topological relations between elements the IFC even provide a model for the topographic and semantic representation of building service elements. The superclass of all building service elements respective of all included distribution systems is represented by the class *IfcDistributionElement*. Those elements are further specialized (see Fig. 4) into flow elements (*IfcDistributionFlowElement*) and controller objects (*IfcDistributionControlElement*) which in turn have further subtypes but do not elaborate further attributes. Those subtypes serve solely for semantically and logically structuring of the model.

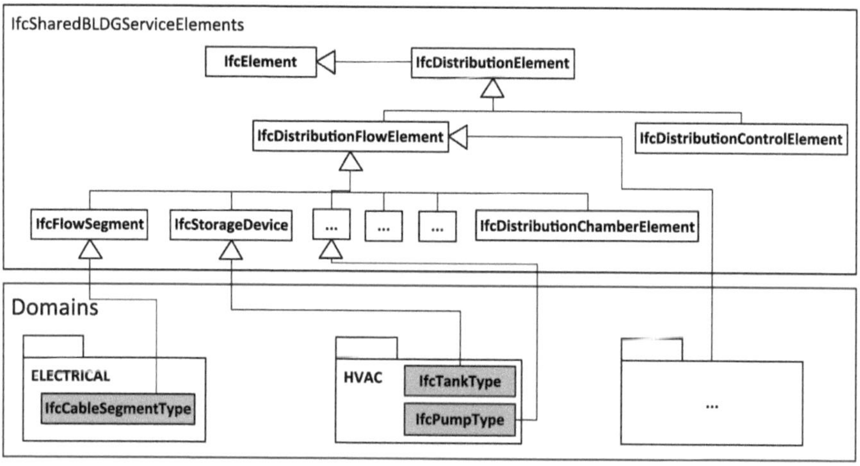

Fig. 4 Principle structure of IFC building service elements (excerpt)

IFC distinguishes flow objects into *IfcFlowFitting*, *IfcFlowSegment*, *Ifc-Flow-Controller*, *IfcFlowTerminal*, *IfcFlowMovingDevice*, *IfcEnergyConversionDevice*, *IfcFlowStorageDevice*, *IfcFlowTreatmentDevice*, and *IfcDistributionChamber-Element*. The *IfcDistributionControlElement* comprises all elements which are necessary to define elements of a building automation control system that are used to impart control over elements of a distribution system (Liebich et al. 2007). Thus, it is possible to control valves, dampers, etc. through explicit actuation (*Ifc-Actuator*). IFC allows both 2D and 3D geometries to represent the real-world shape and extent of network entities. The geometry is given in a local engineering reference frame which is valid for a single building but which lacks the possibility to evaluate the building utilities in an urban or regional context. To summarize this (see Fig. 4) the layer *SharedBLDGServiceElements* forms an intermediate layer that classifies the objects and elements of a buildings service system according to their functionality within the building system. Thus, every building system may consist of objects that move (*IfcFlowMovingDevice*), store (*IfcFlowStorageDe-vice*), distribute (*IfcFlowSegment*), etc. the carried medium. A more precise classification respectively definition of building service system elements is being done within the domain layers *IfcHvacDomain*, *IfcPlumbingFireProtectionDo-main*, *Ifc-ElectricalDomain* and *IfcBuildingControlsDomain*. The IFC utility model intentionally is tailored to the modeling of utility structures within buildings. The integration into citywide utility networks on a larger scale is not supported. Visualizations and analyses are hence restricted to the building scale.

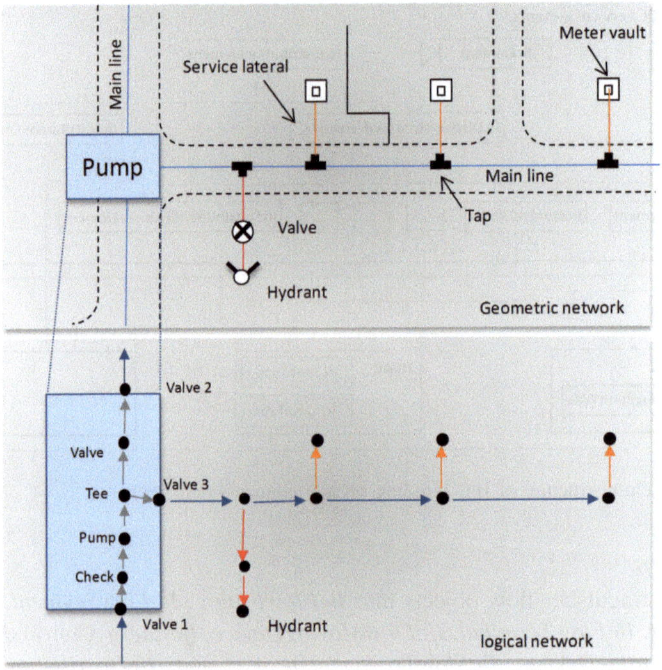

Fig. 5 Example of a feature dataset containing a geometric and logical network

2.3 ArcGIS Network Model

In ArcGIS different types of utility networks are defined based on a core concept called Geometric Networks. This core technology represents the basic structure for all kind of utility networks. A network is constructed from edges and junctions as 2D line and 2D point features. Each real world utility object can be represented as one feature in the network whereas same kinds of features can be represented by a feature class (ESRI 2003, 2007; Grise et al. 2001; Meehan 2007). The geometric part of a utility network (see Fig. 5) is a single graph structure consisting of edge and junction elements with an embedding into 2D space. The graph is composed of features from one or more feature classes in a feature data set (ESRI 2003). It binds together feature classes that form a network and contains all attributes, relationships, and validation rules. The logical network (see Fig. 5) is a special data structure to store the connectivity between features of the network and is implemented by a set of tables.

An ArcGIS utility network may consist of four network feature types: simple edge, simple junction, complex edge, and complex junction and thus provides a more flexible way to create network topology as pure edge-node topology. Whereas a simple edge has a one-to-one relation between the feature and the edge element and connects always two junctions, a complex edge may have more than two connected junctions on their length and may represent a sequence of edge elements divided by junctions. Therefore, those complex edges realize a logically connected sequence of edges, but geometrically they represent a single feature and in this respect, it is a kind of feature hierarchy. Similarly, to the edge concept, a simple junction is a one-to-one relation between a feature and its corresponding junction element and is represented by one point in the network. A complex junction, however, is represented by one object within the geometric network and multiple objects including junctions and edges within the logical network. This is illustrated in Fig. 5 where the pump from the upper part of the picture is represented as a graph like structure in the lower part.

Based on ESRI's geometric network the data models for electrical power distribution, gas distribution, and water distribution were developed. All data models are especially designed for the daily work in utility companies and thus they represent features and relations for as-built documentation in distribution systems. These models include essential sets of object classes for water, gas, or electricity supply networks and properties as well as rules and relationships that define object behavior and provide "…an implementation that focuses on operations and maintenance portions of the facility life cycle" (ESRI 2007, p. 3 Sect. 1). These models extend or address the ArcGIS Geometric Network Model and distinguish between objects that are building the network topology and those which are used for documentation or controlling purposes. Network-forming elements are specializations of *SimpleJunctionFeature/ComplexJunctionFeature* or *Simple-EdgeFeature/ComplexEdgeFeature*. Elements which do not participate in network-forming process are handled as "simple" *Point*, *Polyline*, or *Polygon* features. Since they do not participate as network features, they are not part of the network topology and thus cannot be traced or analyzed by any kind of ArcGIS network tools.

All investigated ESRI utility data models can be structured into 2–3 stages. In a first step a superclass for entities with similar semantics is built and is then associated according to its semantics and geometry as a subclass of either *SimpleJunctionFeature*, *ComplexJunctionFeature*, *SimpleEdgeFeature*, *ComplexEdgeFeature* (see Fig. 6, *ElectricComplexEdge*). All of these superclasses are container classes, which inherit relevant attributes to subclasses which are further specializations of these superclasses (*ElectricLineSegment*, *BusBar*). The last step is a further specialization from the second stage classes in order to further distinguish the entities of the upper class. In case of electricity, the line segments are further specialized into Overhead (*OH*) and Underground structures (*UG*) as well as Primary (*Pri*) and Secondary (*Sec*) lines.

The ArcGIS utility model serves for documentation and planning support in utility companies and city administrations. The models are semantically rich and complex, but do only represent the 2D topography of the network besides the logical network connectivity information. A 3D geometry representation could be associated by using multipath features but these would be uncoupled from the network modeling; just pure 3D visualization. Each of the domain data models is representing a commodity-specific 2D GIS-based abstraction of the respective

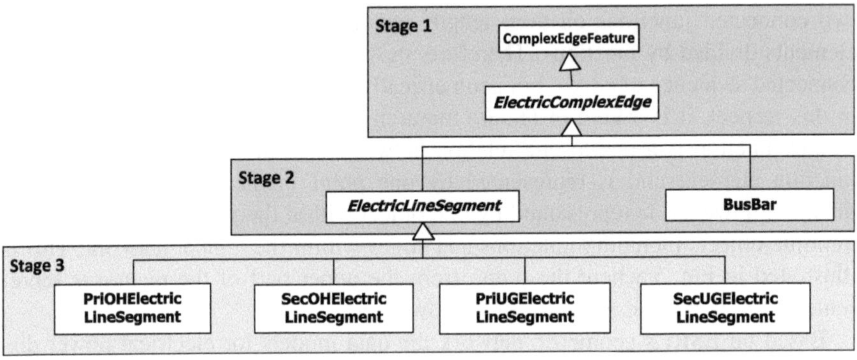

Fig. 6 Example of stepwise refinement of the ArcGIS utility data model shown on ESRI'S electricity data model (excerpt)

utility network in the real-world. It cannot easily be used for a different type of utility network, and thus no common model/database integrating different utility models is being provided. The only ArcGIS component that is shared by all of these data models is the Geometric Network Model that forms the common denominator of all ArcGIS utility networks. Nevertheless, the GNM allows for network tracing and other types of network analysis.

3 Short Introduction to the Utility Network Core Model for CityGML: Utility Network ADE

The *UtilityNetworkCore* as proposed by Becker et al. (2011a, b) is a CityGML application domain extension (ADE) offering the possibility to integrate network structures into the urban environment. In general, it provides classes and concepts to model multiple different infrastructure networks, to embedded the 3D multi-utility networks into the 3D virtual urban environment and to relate them to each other. The base class of the *NetworkCore* model is the abstract class *_Network-Feature* (see Fig. 7). It is a subclass of the CityGML class *_CityObject* and establishes the link between aboveground city objects and network structures located above and below ground. *_NetworkFeature* is the conceptual head for the further semantic and thematic classification and description of network entities. Collections of *_NetworkFeature* instances of one transported medium/commodity can be grouped to *Networks*, which themselves might be structured into *subNet-works*, expressed by a self-association of *Network*. Thus, network hierarchies as they exist in power or gas networks can easily be represented. A similar concept is used to represent component hierarchies. Each *_NetworkFeature* might contain other *_NetworkFeatures*, expressed by a self-association named *consistOf* and, thus, enabling on the one hand a very flexible and on the other hand a very detailed

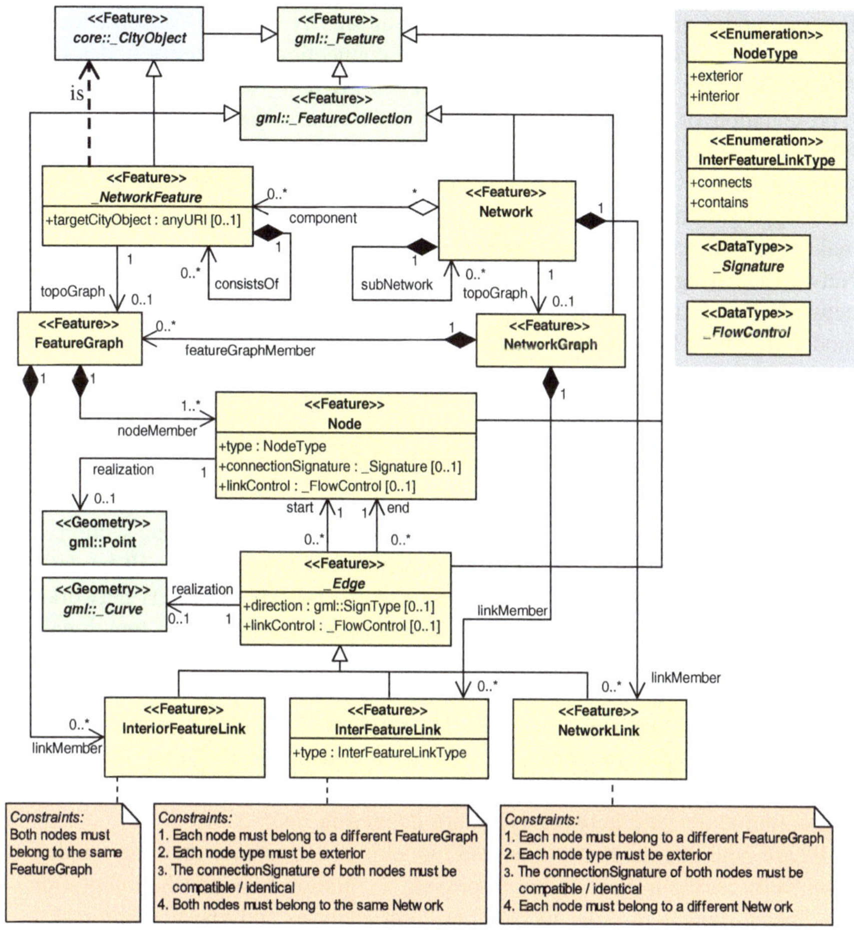

Fig. 7 UML class diagram of the UtilityNeworkADE core model

way to model the hierarchies and affiliation of features, between features, and inside of network features.

Network and *_NetworkFeature* are used for the topographic representation of utility networks. The representation of network topology and connectivity is achieved within the network core model by providing a dual concept for the representation of features where each network component can be represented both by its topography and by means of a complementary graph structure.

The *FeatureGraph* is representing a separate graph structure for each utility element reflecting the functional, structural as well as the topological aspects of each element. Following the general principles of graph theory the *FeatureGraph* may consist of *Nodes* and *InteriorFeatureEdges* (cf. Diestel 2010). Thereby a differentiation into interior and exterior nodes is done. Interior nodes represent structural, functional, logical, or physical internal aspects within a network feature.

Exterior nodes are used to establish the connectivity to other *NetworkFeatures*. Different *NetworkFeatures* can be linked by connecting the exterior nodes via the *InterFeatureLink* forming a complete *NetworkGraph*, which itself is the dual representation of the collection of *NetworkFeatures*—the *Network*. In order to make mutual relations between networks or network elements explicit, the edge subclass *NetworkLink* can be used. Hence, besides the feature aggregation, network aggregation, internal, and external connectivity of features it is possible to make dependencies between networks of different commodity types explicit. Networks sharing the same urban space can therefore be modeled as inter modal networks by having explicitly modeled relations. More details on the *NetworkCore* model are given in Becker et al. (2011a, b).

4 Integration of Network Components

Based on the abstract *NetworkCore* model described in the previous section, now the concrete representation of the entities within utility networks will be defined. Since we are interested in the representation of diverse types of utility networks for different commodities, we first identify the common elements and functionalities over the different types of utility networks. These entities are then further specialized to represent the distinct properties and characteristics of the components of the different utility network types. The representation of the commodities and the general network properties are modeled independently from the network components, because each utility network can transport different commodities. The commodities and network types are defined in Sect. 5.

Please note that the aim of the data model is not to replace the other models or systems discussed in Sect. 2, but to provide a common basis for the integration of the diverse models in order to facilitate joint analyses and visualization tasks.

The represented degree of detail, i.e. object classifications and their attributes, was determined mostly by the use case of the simulation of the propagation of failures of critical infrastructures across different utility networks in the context of disaster management. However, first investigations show that the model is also suitable for supporting strategic energy planning. Although utility networks differ substantially with regard to transported goods/commodities, they have the following elements in common (see Fig. 8): *DistributionElements* are used to transport the commodity from producer to consumer or to connect different network users. *FunctionalElements* are elements which play a role in the operation or maintenance of the network, but which themselves are not elements of the network. For example, manholes are required to access components of a utility network but are not elements of the related network graph. *Devices* represent network elements playing an active role in the operation of networks like controlling, measuring, storing, transforming, or amplifying. *Terminal-Elements* mark end points of the network where the goods/commodities “leave” or “enter” the

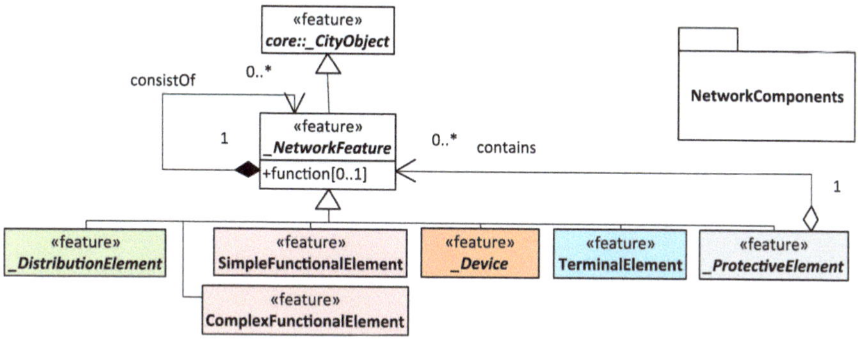

Fig. 8 Network components are modeled as specializations of *NetworkFeature* into 6 main subclasses. For further details see (NetworkComponents Model 2012)

modeled network, e.g. a hydrant or house service connection. *ProtectiveElements* represent shielding cases or beddings of network elements.

The concrete realization of *DistributionElements* is highly dependent on the transported material (gas, liquid, electricity, light, solid medium) and thus is done by cables, pipes, or canals. Pipes and canals can be further specialized according to the shape of the cross section or construction type respectively. Cables do not have to be specialized any further; they are just characterized by diameter/cross section, material type, and transmission type. Therefore, the modeling of utility network entities must take into consideration the functional level of network features as well as the transported commodity type of those features.

Existing infrastructure assets dedicated for the distribution of commodity can be differentiated into *RoundPipe*, used to transport liquids like water, wastewater, domestic hot water, and *RectangularPipes* used to transport medium such as air for cooling systems, exhaust systems, and so on (see the two pictures on the left in Fig. 9). According to this distinction useful attributes can easily be defined which further specify the shape or general state of those pipes, such as *interiorDiameter* and *exteriorDiameter* for *RoundPipes* and *exteriorHeight*, *exteriorWidth*, *interiorHeight*, and *interiorWidth* for *RectangularPipes* (cf. Fig. 10).

Canals are typically used to transport storm and wastewater and can be built as walkable, open top, closed, or as multi-utility system depending on mounting type, model, and size (see the two pictures on the right in Fig. 9). Therefore, canals are specialized into *ClosedCanal* and *Semi-OpenCanal* (cf. Fig. 10). Using additional attributes such as *height*, *width*, and cross section shape further specifies the interior profile. Semi open structures (*SemiOpenCanal*) can be used to represent Storm water systems that exist in urban space as U-shaped features, often along streets or walkways.

Cables are used for energy supply or signal transmission which determines the composition and material type of the cable, but does not affect the exterior appearance (shape) of those. This eliminates the need for further specialization of

Fig. 9 Some examples of DistributionElements (pipes, closed and semi-open canals)

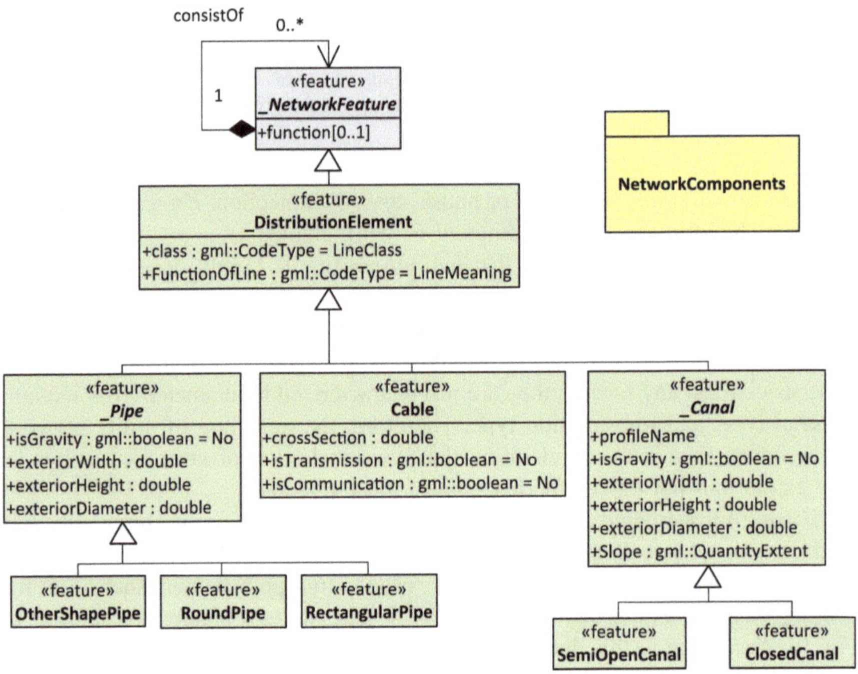

Fig. 10 Modeling approach for *DistributionElements*. Further details are given in (Network-Components Model 2012)

the class *Cable*. The proposed data model for *DistributionElements* and their specific subclasses as explained above are depicted in Fig. 10.

Besides the *DistributionElements* a utility network consists of elements of feeding, elements of abstraction, and elements of control in order to build a working infrastructure. Sewage treatment plants, water treatment plants, and relay stations are feeding elements. Pressure increase stations, pressure decrease stations, and transformer stations are likewise essential components of those networks that do not feed into the network but alter the transported commodity by suitable devices. While such devices obviously are components of the logical network, the entire treatment plant or transformer station is to be seen as an entity which contains these devices. Since entire stations also need to be represented explicitly by a specific entity, the class *FunctionalElement* is introduced (see Fig. 8). *FunctionalElements* are further differentiated in order to fulfill the requirement to represent the main elements of utility networks. *ComplexFunctional-Elements* aggregate *_Network-Features* which build a functional unit such as a water treatment plant. Thus, they include further network entities such as pumps, valves, switches, and generators. *SimpleFunctionalElements* must not contain other network entities since they represent objects useful for maintenance and inspection of the transported commodity. Manholes or inspection chambers are examples for *SimpleFunctionalElements*.

The next major subclass of *_NetworkFeatures* is *Device*. The amount of devices within gas, power, water, and wastewater networks is huge and the definition of super classes (generalization classes) is difficult due to the fact that the specialization within existing utility networks, GIS based as-built documentations is very fine granular. However, the classification of those devices according to their main functionalities provides a feasible approach for a distinction into subsets. Each utility network contains—according to their functionality—*StorageDevices*, *ControllerDevices*, *MeasurementDevices*, *TechDevices*, and *AnyDevices* (the latter representing features with unspecific functionality such as a blind flange).

A *StorageDevice* can be a battery, reservoir, underground storage, or any other device that is used to buffer or put aside the commodity for future use. *ControllerDevices* are devices such as valves, switches, gate valves, etc. that are used to control, limit or influence the flow of commodity. *MeasurementDevices* serve as entities for the quantification of commodity flow, commodity quality, or distribution, e.g. pressure sensor, meter, volumetric flow rate sensor, etc. *TechDevices* might have the same functionality as controller devices and measurement devices but have a clear dependency on power supply, such as cathodic corrosion protection, electrical driven valves, gauges, and slider. Thus, an additional classification of relevant network features is made and is depicted in Fig. 11.

Finally, two more feature classes have to be taken into account. Entities like water-tap, street light, gas lamp, hydrant, anything being or situated at an end of a line can be seen as a sink and, thus, be represented by using the class *TerminalElement* (see Fig. 12). *TerminalElements* represent interfaces between the utility network model and the environment, where goods/commodities "enter" or "leave" the network.

The final subclass of *_NetworkFeature* is *ProtectiveElement* (see Fig. 12). All types of elements intended or used to provide protection of some kind, such as duct work, cable lines, cable protection packages, or cladding tube are covered by this class. *ProtectiveElements* are further differentiated into *_ProtectionShell* and *Bedding*, the latter representing cable lines or tracks that build the near surrounding of a utility line. According to the profile shape *_ProtectionShell* is further distinguished into *RoundShell*, *RectangularShell*, and *OtherShell*. Since the *ProtectiveElement* requires objects to be protected, it can contain any other *_NetworkFeature* including more *_ProtectiveElements*. As already mentioned in Sect. 3 the network core model supports a very flexible modeling of network features,

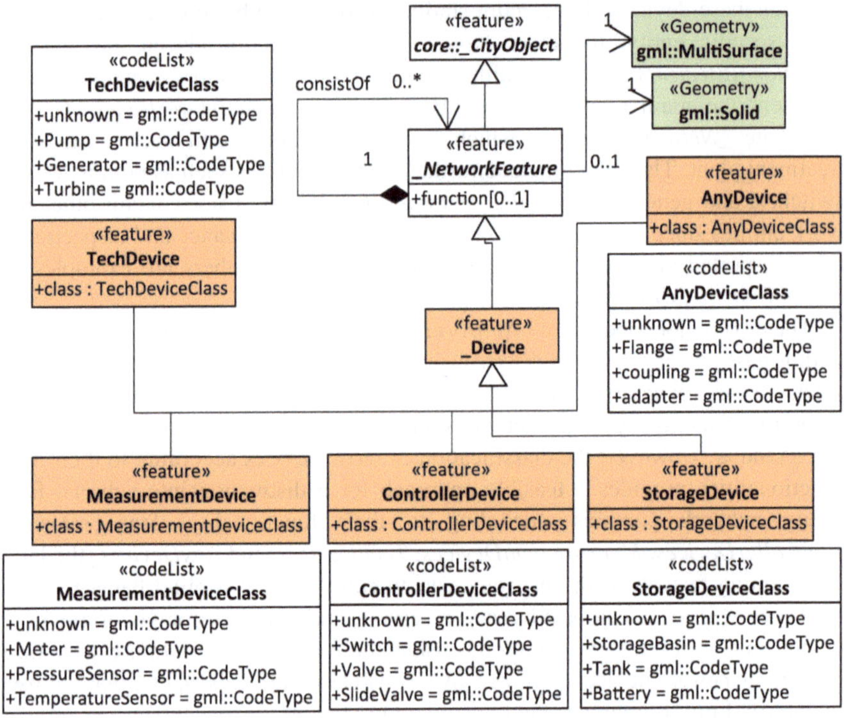

Fig. 11 Inclusion of network entities suitable for measuring, controlling, storing and manipulating the transported material. Further details are given in (NetworkComponents Model 2012)

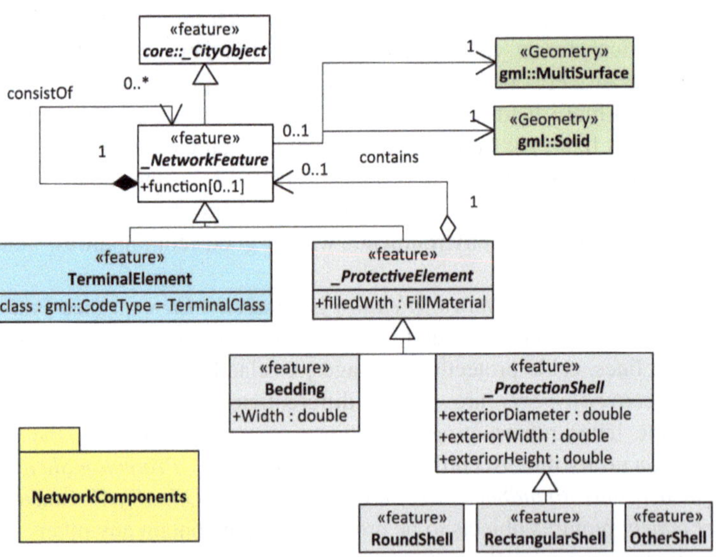

Fig. 12 Integration of *TerminalElement* and *ProtectiveElement* into the model

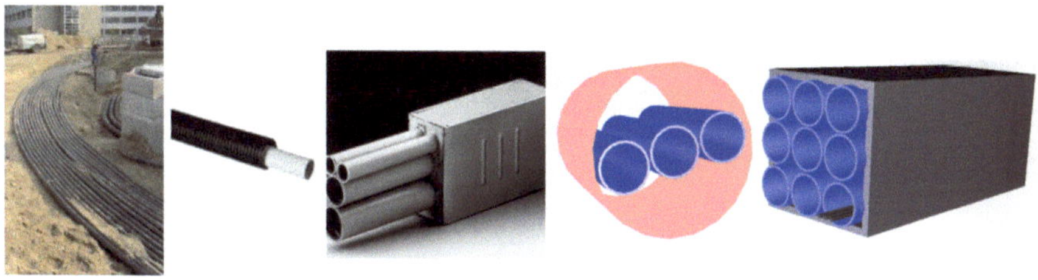

Fig. 13 Typical *ProtectiveElements* from *left* to *right*; cable line, cladding, cable protection package, cladding tube (i.e. district heating), cable protection package (i.e. power supply)

especially the aggregation of network objects using the *consistOf* composition. *ProtectiveElements,* however, might be an aggregate of different protection entities and thus each network entity might exist without the parent elements, i.e. a power cable can exist without the parent protection element surrounding it. A switch in a switchgear cabinet, however, cannot exist without the surrounding cabinet. Using both association types allows for simple (cladding tube) and complex (cable protection package) feature modeling as illustrated in Fig. 13.

5 Modeling Network Properties and Commodities

As already mentioned in the previous section the classification of network objects is not only done with respect to the functionality of elements in the network, but also to the transported material. Elements as cable, pipe, and so forth are used for water, gas, and electricity supply as well as in the field of industrial manufacturing of goods of any type. They are even used in production facilities and buildings. Hence, a simple characterization of network elements into water, gas, or electricity related objects is not appropriate and a more specific way to describe the network properties has to be found in order to differentiate all elements in a thematic and semantic manner that will cover all application fields of utility networks.

The abstract class *_CommodityType* serves as a container for the chemical and hazard classification of the transported material as well as the material classification of the transported commodity and, thus, for the description of its material properties in particular. As mentioned in Sect. 3 and further described in Becker et al. (2011a, b) a collection of *_NetworkFeatures* transporting the same commodity are assigned to one network, which itself might consist of sub networks (network hierarchy) which all transport the same commodity. The commodity type therefore is related to the *Network* and not to the individual network features.

Utility networks transport all types of material in every type of physical condition such as gas, electricity, light, and water. According to this a classification into liquid (*LiquidMedium*), gaseous (*GasMedium*), and solid (*SolidMedium*) material is required (see Fig. 14). Each physical condition, respectively each commodity type possesses its own property set which is needed to describe the

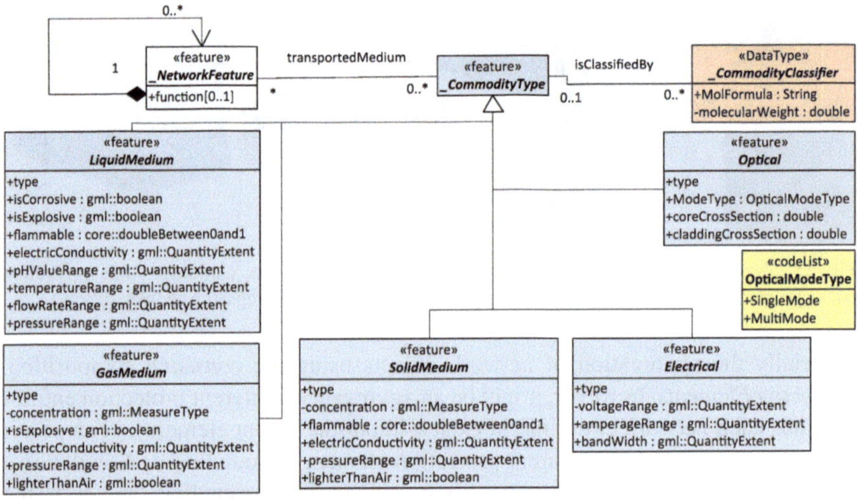

Fig. 14 Classifications of transported medium according to physical condition. Further details are given in (NetworkProperties Model 2012)

transported material adequately. Important properties are for example whether a transported material *isExplosive* or flammable. Also the *electricConductivity* besides the *Flow-Rate*, *Temperature*, *concentration*, and *pHValue* of a commodity can be specified and are needed to inform users of a system or city model about the transported material.

As depicted in Fig. 14 a network might also be used to transport goods that cannot be described by a physical aggregate state. In fact, another commodity type is needed that allows for representing electrical energy or signal transmission. As Fig. 14 indicates a differentiation between electrical and optical transmission was made due to the fact that the transmission is different and, thus, the needed properties. Whereas the general principle of optical signal transmission is based on total reflection and the needed properties are core, cladding cross section, and mode type, the electrical energy or signal transmission is based on the movement of electrons and the relevant properties are *frequency bandwidth*, *voltageRange*, and *amperageRange*. Thus, every commodity type can be expressed by its respective physical condition or transmission type.

Further classification of commodity types can be done using well-defined and standardized *_CommodityClassifiers*. These are explained in detail in (Network-Properties Model 2012).

6 Implementation and Realization of the Model

The developed data models were brought into practice first within the project SIMKAS-3D (www.simkas-3d.de). The aim of that project was to develop methods for the identification and analysis of the mutual interdependencies of

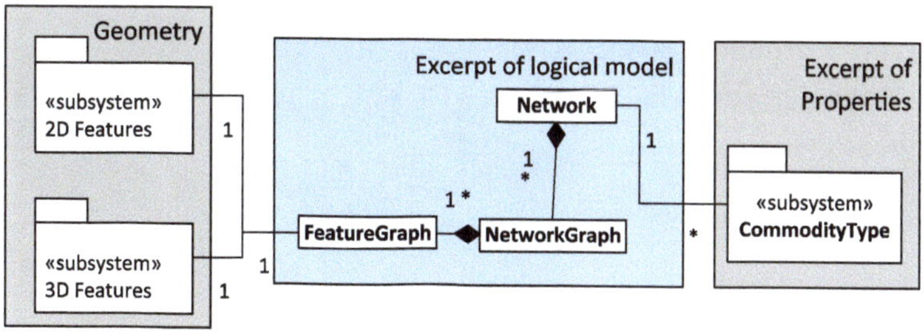

Fig. 15 Implementation of the CityGML Utility Network ADE as a geodatabase model for ESRI ArcGIS

critical infrastructures including the simulation of cascading effects in the failure of supply infrastructures (see Becker et al. (2011a, b) for more details on the background).

In order to achieve the project goals a data model and geodatabase for the homogeneous representation of different utility networks such as water, gas, long-distance heating, and power supply had to be developed. The integrated database should facilitate the common operational picture (COP) for disaster management as well as for the simulation of cascading effects in case of network failures.

The *NetworkCore* model, the *NetworkComponents* model, and the *Network-Properties* model have been mapped to a relational database schema and are stored using the ESRI File-Geodatabase format. According to the developed three data models the database schema is partitioned into three major parts (see Fig. 15) as well. One is representing the geometry of the network components in 2D (poly-line, point) and 3D (multipath), one is representing the logical model—the core model, in tables and the last one is representing the network properties (commodity types) as a relation to the networks. The utility networks of the supplier companies were converted into the created geodatabase by customized FME workbench processes. The proprietary GIS systems were the data source for the process and the created geodatabase has defined the destination writer type and schema.

The interdependencies between networks, network objects, as well as city objects were identified and added. Figure 16 shows 3D visualizations of the available data and its embedding into the urban space. Each building of the dataset is logically connected to the available network. Thus, the possibility is given to perform complex analysis and simulations from producer (treatment plant) to the utility client (building) with respect to cascading effects, network tracing, and more.

7 Conclusions and Outlook

In this paper a new geospatial information model for multi-utility networks was proposed. It specializes the UtilityNetworkADE core model for CityGML as previously presented in Becker et al. (2011a, b) by concrete classes and

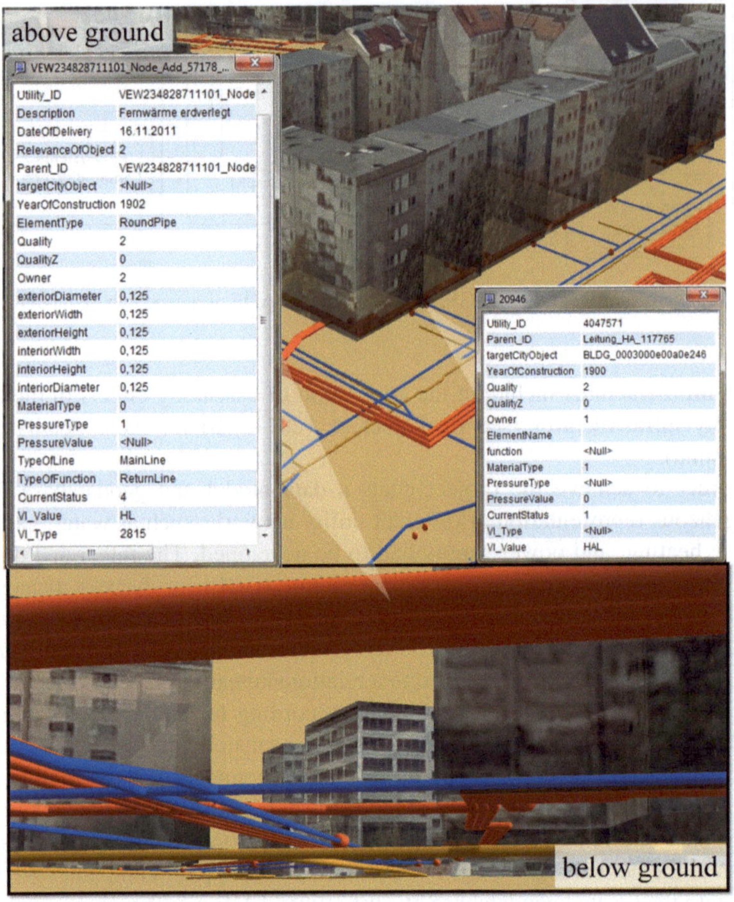

Fig. 16 Embedded multi-utilities into 3D urban space in a perspective view (*top* = above ground; *bottom* = view from below ground)

relationships for the representation of the network entities used in the different types of utility networks and commodities. The semantic classification of network components into the essential entity types was based on empirical studies carried out at different utility providers. A comparison with already existing models and approaches was given in the paper. The integration with the CityGML standard facilitates the integration of multi-utility networks into the urban space respectively into 3D city models for joint visualization and analysis tasks. Furthermore, interoperable exchange of and access to 3D multi-utility networks is enabled.

The data model represents 3D topography, 3D topology, and functional properties and interdependencies of the networks and their components. Hierarchical representations for both networks and components are supported as well. These characteristics allow to perform geospatial analyses in order to determine the implicit interdependencies between network components within the same or different infrastructures or between network features and other city objects based on spatial relations like proximity. For example, collision detection would prevent many pipeline ruptures caused by excavation works. The logical representation of networks and their interdependencies support complex analyses and simulations as needed in the fields of disaster management, critical infrastructure analysis, strategic energy planning, and simulation of power grids.

Concerning the previously existing utility network models, the suggested model can be considered a superset with regard to model expressivity. This will improve the possibilities to exchange or link data between different systems (INSPIRE, ESRI Utility Networks, IFC ⇔ CityGML). However, it will be a task of future work to show that datasets represented according to these frameworks can be cast into the proposed model without information loss. For this purpose we intend to follow the line of Hijazi et al. (2011) where the lossless mapping of IFC utility networks onto CityGML utility networks and vice versa was shown.

In the current version the model does not distinguish different levels-of-detail (LOD). Also the 3D geometry model is limited to the usage of Multisurfaces, Solids, and graph structures with a 3D embedding. The investigation of multiple LODs and alternative geometry representations like sweep geometries is subject of ongoing work. Of course, the data model can also be extended to create a more fine-grained and semantically further enriched model, e.g. for the representation of material properties of network components and cross sections of pipes and canals.

Further research is currently being undertaken on the cartographic visualization of multi-utility networks and their operating status meeting the requirements of disaster management and stressful situations. The cartographic representation will focus on the functional aspects of network entities and a geometric simplification will be done in order to provide a common operational picture (COP) of the critical infrastructures within a city.

Acknowledgments The presented work was mainly carried out within the collaboration project SIMKAS-3D funded by the Federal Ministry of Education and Research of Germany (BMBF). We are grateful to the city of Berlin for providing access to the city model of Berlin, to all members of the SIMKAS-3D project and to all members of the modeling group of the SIG-3D for the cooperation and fruitful discussions.

References

Becker T, Nagel C, Kolbe TH (2011a) Integrated 3D modeling of multi-utility networks and their interdependencies for critical infrastructure analysis. In: Kolbe TH, König G, Nagel C (eds) Advances in 3D Geo-information sciences, Springer, Berlin

Becker T, Nagel C, Kolbe TH (2011b) UtilityNetworkADE—core model. Online available at http://www.citygmlwiki.org/index.php/CityGML_UtilityNetworkADE. Accessed 9 Mar 2012

Diestel R (2010) Graph theory, 3rd edn. Series on graduate texts in mathematics 173. Springer, Berlin

ESRI (2003) ArcGIS water utility data model. ESRI, Redlands, CA. Online available at http://www.downloads2.esri.com/resources/datamodels/ArcGISWaterUtilityDataModel.pdf. Accessed 27 Mar 2012

ESRI (2007) GIS Technology for water, wastewater, and storm water utilities. Environmental systems research institute. Redlands, CA. Online available at http://www.esri.com/library/brochures/pdfs/water-wastewater.pdf. Accessed 27 Mar 2012

Grise S, Idolyantes E, Brinton E, Booth B, Zeiler M (2001) Water utilities. ArcGIS™ data models. Environmental systems research institute, Online available at http://support.esri.com/en/downloads/datamodel/detail/16. Accessed 27 Mar 2012

Gröger G, Kolbe T H, Czerwinski A, Nagel C (2008): OpenGIS® city geography markup language (CityGML) encoding standard. Version: 1.0.0, OGC 08-007r1, http://www.opengeospatial.org/standards/citygml. Accessed 27 Mar 2012

Hijazi I, Ehlers M, Zlatanova S, Becker T, Berlo L (2011) Initial investigations for modeling interior utilities within 3D geo context: transforming IFC-interior utility to CityGML/UtilityNetworkAD. In: Kolbe TH, König G, Nagel C (eds) Advances in 3D geo-information sciences. Springer, Berlin

INSPIRE Consolidated UML Model (2011) INSPIRE consolidated UML model. Online available at https://inspire-twg.jrc.ec.europa.eu/annexII+III/inspire-model/. Accessed 13 Dec 2011

INSPIRE Data Specifications Drafting Team (2010a) D2.5: Generic conceptual model, Version 3.3. Online available at http://inspire.jrc.ec.europa.eu/documents/Data_Specifications/D2.5_v3_3.pdf. Accessed 13 Dec 2011

INSPIRE Thematic Working Group Transport Networks (2010b) D2.8.I.7 INSPIRE data specification on transport networks—guidelines. Online available at http://inspire.jrc.ec.europa.eu/documents/Data_Specifications/INSPIRE_DataSpecification_TN_v3.1.pdf. Accessed 13 Dec 2011

INSPIRE Thematic Working Group Utility and governmental services (2010c) INSPIRE data specification on utility and governmental services—draft guidelines, 2011-06-17. Online available at http://inspire.jrc.ec.europa.eu/documents/Data_Specifications/INSPIRE_Data Specification_US_v2.0.pdf. Accessed 13 Dec 2011

ISO 19109 (2005) ISO 19109 geographic information—rules for application schema. Online available at http://www.iso.org/iso/iso_catalogue/catalogue_tc/catalogue_detail.htm?csnumber=9891. Accessed 27 Mar 2012

Liebich T, Adachi Y, Forester J, Hyvarinen J, Karstila K, Reed K, Richter S, Wix J (2007) Industry foundation classes—IFC2x edition 3 technical corrigendum 1. Online available at http://buildingsmart-tech.org/ifc/IFC2x3/TC1/html/index.htm. Accessed 27 Mar 2012

Liebich T (2009) IFC 2x edition 3. Model implementation guide, Version 2.0. AEC3 Ltd. Online available from http://buildingsmart-tech.org. Accessed 27 Mar 2012

Meehan B (2007) Empowering electric and gas utilities with GIS. Series on case studies in GIS. ESRI Press, Redlands CA

NetworkComponents—Model (2012) Online available at http://www.citygmlwiki.org/upload/b/b1/NetworkComponents_-_UtilityNetworkADE.pdf. Accessed 13 Mar 2012

NetworkProperties-Model (2012) Online available at http://www.citygmlwiki.org/upload/d/d7/NetworkProperties_-_UtilityNetworkADE.pdf. Accessed 13 Mar 2012

Generalization and Visualization of 3D Building Models in CityGML

Siddique Ullah Baig and Alias Abdul Rahman

Abstract Generally, cities are expanding due to rapid population growth and require 3D city models for effective town planning, communication and disaster management. Rendering of 3D scenes directly is not so much appropriate as appearance properties, textures and materials attached with city models drastically increase the loading time for visualization and spatial analysis. Additionally, different applications or users demand different Level of Detail (LoDs), thus one of the questions arises—how different LoDs can be made available to these applications? Generation of lower LoDs given by OGC standard CityGML from higher LoDs to reduce data volume is a generalization problem. Relying only on existing geometric-based generalization approaches can result in the elimination or merging of important features, hence, semantic information can be considered. A review of pertinent generalization algorithms proposed by several researchers is presented. Additionally, this paper provides a method for generalization of 3D structures with the aim to derive multiple LoDs keeping semantic information into account. For this purpose, height and positional accuracy of objects at different LoDs provided by CityGML are considered. Initially, building parts and installations are removed. 2D footprints of remaining 3D structures are projected onto ground and simplified to derive LoDs building geometry. An adoption of methods of Sester and Brenner (Continuous generalisation for visualisation on small mobile devices. Heidelberg, pp. 355–368, 2004) extended by Fan et al. (Lecture notes in geoinformation and cartography, advances in giscience. Springer, Heidelberg, pp. 387–405, 2009) are applied for simplification and aggregation of projected

S. U. Baig (✉) · A. A. Rahman
3D GIS Research Lab, Faculty of Geoinformation Science and Real Estate,
Universiti Teknologi Malaysia (UTM), Johor Bahru, Malaysia
e-mail: subaig2@live.utm.my

A. A. Rahman
e-mail: alias@utm.my

J. Pouliot et al. (eds), *Progress and New Trends in 3D Geoinformation Sciences*,
Lecture Notes in Geoinformation and Cartography, DOI: 10.1007/978-3-642-29793-9_4,

footprints. The experiments showed that due to repetition of coordinates of connected nodes in CityGML increase both the rendering time and memory space. However, elimination of important smaller features can be avoided by taking semantic information into account while performing generalization operations.

Keywords 3D building modeling · Generalization · Simplification · Aggregation · Level of details (LoDs)

1 Introduction

The demand for visualization, analysis and simulation of 3D city models is increasing as 3D geospatial data is commonly available as cities are expanding due to rapid population growth and require 3D models for planning effective communication, town planning and disaster management. Specific applications of 3D city models include simulation of air pollutants, noise emission simulations, and 3D navigation systems (Gröger et al. 2008). Currently, popular 3D visualization tools e.g. Digital City and Earth Viewer is few of commercially available based on 3D visualization standards like X3D and KML for online visualization of 3D city models. These standards are only useful for visualization but not for representation of 3D city models so spatial analysis could not be carried out due to missing semantic information attached with city models. Alternatively, an OGC approved standard like CityGML (OGC 2009) defines not only the shape and photo-realistic appearance of 3D building objects but also thematic properties, attached with rich semantic information, classification and aggregation of 3D city models Kolbe (2008).

Rendering of 3D scenes directly from CityGML is not so much appropriate as appearance properties of 3D models depends on both the textures and materials (Mao et al. 2010) hence drastically increase loading time in visualization and spatial analysis. Therefore, different task-specific applications demand different abstractions (generalized) 3D scenes of 3D models. Therefore, lower LoDs are generated from higher LoDs of 3D models by reducing data volume (depends upon application's need) by applying a number of operations but main characteristics of the building are preserved. Reduction of data is one of the components of generalization process. Technically, a set of generalization requirements such as automatic generation of different LoDs, multiple representation of generated LoDs; and functionalities to automatically provide the required LoD to a given application or user, have to be met to automate the generalization process. Abstract models as a result of this generalization process can be visualized and analyzed efficiently in different scales.

Demands of different applications vary so generalization process should be tailored to be able to fulfill the requirement of a specific application. Components of generalization process can be altered to be made suited for a specific application. For example, some disaster-related applications may require to keep outer building installations, openings (doors, window) while deriving LoD2 from LoD3 to assess potential affected features of a building. Similarly, application specific attributes e.g. outer building installations and openings are not required for navigation systems so can be removed. Tourism-specific application may require semantic information about models to be preserved e.g. type of building models and their components such as entrances of a museum for visitors, exit points of sport complex, castle etc. Therefore, parts of the generalization process can be customized to entertain queries from users or applications. The queries can be e.g. "maintain exit points", "remove windows", "preserve bigger walls" etc. Similarly, during simplification operation, generalization engine can ask user to enter the threshold (40 or 50 m) to see a height of a building if a user wants to mount antennas for communication purposes. Another benefit of generalization concept in regard with 3D visualization is, building models closer to the observer should be clear and in detailed compared to the buildings residing far-away as observers are more interested in closer models.

Another method for generalization of 3D structures based on semantic information is proposed in this paper. Multiple LoDs are derived based on height and positional accuracy of points at different LoDs provided by CityGML. 3D objects to be simplified are projected onto ground as footprints. A test of geometry for details and possible aggregation of projected footprints is achieved. Initially, CityGML file which is XML-based is parsed and stored in C++ class objects. The resulting objects contained both geometry and semantic information from the input CityGML dataset. An adoption of algorithm of Sester and Brenner (2004) extended by Fan et al. (2009) is used to simplify footprints. Libraries of OpenGL are used to visualize the generalized LoDs.

Existing geometric and semantic-based generalization algorithms are discussed in Sect. 2. Another method for generalization of 3D buildings modeled in CityGML to derive multiple LoDs is presented in Sect. 3. Implementation and results are described in Sect. 4. A comparison of results for evaluation is presented in Sect. 5. Finally, the conclusion and outcome of the work is highlighted in Sect. 6.

2 Related Works

2.1 Geometric-Based Generalization Approaches

Generalization method proposed in Thiemann (2002) is useful for single buildings. Small features e.g. chimneys, balconies, windows and doors are extracted and separated from main body. Segmentation is carried out on main features and the sequence of operation is stored in a Constructive Solid Geometry (CSG) tree. Thiemann and Sester (2005) proposed a method for segmentation of boundary of building surface for generating a hierarchical generalization hierarchy tree. Protrusions are removed and holes are detected resulting finite lump. Elements of hierarchy tree are removed or recognized through elementary generalization operations.

Building models are categorized into a limited number of classes keeping their characteristic shapes and made similar to available 3D templates in Thiemann and Sester (2006). Thus, an intended building model is substituted with a similar 3D template, which is best fit to the real object model. Automatic generalization based on minimal constraints is implemented.

A method proposed in Forberg (2007) is based on scale-space approach, which provides morphology and curvature space operators to be used for splitting and merging of 3D building parts. Different representations of models at different scales from an image are derived to obtain scale-space. The resulted scale-space has inherited capability of abstraction to be used to initiate generalization process for 3D building models. Therefore, scale-space is used to generate different LoDs or coarser representations of 3D city models. For this, parallel facades are moved to each other to eliminate the gap between them. Smaller features between two facades are removed by dragging facades away from each other resulted a split between them. Squaring technique is applied to generalize non-orthogonal structure like roofs, etc.

2.1.1 Simplification of 2D ground plans

A set of generalization actions were proposed in Staufenbiel (1973) for simplification of 2D footprints based on human interaction. Powitz (1992), Regnauld (2001) and Harrie (1973) developed algorithms to remove line segments of ground plan by extending and crossing their neighbor segments based on a predefined threshold of length by taking angles and minimum distances into account. Mayer (2005) and Rainsford and Mackaness (2002) use vector templates for generalization 2D ground plan. An adoption of approach developed by Sester and Brenner (2004) and extended by Fan et al. (2009) is applied for simplification of ground plan in Mao et al. (2010).

An algorithm proposed in Sester (2005) using Least Square Adjustment (LSA) removes irrelevant facades of a building. A number of parameters need to be changed during the process to generalize a building model, so, direct application of LSA become irrelevant. It become two-steps approach: smaller facades are removed or replaced during simplification process to generate an estimated ground plan; then a parametric model from the resulting simplified ground plan is derived. Subsequently, adjustment process uses the resulting parametric model to be adapted to the new shape.

The algorithm by Sester and Brenner (2004) is applied to simplify ground plan and intrusions and extrusions are eliminated. Roof structures are adjusted with respect to ground plan.

2.2 Semantic-Based Generalization Approaches

A semantic-based method is proposed in Fan et al. (2009) for deriving different lower LoDs from higher LoDs of buildings modeled in CityGML. Generalized LoDs are

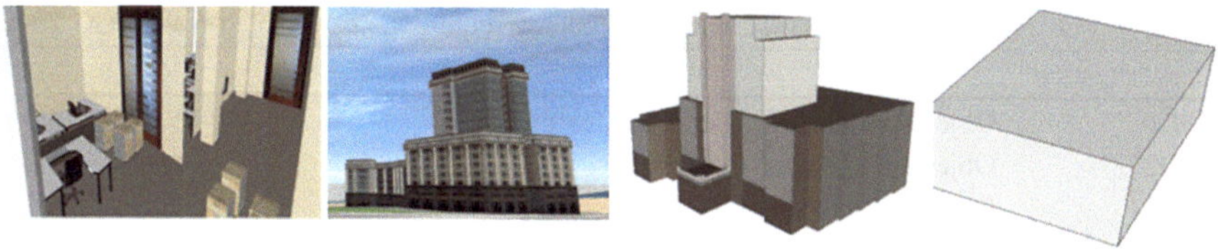

Fig. 1 Four buildings of Putrajaya city (Malaysia) in different LoDs, from *left* to *right*: (LoD4–LoD1)

stored in multi-scale representation as a CityTree. An intermediate level of detail (LoD2.5) was created to reduce the gap between LoD3 and LoD2 for more efficient visualization of 3D models. The geometric and semantic information of LoD2 is more generalized than LoD3 hence the process for deriving LoD2 from LoD3 will be relatively more complicated than deriving LoD1 from LoD2. A novel for multiple representation data structure for dynamic visualization of 3D city models is proposed in Mao et al. (2010) and produced CityTree to store different LoDs of generalized models.

3 3D Modeling and Generalization

3.1 CityGML

The building model is the most detailed thematic concept of CityGML. It allows for the representation of thematic and spatial aspects of buildings, building parts and installations at four levels of detail (LoD4–LoD1) illustrated in Fig. 1. At LoD1, 3D buildings are represented by block model with flat roofs. At LoD2, buildings have differentiated roof structures and thematically differentiated surfaces. LoD3 denotes architectural models with detailed wall and roof structures, balconies, bays and projections. LoD4 completes a LoD3 model by adding interior structures.

On the other hand, the LoDs in CityGML are also characterized by differing accuracies and minimal dimensions of generalized objects. Table 1 shows all object blocks as generalized features with a footprint of at least 6 m by 6 m have to be considered in LoD1 as compared to 4 m by 4 m in LoD2. In the detailed model at LoD3, the minimal footprint should be 2 m by 2 m. Moreover, CityGML supports the aggregation/decomposition by providing an explicit generalization association between any City Objects (Gröger et al. 2008). In this sense, buildings at a certain LoD could be generalized to be represented by an aggregate building at a lower LoD.

3.2 Generalization

The proposed method for derivation of multiple LoDs is based on semantic information and height and positional accuracy of LoDs provided by CityGML. LoD3 include outer building installations, opening objects e.g. windows, doors and

Table 1 (LoD1–LoD4) of CityGML with its accuracy requirements (adopted from OGC 2009)

Generalization		
LoD1	Objects blocks as generalized features	>6 × 6 m/3 m
LoD2	Objects as generalized features	>4 × 4 m/2 m
LoD3	Objects as real features	>2 × 2 m/1 m
LoD4	Constructive elements and openings are represented	

different types are roofs. Opening objects are removed and remaining objects are projected on ground as footprints to be simplified. Roof structures are projected and simplified similar to footprints. If side of feature is less than 4 m are preserved otherwise deleted. Similarly, the remaining features e.g. outer building installations are simplified resulting LoD2. Outer building installations are eliminated resulting LoD1 as a process of aggregation and simplification of geometries of building at LoD2. Following section explains the simplification process of projected footprints.

The following pseudocode shows the workflow of generalization process.

1. Declaration of C++ objects, which contain both geometry (coordinates of vertices) and semantic information about components of LoD3
2. Parsed LoD3 model to C++ objects
3. Displayed LoD3
4. Removed "Opening objects" from objects
5. Removed sides of remaining footprints of components of LoD3 resulting LoD2 based on specification of CityGML
6. Displayed LoD2
7. Repeated (5) for simplification and aggregation of footprints at LoD2 resulting LoD1
8. Displayed LoD1

3.2.1 Simplification of Footprints

Involvement of human interaction for simplification of 2D footprints is an old method for generalization of 2D geospatial data. Similarly, another ancient method is extending and crossing neighboring segments to remove line segments of ground plan on a given both thresholds the angles and distances. Here, in study, an approach developed by Sester and Brenner (2004) and extended by Fan et al. (2009) is applied for simplification of ground plan. The proposed method is an adaption of their methods and has been modified due to different scenarios like if building contains sharp edges in case of dome of a mosque, etc. or if smaller features make a special pattern. Simplification methods adopted in our study to simplify footprints based on neighboring edges.

The following semantic-based rules can be imposed to avoid removal/addition of important parts of building model:

- Building side smaller than a threshold should be removed but in certain cases similar to Fan et al. (2009), an important object, which lies on the side should not be removed.

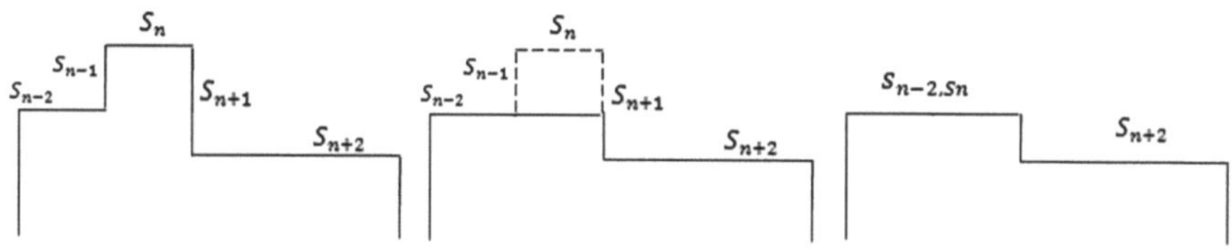

Fig. 2 Removal of Offset taking neighboring edges into account for simplification of ground plan (redrawn from Mao et al. 2010)

- If more than one smaller parts consecutively exists which are smaller than threshold and make a pattern, then all of them needs to be treated collectively not individually.
- If removal or modification affects the neighboring feature then or if these is an overlapping between two smaller features should be treated together.

3.2.2 Removal of Intrusion/Extrusion and Offset

Figure 2 illustrates removal of offset taking neighboring edges into account for simplification of foot print similar to Sester and Brenner (2004). Before simplification each side of building needs to be checked to determine if the side of building is smaller than threshold. Here in our case, building side S_n is smaller than threshold and there is no important part of building lies on it. Other sides connecting with S_n are S_{n+1} and S_{n-1} which are larger and parallel to each other while S_{n-2} share an edge with S_{n-1}. Larger side S_{n-2} is extended and vertical side S_{n+1} is reduced resulted a new side S_{n-2}, S_n.

3.2.3 Removal of Corners

Case 1 (Fig. 3a): Corner of a building side can be reduced without disturbing shape of sides as both sides (S_{n+1} and S_{n-1}) of corner are not parallel but the distance of each side is same as illustrated in Fig. 3a. A new point P is produced by intersecting perpendiculars of both sides. The resulting sides become (S_{n+1}, P) and (S_{n-1}, P) and previous longer sides (S_{n+1}) and (S_{n-1}) an be deleted but shape is preserved.

Case 2 (Fig. 3b): The side of a corner needs to be checked if it is shorter than two longer sides and threshold. If an important feature like door or window exists on the wall connected with this side then the smaller side (S_n) should not be altered. Otherwise, (S_{n-1}) and (S_{n+1}) which are neighbors of (S_n) are extended until they interest with each other at point P. Then the smaller side (S_n) is removed resulting two longer sides (S_{n-1}, P) and (S_{n+1}, P).

Case 3 (Fig. 3c): This case is reverse of Case 2 where a point P is deleted by reducing both sides (S_{n-1}, P) and (S_{n+1}, P) inwards resulting a new side (S_n, P).

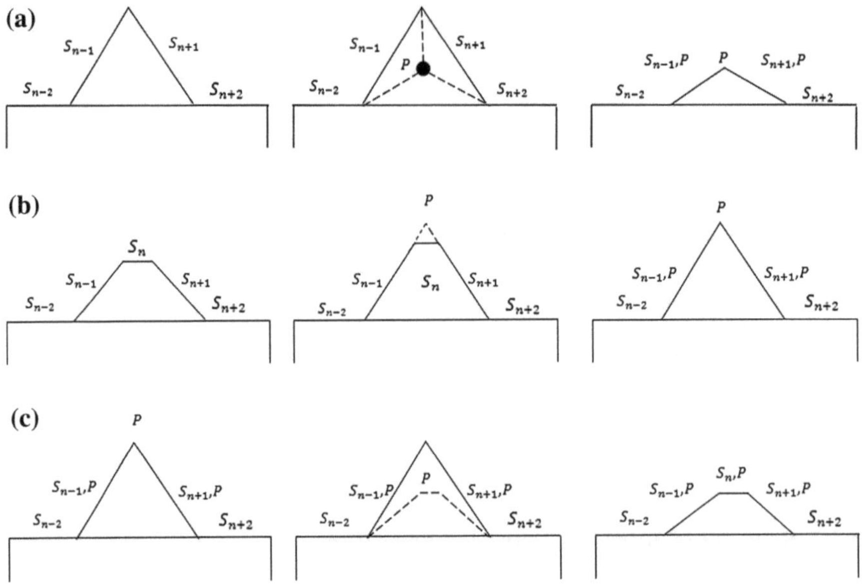

Fig. 3 Removal of corners taking neighboring edges into account **a** introduction of new point at center, **b** extension of longer faces until their intersection, **c** removal of point P and reducing face inward. (redrawn from Fan et al. 2009)

Sometime we encounter with such buildings which contains intrusions/extrusions, offsets and corners together and make a pattern on each side of building part. Figure 4a illustrates three portions (A, B and C) of a larger surface. The pattern of two portions A and B are similar with two corners and two flat intrusions but the third portion (C) is slightly different form rest of portions A and B and an intrusion is missing.

Case 4: In this case (Fig. 4b), we need to maintain pattern before reducing data. It can be said alternatively, we have to increase data before reducing. Therefore, missing intrusions are produced by extending larger sides vertically resulting two parallel sides. As next step, edges of both newly formed parallel sides are connected resulting new intrusion. Now, pattern of portion A, B and C is maintained. Process of insertion of new intrusion is illustrated in Fig. 4b. Identified portion (C) contains two larger sides ($S_{n\text{-}2}$) and (S_{n+2}). An additional intrusion having smaller sides (S_n), (S_{n+1}) and ($S_{n\text{-}1}$) is produced by reducing a smaller side resulting three smaller sides ($S_{n,New}$), ($S_{n,New\text{-}1}$) and ($S_{n,New+1}$). The number of intrusions/Offset in all three portions A, B and C becomes same and pattern is maintained.

Case 5: (Fig. 4c), as shown in (b), smaller sides ($S_{n,New+1}$) and ($S_{n\text{-}1}$) which are smaller than threshold need to be reduced. They are reduced and a new side($S_{n.New}$, S_n, $S_{n,New+2}$) is produced. Similarly, the process is repeated similar to Fig. 3 for Offset and sides of smaller sides are reduced and deleted. The resulting sides larger than threshold become ($S(_{NewLine})_1$), ($S(_{NewLine})_2$) and ($S(_{NewLine})_3$).

The above-mentioned method is applied to simplification of 3D buildings in Putrajaya city of Malaysia and presented the results in last section.

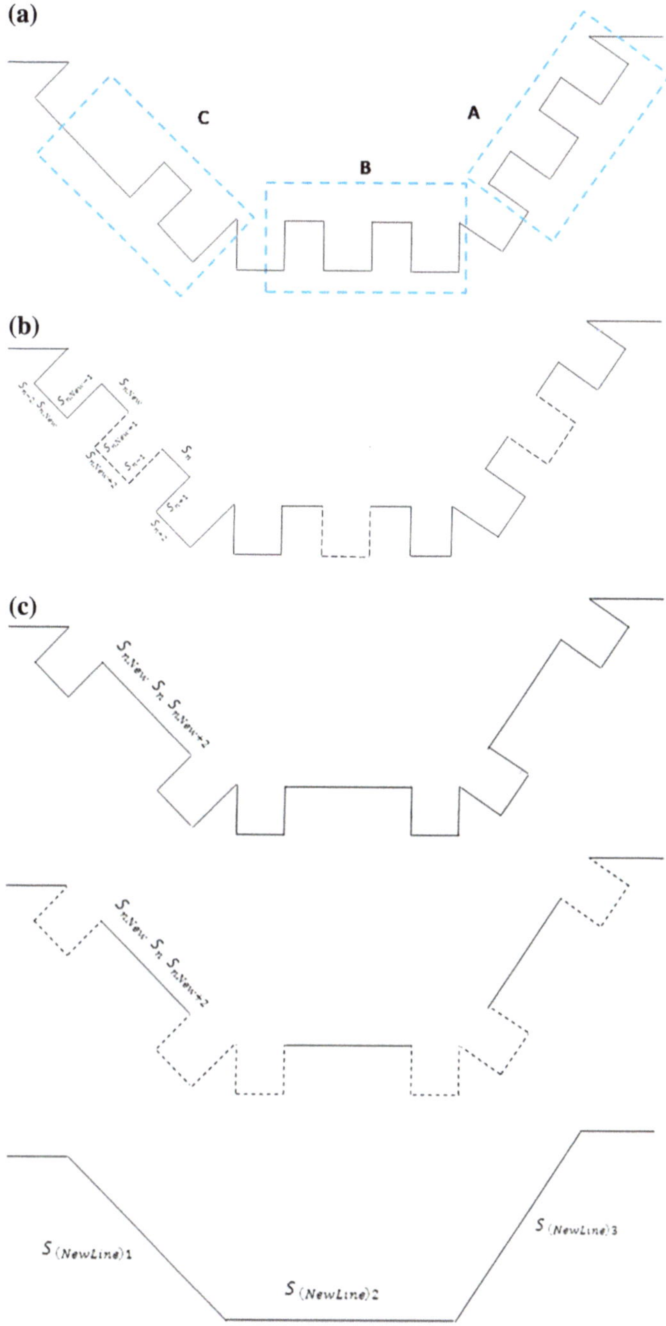

Fig. 4 Creation of an intrusion to maintain pattern of features of footprint: **a** three portions (*A*, *B* and *C*) of a larger surface, **b** insertion of new intrusion, **c** reduction of all intrusions together

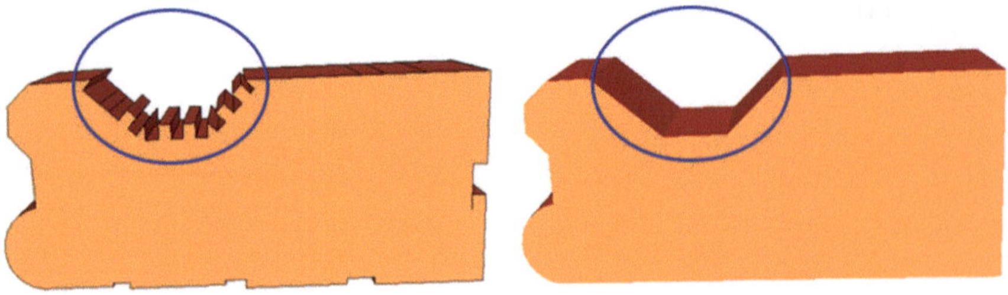

Fig. 5 Result of removal of intrusion, corners of the Jabatan Pendaftaran Negara (JPN), Putrajaya, Malaysia

Building model presented in Fig. 5 is of the *Jabatan Pendaftaran Negara (JPN)*, Putrajaya city of Malaysia having similar characteristics mentioned. The model is simplified by applying methods presented in different cases.

3.2.4 Aggregation of Footprints

Aggregation methods developed by a number of researchers for 2D cartographic objects can be used for 3D objects. Glander and Döllner (2007) introduced a method for aggregation by cell-based clustering to merge building blocks. "Direct-merge" and "Snap-merge" are two aggregation operators introduced by Bundy et al. (1995) to move two objects near to each other for aggregation purpose. A method developed by Anders (2005) is quite suitable for aggregation of footprints located on a straight line with same orientation but both simplification and aggregation of footprints are processed simultaneously. In this case, this method can preserve the location of original footprints, and creates the new by prolonging the existing ones but it is not necessary to club buildings horizontally.

In our approach, application of aggregation methods depends upon certain conditions which include Direct-merge and Snap-merge by Bundy et al. (1995) and triangulation for aggregation of footprints of building models illustrated in Fig. 6. Following conditions need to be checked first:

If one or more than one important feature lies in the way between two large features to be aggregated then important features need to be aggregated first followed by aggregation of both larges features.

- If the distance between two separate footprints is smaller than the size of one of smaller object then Snap-merge method of Bundy et al. (1995) should be applied otherwise triangulation method similar to Mao (2010) should be applied.
- Smaller part of ground plan should move to larger one in case of Snap-merge or direct-merge.
- In case of tilting of surfaces which cannot be identified in footprints need to be addressed taking height of footprint of wall or surface into account.
- All vertices of footprints need to be taken into account for merging.

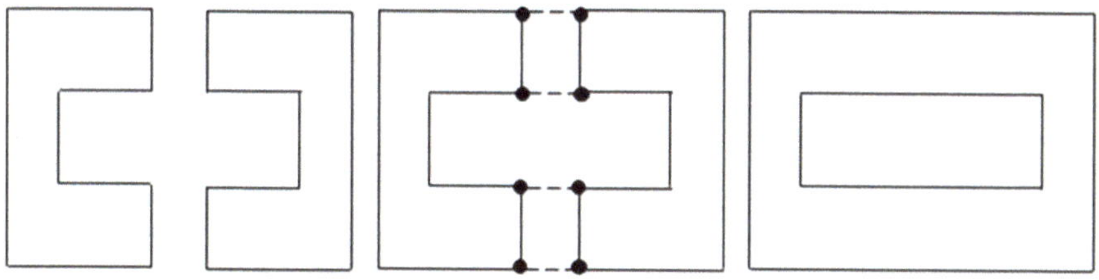

Fig. 6 Aggregation of footprints connecting neighboring edges of building model

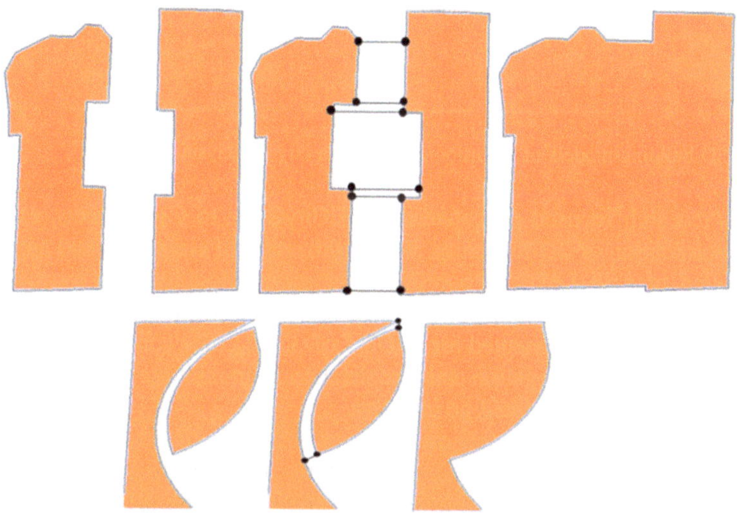

Fig. 7 Aggregation of footprints of Putra Mosque and Jabatan Pendaftaran Negara (JPN) in Putrajaya, Malaysia: original model (*left*), triangulation (*middle*) and aggregated (*right*)

The Fig. 7 illustrates aggregation of footprints of Putra Mosque and *Jabatan Pendaftaran Negara* (*JPN*) in Putrajaya, Malaysia using triangulation method. Results of footprints of two buildings of *Jabatan Pendaftaran Negara* (*JPN*) and Putra mosque in Putrajaya city of Putrajaya, Malaysia are presented.

4 Implementation and Results

Generalization and visualization of generalized models were implemented in a C++ application. The platform is Microsoft Visual Studios 2008 runs on a PC Pentium (R) Dual-Core CPU 2.10 GHz, 2.00 GB RAM, and Microsoft Windows 2007. The CityGML file was parsed using CMarkup 10.1. 3D building models of Putrajaya city of Malaysia were used for this research.

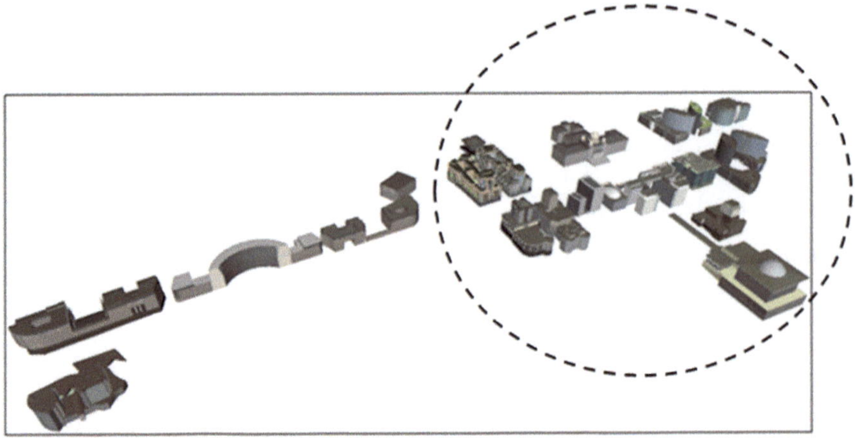

Fig. 8 3D building models of a portion of Putrajaya city (Malaysia)

This research focused on conversion of CityGML to C++ objects, extraction and simplification of 2D footprints and visualization of simplified and aggregated 2D ground plan in LoD1 in our implementation. Modifications were made in CityGML file before conversion and unnecessary data was removed. The proposed methods were not implemented for LoD2, LoD3 and LoD4 completely as methods to merge nodes for aggregation and visualization from multiple representation data structure is not yet completed as intended.

Initially, 3D building models are converted into C++ objects after CityGML is parsed using CMarkup. We used some of the methods like Load, FindElem, FindChildElem, FindNode, etc. to populate the CMarkup object from CityGML file and to parse it. FindElem method is used to locate next element, optionally matching name of the tag. Similarly, IntoElem method takes control into current main position element that becomes like the current position as a parent.

For visualization, initially, vertices of 2D faces as points are plotted using various commands of OpenGL. Similarly, other commands are used to define a three-dimensional vertex using three floats. There are other variants ranging from four doubles to an array of two shorts. Figure 8 shows the visualization of 3D building models of Putrajaya city of Malaysia around MaCGDI (Malaysian Center for Geospatial Data Infrastructure) building without generalization and the results of generalization processes using the above-mentioned proposed methods is presented in Fig. 9a, b, Fig. 10 and Fig. 11.

Figure 9a shows the simplified models of a small portion highlighted in Fig. 8 while Fig. 9b illustrates simplified and aggregated building models. Figure 10 shows the results in level as street views. It can be visualized that the buildings located closer to viewing point of a user are less generalized as compared to those locating far from user illustrated in Fig. 11. Far-away buildings are aggregated compared to closer buildings.

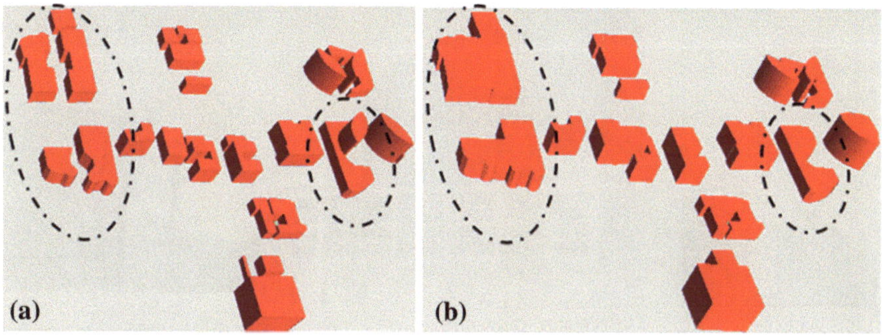

Fig. 9 Experimental results. **a** Simplified models. **b** Simplified and aggregated building models

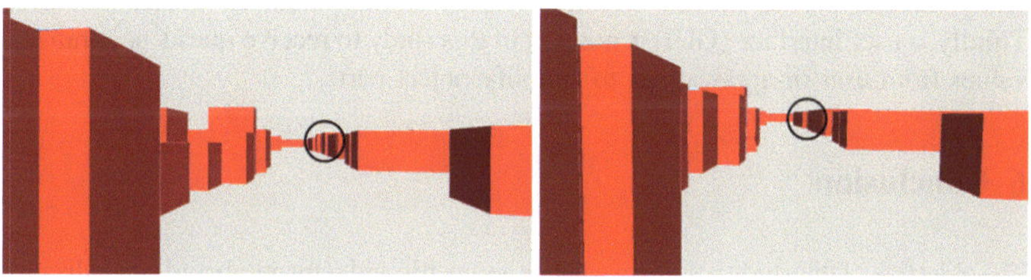

Fig. 10 3D building models (*street view*)

5 The Evaluation

The evaluation plays a vital role if it is necessary to choose between alternative generalization results. A comparison of results of two algorithms in terms of geometrical characteristics of two objects is evaluated to determine the level of difference of their appearance. For this purpose, results of simplification algorithm of Sester and Brenner (2004) are compared with our proposed algorithm as different algorithms affect characteristics of these objects. Different characteristics of an object as a whole is taken into account while determining total effect of both algorithms to evaluate its quality. During simplification process proposed in the algorithm of Sester and Brenner (2004), intrusions and extrusions are eliminated and roof structures are adjusted with respect to ground plan taking length of part into account. In this paper, pattern as well as length of intrusions and extrusions are taken into account and enhanced initially rather than removed completely. These intrusions/extrusions should be enhanced first rather than removed directly because a larger value than the minimal simplification distance can indicate that essential components of the city model have been removed, illustrated in Fig. 4.

Semantic-based removal of smaller parts of building is one of the strength of algorithm as different applications demand maintenance of important parts based on their priority and feature type. Desktop visualization of generalized 3D building models is one of the drawbacks of this study as evaluation of results based on time

Fig. 11 Models closer to observer are in detailed (*street view*)

and consumption of memory space could not be made as part of this work. Secondly, results are visualized directly without storing them in a data structure. Thirdly, a user interface (GUI) is not part of this study to receive quests or parameter values from user or applications to simplify object parts.

6 Conclusion

The objective of this study was to investigate geometric and semantic-based generalization of 3D building models approaches proposed by different researchers. Currently available methods applied in different scenarios or cases for simplification and aggregation of buildings were discussed and extended some of the methods to make them useful.

Building models represented in CityGML which is OGC standard is used for this study. To process the data from CityGML, we used CMarkup class as a parser to retrieve contents of XML-based CityGML file. Components of buildings were loaded and populated C++ objects as a tree to be able to process or to display host information. CMarkup class used *Markup.h* as an interface to initialize, produce output. The resulting C++ objects contained data related to components of building. From C++ objects, OpenGL routines are used to extract footprints by adopting methods of Mao et al. (2010) followed by simplification. Simplification methods were adopted from algorithm of Sester and Brenner (2004) extended by Fan et al. (2009).

The experiments showed that coordinates of vertices in CityGML are repeating for connected polygons hence increase memory space and rendering time. However, generalization of models keeping semantic information into account help avoid elimination of important smaller features.

Further research will be conducted to generate *BuildingTree* as multiple data structure to represent multiple LoDs as a result from generalization process. A Graphical User Interface (GUI) for the proposed generalization is also part of the future work.

Acknowledgments We would like to convey our deepest acknowledgement firstly to Universiti Teknologi Malaysia (UTM) for providing research grant Vote No. Q.J130000.7127.04J81, and also to Mr. Muhammad Imzan Hassan for managing the funding. Last, but not least, our sincere appreciations to Research Management Centre (RMC) of UTM and Ministry of Higher Education (MOHE), Malaysia for enabling us to carry out this research project.

References

Anders K (2005) Level of detail generation of 3D building groups by aggregation and typification. In: Proceedings of the XXII international cartographic conference, La Coruna, Spain

Bundy GL, Jones CB, Furse E (1995) Holistic generalisation of large-scale cartographic data. In: Müller JC, Lagrange JP, Weibel R (eds) GIS and generalisation, Gisdata 1. Taylor & Francis, London, pp 106–119

Fan H, Meng L, Jahnke M (2009) Generalisation of 3D buildings modeled by CityGML. In: Paelke V, Sester M, Bernard L (eds) Lecture Notes in Geoinformation and Cartography, Advances in GIScience. Springer, Berlin, Heidelberg, pp 387–405

Forberg A (2007) Generalisation of 3D building data based on scale-space approach. ISPRS J Photogrammetry Remote Sens 62(2):104–111

Glander T, Döllner J (2007) Cell-Based Generalization of 3D Building Groups with Outlier Management. In: Proceedings of the 15th international symposium on advances in geographic information systems, ACMGIS

Gröger G, Kolbe TH, Czerwinski A, Nagel C (2008) OpenGIS® City Geography Markup Language (CityGML) Implementation specification. ClinMed NetPrints. http://www.opengeospatial.org/legal/. Accessed 10 Oct 2011

Harrie L (1973) An optimisation approach to cartographic generalization. Doctor thesis, Lund University

Kolbe TH (2008) Representing and exchanging 3D city models with CityGML. In: Zlatanova S, Lee J (eds) 3D geo-information sciences. Seoul, South Korea. Springer, Berlin, Heidelberg

Mao B (2010) Visualization and generalization of 3D city models. Doctoral thesis, Lund, Sweden

Mao B, Ban Y, Harrie L (2010) A multiple representation data structure for dynamic visualisation of generalised 3D city models. ISPRS J Photogrammetry Remote Sens, J Mol Med. doi:10.1016/j.isprsjprs.2010.08.001

Mayer H (2005) Scale-spaces for generalisation of 3D buildings. Int J Geogr Inf Sci 19(8–9):975–997

OGC, (2009). CityGML specification. ClinMed NetPrints. http://www.opengeospatial.org/standards/citygml. Accessed 25 Nov 2011

Powitz B (1992) Kartographische generalisierung topographischer daten in GIS. Kartographische Nachrichten 43(6):229–233

Rainsford D, Mackaness W (2002) Template matching in support of generalization of rural buildings. In: Joint international symposium on "GeoSpatial theory, processing and applications" (ISPRS/Commission IV/SDH2002), Ottawa, Canada

Regnauld N (2001) Contextual building typification in automated map generalisation. Algorithmica 30(2):312–333

Sester M (2005) Optimization approaches for generalization and data abstraction. Int J Geogr Inf Sci 19(8):871–897

Sester M, Brenner C (2004) Continuous generalisation for visualisation on small mobile devices. In: 11th international symposium on spatial data handling, Heidelberg, p 355–368

Staufenbiel W (1973) Zur Automation der Generalisierung topographischer Karten mit besonderer Berücksichtigung grossmasstäbiger Gebäudedarstellungen. PhD Thesis, Fachrichtung Vermessungswesen, Universitate Hanover, Hanover

Thiemann F (2002) Generalization of 3D building data. The international archives of the photogrammetry, remote sensing and spatial information science, 34 (Part 4)

Thiemann F, Sester M (2005) Interpretation of building parts from boundary representation, workshop on next generation 3D city models, Bonn

Thiemann F, Sester M (2006) 3D symbolization using adaptive templates. In: ISPRS technical commission II symposium, Vienna, Austria

From the Volumetric Algorithm for Single-Tree Delineation Towards a Fully-Automated Process for the Generation of "Virtual Forests"

Arno Buecken and Juergen Rossmann

Abstract When we introduced the volumetric algorithm for single-tree delineation at the 3D GeoInfo 07, it was already a powerful algorithm with a high detection rate and the capability to generate trees for forestry units with only minimal user interaction. For the first test-area of 82 km^2 this was acceptable, but as the test-areas grew, it showed that even the little user interaction does make the process laborious and strenuous. For currently envisaged test-areas of more than 1,000 km^2, it is essential to further limit the required user interaction. In this paper we will show how to reduce the computational complexity of the volumetric algorithm and how to automatically calculate the free parameter that had to be set interactively in the earlier implementation. We will use the so called Receiver Operator Characteristic (ROC), an approach that is being used to model and imitate the human decision process when it comes to making a parameter decision in statistical processes. It turns out that this method, which is commonly used in other fields of scientific decision making, is also valuable for many other geo-information processes.

Keywords Remote sensing · LIDAR · Single tree delineation · Forest · 3D modeling

A. Buecken (✉) · J. Rossmann
Institute for Man-Machine Interaction, RWTH Aachen University, Aachen, Germany
e-mail: buecken@mmi.rwth-aachen.de

J. Rossmann
e-mail: rossmann@mmi.rwth-aachen.de

J. Pouliot et al. (eds), *Progress and New Trends in 3D Geoinformation Sciences*, Lecture Notes in Geoinformation and Cartography, DOI: 10.1007/978-3-642-29793-9_5,

1 Introduction

In the "Virtual Forest" project we develop a next-generation geo-information-system which is no longer limited to two dimensions. The "4D-GIS" of the Virtual Forest supports three spatial dimensions and keeps track of changes over time. The final goal of the project is to provide a forest inventory database for the entire forest in North-Rhine Westphalia, Germany (approx. 9.000 km^2 and 240 million trees, whose diameter breast height is bigger than 20 cm), resulting in a new 4D-geo-information-system for the forest. We therefore had to consider available algorithms and developed new approaches for single tree delineation.

Many approaches for single tree delineation from LIDAR-data were published within the last 15 years. Hyyppä and Inkinen published a variant of the watershed algorithm (Hyyppä and Inkinen 1999). Persson, Holmgren and Södermann used Gaussian filtering and a clever heuristic to delineate trees from high resolution laser scanner data (Persson et al. 2002). They observed a hit-rate of 71 %. Erikson described an algorithm based on particle simulation (Erikson 2003). Diedershagen, Koch, Weinacker and Schütt extendend the Watershed algorithm with a heuristic (Diedershagen et al. 2003). Gougeon developed a valley following algorithm that was also used for LIDAR data sets (Gougeon 1998, 2010). Popescu used a filter window, which size was adapted to the yield classes of the stand (Popescu and Wynne 2004). Garcia, Suarez and Patenaude did a comparison of the algorithms of Gougeon, Popescu and Weinacker (Garcia et al. 2007). They stated, that the Gougeon algorithm was most suitable to estimate crown and stem diameter, while Popescu's approach delivered the best top height and Weinacker provided the best estimate for basal area and volume of the stand. Garcia et al. observed a hit-rate of 71.8 % for linked trees in the Popescu algorithm, 76.0 % for Weinacker and 60.2 % for Gougeon on a LIDAR dataset with 3–4 sample points per square meter.

The Volumetric Algorithm was already introduced at the 3D GeoInfo 07 in Delft (Rossmann and Buecken 2007). This algorithm provides high accuracy and good detection results, but in the presented version it was rather slow and it was still necessary to manually adjust a free parameter for each forestry unit. For a state-wide application, we needed a fast, fully automated approach, which works on a cluster of computers without user interaction.

In this paper we will first analyse the computational complexity of the original algorithm and then improve this to a linear complexity. Compared to the complexity of the original implementation, which is $O(n^2)$, this is a major step ahead. It is not only a speed-up, that one could also accomplish by using a faster computer, but it also makes the time, which is required for the calculation, more predictable, because the time does no longer depend on the size of the individual forestry units. Only the total size is now the criterion. Afterwards we will show how to estimate the free parameter by means of the Receiver Operator Characteristic (Fawcett 2003).

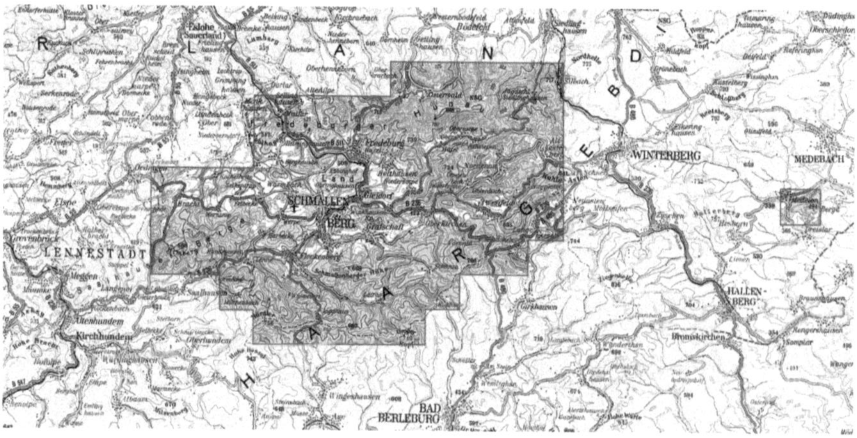

Fig. 1 The test-site in Schmallenberg in the Sauerland, North-Rhine Westphalia, Germany

It turns out that there is a correlation between the dominant height of a unit and the threshold. The result is a curve that can be used to calculate the free parameter of the algorithm based on the dominant height of the forestry unit.

The results of this automated approach have been tested and were compared to the detection results of an algorithm that was interactively tuned by a human operator. We observed that the automated approach performs in a similar quality as the human operator.

2 Geo-Data

We currently have recorded LIDAR data for test-areas of about 1400 km^2. The examples used in this paper were taken from the 2007 flight campaign in Schmallenberg (Fig. 1).

The LIDAR data was recorded with a Riegl LMS-Q560 with a density of six points per square meter and with full wave-form information. The data was then filtered and converted into a digital terrain (DTM) and digital surface model (DSM) with a 40 cm grid (6.25 grid-points per square meter). The elevation difference between the DSM and the DTM was calculated and serves as the foundation for the single tree delineation process. This model is known as normalized digital surface model (nDSM) and is sometimes also referred to as canopy height model (CHM) or differential model (DM). It defines the height of the vegetation at a certain coordinate. These steps were completed by different external companies.

An additional source of information, that is required to calculate the single tree layer, is a species map which has been generated from spectral and LIDAR data (Rossmann and Krahwinkler 2009). This data is used to facilitate the setting of different parameters of the algorithm for different species and to annotate the generated tree with its correct species and with further attributes like diameter at breast height, which cannot be detected in remote sensing data, but can be statistically derived from other parameters like crown diameter und tree height.

3 The Volumetric Algorithm

In this section we will first motivate the idea of the Volumetric Algorithm for single tree delineation and then optimize the original implementation by using a geometrical algorithm.

3.1 The Idea of the Algorithm

The Volumetric Algorithm (Rossmann and Buecken 2007) provides an effective solution to delineate individual trees in a forest that is described by a nDSM. This is equal to the task of deciding whether a local maximum is caused by a tree-top or a lateral branch. While the Watershed Algorithm only uses the basal area (i.e. a 2D projection) of a potential tree-top for its decision, the Volumetric Algorithm adds the third dimension and relies on the volume of a peak pointing out of the canopy. Figure 2 illustrates the workflow of this approach starting with the tree and its nDSM (a). In the subsequent images we flipped the nDSM upside down with the most significant points—the maximum heights in the original data that may represent tree-tops—as local minima of the graph. The inverted graph makes it easier to imagine the volumetric algorithm as a simulation of a water-flow. To get volumetric information we, figuratively spoken, fill the nDSM with water (b). Then in each cycle the point with the highest water-level is "opened", the flow of the water is simulated and the amount of water that drains out of the opening is measured (d). The measured volume is composed of the volume of the peak itself and a basal volume below the peak and the adjacent peaks which are dominated by this peak. The interesting feature is that the resulting volume emphasizes peaks that are dominant in the surrounding. For each volume that is higher than a threshold t a tree is generated. In the example only one tree was detected for the point with a volume of 495 m^3.

This threshold t differs for each forestry unit. In the implementation of 2007, this parameter was set interactively by an operator. This process was supported by a graphical user interface (GUI) which visualizes the effect of the current threshold setting. In this GUI the operator moved a slider until the generated trees match the scenery best.

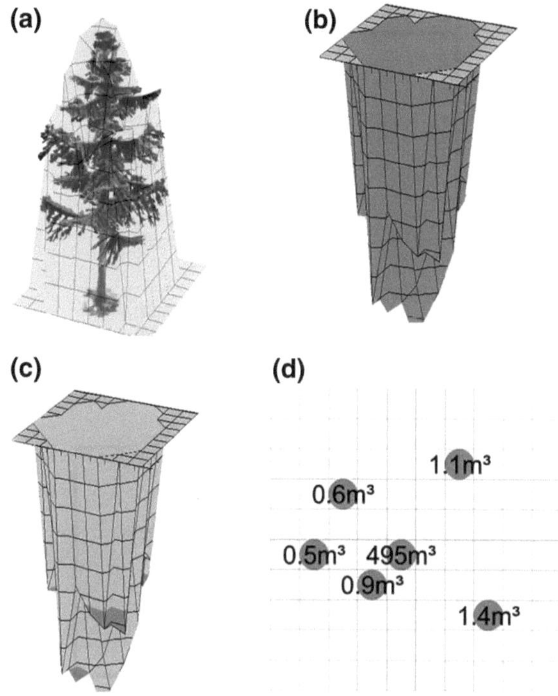

Fig. 2 The volumetric algorithm: a tree (**a**), its flipped nDSM filled with water (**b**), the first cycle of the algorithm (**c**) and the calculated volumes (**d**)

3.2 The Original Implementation

In this section we will discuss the complexity of the flow simulation on the nDSM for a grid with the same number of grid cells in both dimensions only ($s\ x\ s$ cells). It can be shown easily that grids with different amounts of cells in the two dimensions ($s\ x\ t$ cells) will lead to the same complexity.

If we speak of flow simulation on a discreet, grid-oriented data-set, the traditional class of algorithms would be cellular automata, which was also the first implementation of the algorithm that we chose. Figure 3 shows how this implementation works. For each local maximum a new flow simulation is started. In the worst case a grid of n cells can have

$$m = \frac{n}{4} \tag{1}$$

local maxima which leads to the same number of executions of the outer loop of the algorithm.

In each loop we see two actions that require more than constant time: the search for the cell with the highest amount of water and the flow simulation. The search has to consider all cells because the water level in each cell could have changed during the last step, which would change the order in which the cells have to be processed.

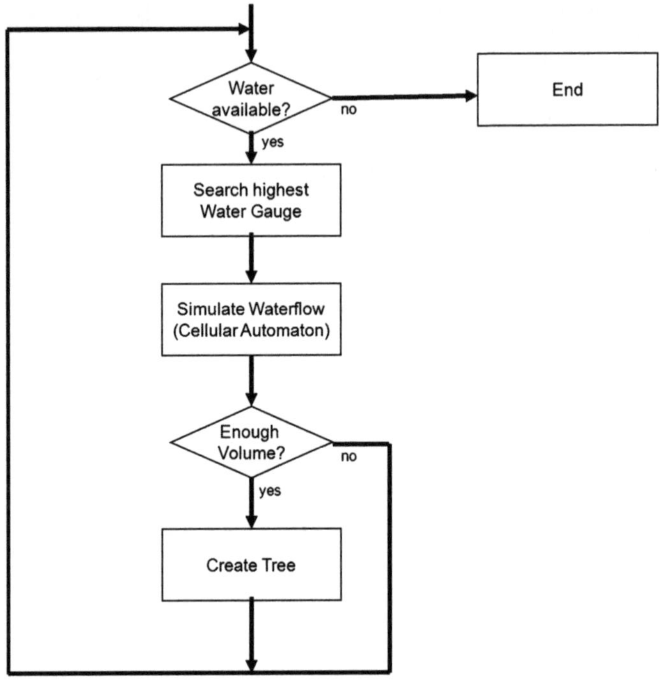

Fig. 3 The volumetric algorithm in its original implementation

Figure 4 shows how the flow simulation on a cellular automaton works. It pictures the first cycle of the example in Fig. 2. At the beginning (a) we see the grid-cells in top-view where the grey-scale-level is equivalent to the amount of water that is currently stored in each cell. In the first step the central cell, which carries the highest amount of water, is emptied. Then all cells adjacent to the first are considered. This is repeated until either the whole grid has been considered or the automaton has not changed for one cycle. The final state is displayed on the image bottom middle. It seems as though no water is left in the grid, but an amplified display of the same situation (g) shows the water that has been left. This simulation considers a single cell in the first step, then the cell and its neighbours and so on. In the worst case—when the local maximum is located in the corner of the grid—the simulation would therefore require

$$s = \sqrt{n} \tag{2}$$

steps on the s x s grid with growing sizes of the considered area (n: number of grid-cells). This would lead to a complexity of

$$\sum_{i=1}^{s} i^2 = \frac{s(s+1)(2s+1)}{6} \approx \frac{s^3}{3} = \frac{n^{1,5}}{3} \tag{3}$$

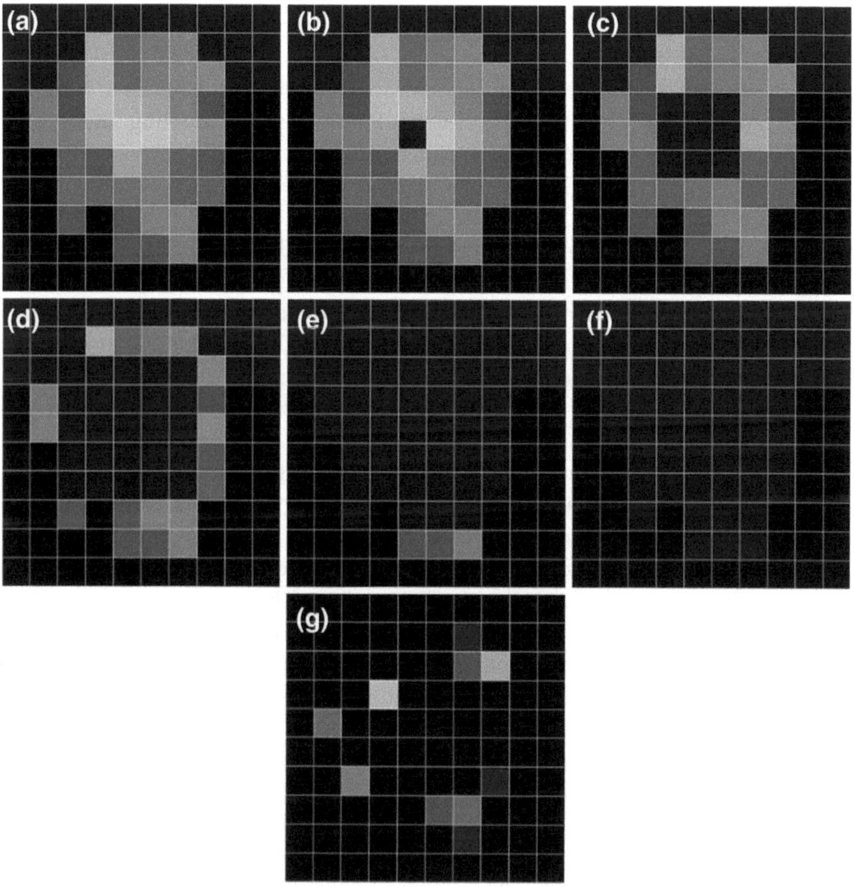

Fig. 4 The flow simulation for the first step of the example in Fig. 2. Volumes are displayed by the level of brightness, a bright cell is equivalent to a high volume. **a** initial situation. **b** First cycle, the local maximum has been processed. **c** Second cycle, the local maximum and the adjacent cells have been considered. **d** Third cycle, the cells next to the previous areas have been processed. **e** Fourth cycle. **f** Fifth and last cycle: all cells have been processed. **g** Changing the scale between brightness and volume reveals that there are still small volumes left that need to be processed in separate passes

There exists also a faster, list-oriented implementation for this step that only considers the borders of the area and does not need to look at the center of the actual peak again. This implementation touches each cell twice which leads to a linear complexity of this step.

In the worst case this algorithm needs n repetitions of the loop where the action within the loop has a linear complexity. This again leads to a total complexity of $O(n^2)$.

3.3 The Sweep-Plane-Implementation

In a data-set with a resolution of 40 cm a forestry unit of 1 ha includes n = 62,500 sample points. This gives an impression of how fast the running time of an algorithm with quadratic complexity will increase. When we started to consider larger units we therefore had to find a more efficient implementation of the Volumetric Algorithm.

While the cellular automaton works with a two dimensional grid, the same dataset can also be displayed as a three dimensional grid that consists of voxels (Fig. 5a). Like a dicom-image of a MRT scan in radiology we can also consider this grid as a pile of layers and can display each of the layers with a sectional view that shows all the voxels on this layer. Each voxel can be assigned to one of the local maxima and the number of voxels is then proportional to the volume that is measured at this peak.

The new implementation of the Volumetric Algorithm uses a geometric approach with a sweep plane that iterates through voxel layers to calculate the volumes of the peaks. In this representation the pile of voxels on each cell has got the height, that is specified in the nDSM for this cell. In the first step of the algorithm the topmost voxels of each pile are sorted by their z-coordinate. This provides an order in which the voxel piles have to be considered when the voxel layers are iterated from top to bottom. The sort operation can be done in linear time with bucket sort because the values are discreet and lie within a limited range.

The sweep-plane contains a grid with the same size as the nDSM. It starts at the top-most layer that contains voxels. These are transferred as points to the sweep-plane grid and afterwards all independent regions within this grid are detected with a run through all cells of the plane and a call of a flood fill algorithm for each new region ("seek and paint"). Each region is given a number and marked on the sweep plane with this number. The number of cells for each region, which equals the number of voxels in this layer for each region, is stored. Then the sweep-plane moves down layer by layer. If a layer does not contain new voxel piles, the situation on the sweep-plane remains unchanged and the same number of voxels that was already used in the previous layer is again added to the counter of each region. If the layer contains new points, they are added to the sweep-plane. After that another seek-and-paint run has to be started. This time it has to consider the already existing regions in the order of their generation. The algorithm starts a flood-fill at the center of each region and counts the new voxels for it. Afterwards the algorithm checks the sweep-plane for still untouched cells. If it finds untouched cells, it starts a new region at this point and again counts the voxels for this region with a flood-fill.

This procedure is repeated until the sweep-plane reaches the ground. The order of the flood-fill-calls ensures that the dominant peak still gets the volume of the peak itself and the basal volume below the peak and the adjacent sub-ordinate peaks. Figure 5b-g shows how the algorithm works on the example. Three different layers are shown and the regions are painted in different colours

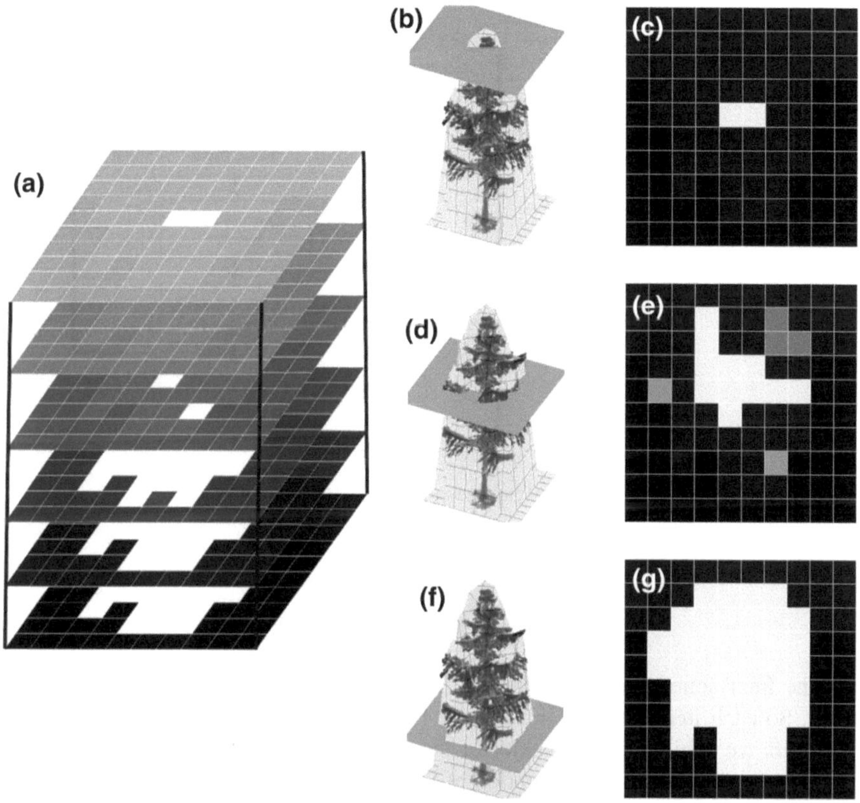

Fig. 5 The 3d-voxel-representation decomposed into layers

according to the region which they were assigned to. Figure 6 shows a diagram of the algorithm.

The number of iterations depends no longer on the number of grid-cells but on the number of layers. Our implementation considers 10,000 layers—a maximum of 100 m in height and 1 cm resolution. Each iteration consists of a maximum of three subsequent actions: insertion of new points, seek-and-paint-step and update of the number of total voxels for all regions. In the worst case all cells are added in one layer. In this case the insertion step will take a time proportional to n. For the seek-and-paint-step it is important to see that all calls of the flood-fill together have a linear complexity. With the 4-connected- or 4-neighbour-flood-fill a grid-cell can be entered from 4 directions only. Note that this cannot happen more than once for each direction during a single iteration of the algorithm. Additionally, there are a maximum of m (maximum number of regions, see formula 1) center-points of regions that have to be checked and where the flood-fill might start and n cells that are checked for a new region. Therefore this step has still got linear complexity:

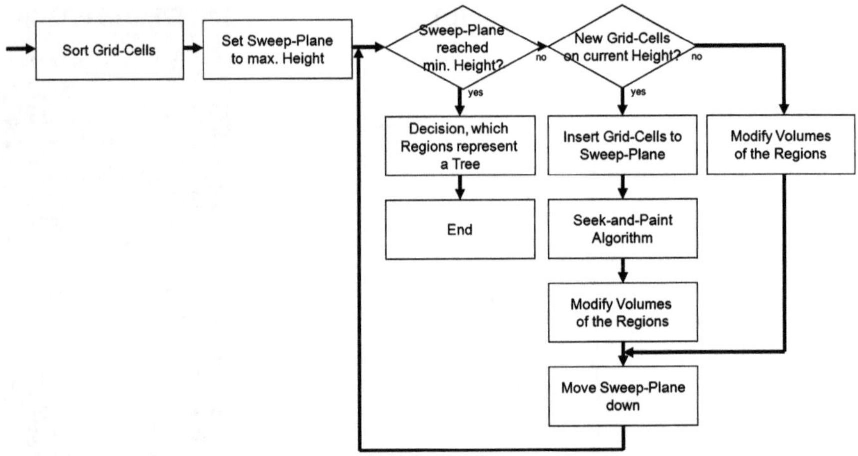

Fig. 6 The volumetric algorithm in the sweep-plane implementation

$$4 * n + m + n = 5\frac{1}{4}n \in O(n) \tag{4}$$

The final action, the update for all regions, can consume a time proportional to m. So each iteration consumes linear time. Together with the constant number of layers we achieve a complexity of the algorithm of O(n).

This sweep-plane implementation of the Volumetric Algorithm features a linear complexity in comparison to the original implementation with a quadratic complexity. This leads to an acceleration of the running time for forestry units of 1 ha and more. For example a test unit with 5.37 ha takes 3:55 min with the original implementation and 2:03 min with the sweep-plane implementation.

There is one more interesting fact about the sweep-plane algorithm: Because all grid-cells have to be touched at least once, there cannot be an algorithm with a complexity better than O(n). This implies that all other algorithms that solve the same problem are at a maximum faster by a constant factor.

4 Elimination of the Free Parameter

The new sweep-plane implementation of the Volumetric Algorithm accelerates the calculation of the volumes for the local maxima, but it does not change the second step: It is still necessary to define a threshold t for each forestry unit, which determines whether a local maximum with its associated volume represents a tree top or a lateral branch. In the following part of this paper we will show how to determine a heuristic for this threshold.

Table 1 The four Sets that are used in the Receiver Operator Characteristic

		Ground-Truth Yes (P)	Ground-Truth No (N)
Detection Result	Yes (D)	True Positive TP	False Positive FP
Detection Result	No (ND)	False Negative FN	True Negative TN

4.1 The Approach via "Receiver Operator Characteristic"

The Receiver Operator Characteristic is a common approach in problems related to signal detection. For example in medical applications it is used to determine thresholds for the feature based derivation of a diagnosis (Obuchowski 2005).

The ROC uses ground truth data and calculates results for different settings of the independent parameter of the algorithm (classifiers)—in our example for different settings of the threshold value. It examines a set of candidates C, which is a superset of the set of the ground-truth elements and the detected elements. In our example it is the set of all local maxima in the normalized digital surface model nDHM. All elements of C are grouped into four subsets TP, FP, FN and TN according to Table 1.

In the example of single tree delineation, the classes in the table contain the following elements:

- P: All local maxima that correspond to a tree in the ground truth dataset.
- N: All local maxima that do not correspond to a tree in the ground truth dataset.
- D: All local maxima that were detected as a tree.
- ND: All local maxima that were neglected as a lateral branch.
- TP: The detected trees that correspond to a tree in the ground truth dataset.
- FP: The detected trees that have no corresponding tree in the ground truth dataset.

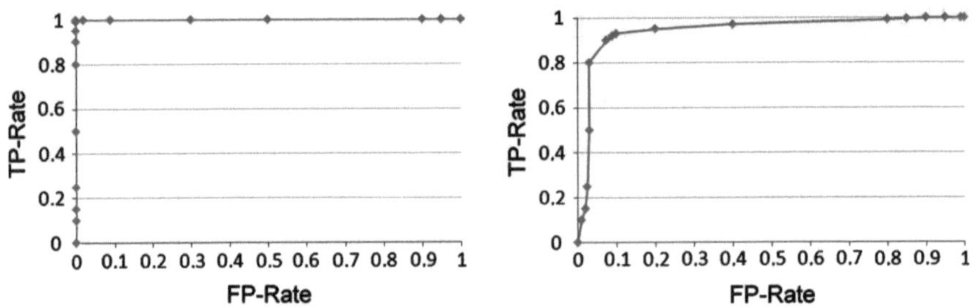

Fig. 7 An ideal (*left*) and a real ROC graph (*right*)

- FN: The local maxima in the nDSM, which were classified as lateral branches but correspond to a tree in the ground truth dataset.
- TN: The local maxima in the nDSM, which were classified as a lateral branch and which have no corresponding tree in the ground truth dataset.

In the Volumetric Algorithm for single tree delineation, a smaller threshold corresponds to a finer segmentation. Thus, we can expect that the number of true positives as well as the number of false positives increases when the threshold is decreased. Once a tree is detected for a certain threshold t_1, it will also be detected for all threshold t_x which are smaller than t_1. This corresponds to a monotonic behaviour of TP and FP.

With the ROC graph it is possible to characterize the change of TP and FP for different classifiers. It plots either the TP rate ("hit-rate")

$$TP - Rate = \frac{|TP|}{|P|} \tag{5}$$

against the FP rate ("false alarm rate")

$$FP - Rate = \frac{|FP|}{|N|} \tag{6}$$

or $|TP|$ against $|FP|$(absolute numbers).

In our example it is reasonable to choose absolute numbers, because $|N|$ is usually much larger than $|P|$ and we want to keep the number of false positives reasonable. In the ROC graph, each possible setting of the algorithm is displayed as a point

$$p = (|FP|; |TP|) \tag{7}$$

In the ideal case, this graph would look like Fig. 7 left. In our example this graph corresponds to the following situation: First the threshold is higher than the maximum volume of a peak that appears in the scene: No tree is generated. When the threshold gets smaller, the first trees are recognized by the algorithm. This continues until all trees in the unit are recognized but no additional tree was

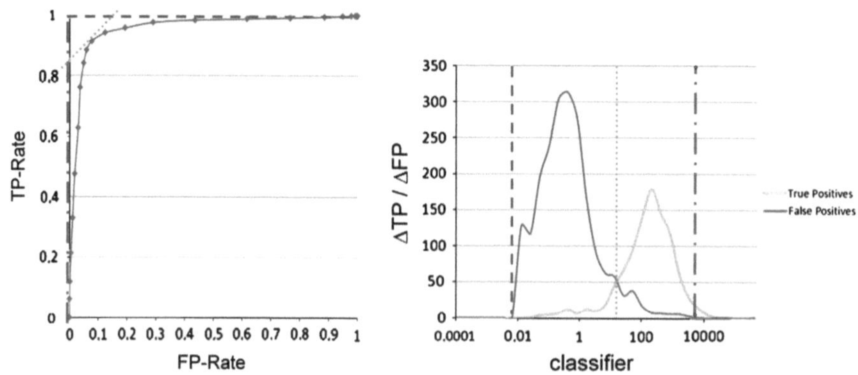

Fig. 8 A ROC graph (*left*) and the corresponding density graph (*right*) with three marks in the graphs

generated. When the threshold further decreases, additional trees (false positives) appear.

For most practical computer vision applications this ideal shape can hardly be achieved. Usually, the graph is shaped like Fig. 7 right. This shows that the first false positives appear before the last true positive was detected. But independent of the exact form of the graph, we can define a rule for the selection of the classifier with it and therewith objectify the selection. In order to visualize the effects of several rules, we use a second kind of graph—the density graph. The density graph plots the change of true positives as well as the change of false positives against the classifier. For each classifier (x-axis) it shows how many additional true positives and false positives (y-axis) appear.

We marked some rules in the density graph (Fig. 8):

- Only true positives (line out of dots and dashes). The yellow line in the right graph marks the minimum classifier where no false positive appeared. In the example this would correspond to a segmentation where all detected trees correspond to a real tree and no additional tree was detected. In the ROC graph this is the point where the graph leaves the y-axis or the highest point where the yellow 90°-tangent contacts the graph.
- All true positives detected (line out of dashes). The blue line in the right graph marks the maximum classifier, where all true positives appeared. In the example this is the situation, when the maximum number of trees is detected, independent of the number of additional trees that appeared at this time. In the ROC graph this is the point where the graph contacts the blue 0° tangent first.
- Trade-off between true-positives and false-positives (line out of dots). In the density graph we look for the point where the number of additional true positives equals the number of additional false positives. In our example this is the break-even point between additional correctly detected trees and misclassified trees. If we decrease the threshold further we will get additional correctly detected trees, but this happens at the price of a larger amount of false positives. In a ROC graph, that plots absolute numbers, this is the point where the green 45° tangent contacts the graph.

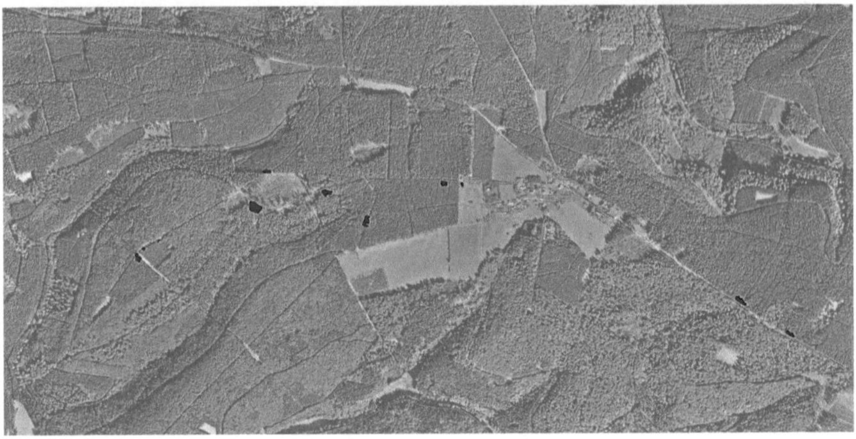

Fig. 9 The test units, that were used for the analysis in the paper

Table 2 Attributes of the test units

Unit	Dominant height (m)	Yield factor	Age (years)	Number of trees
1	21.11	0.75	47	26
2	16.37	0.98	39	16
3	12	0.87	30	14
4	31.8	0.66	91	35
5	19.86	1.02	46	26
6	30.89	0.7	86	19
7	25	0.74	60	26
8	19.19	0.9	45	29
9	16.66	0.88	39	14
10	27.5	0.84	69	21
11	28	0.86	71	13

Each of these rules defines an objective criterion for the choice of the classifier for a given example.

4.2 Parameterization of the Volumetric Algorithm

In this section, we will show how the concept of the Receiver Operator was used to find a formula that specifies the threshold of the Volumetric Algorithm in relation to parameters of the forestry unit that can be monitored from remote sensing data.

We first collected ground truth data for eleven forestry units in the area of Schmallenberg (North Rhine-Westphalia, Germany) (Fig. 9 and Table 2). The test units contain 237 trees and represent a broad spectrum in the attributes dominant height (12–30.89 m),

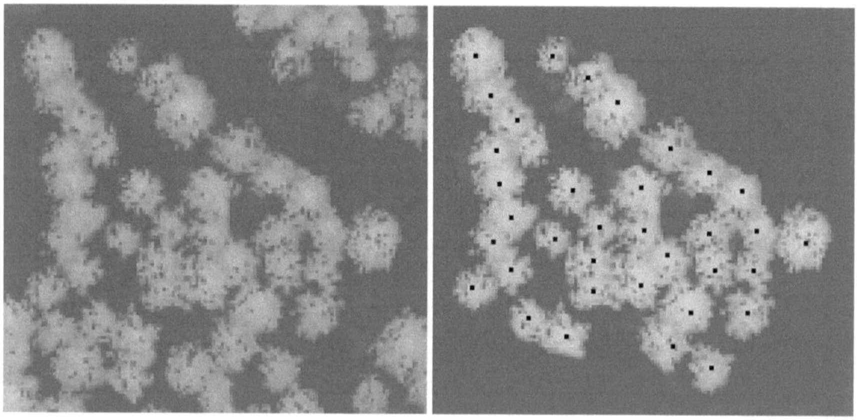

Fig. 10 The nDSM (*left*) and the ground truth map (*right*) for unit 4

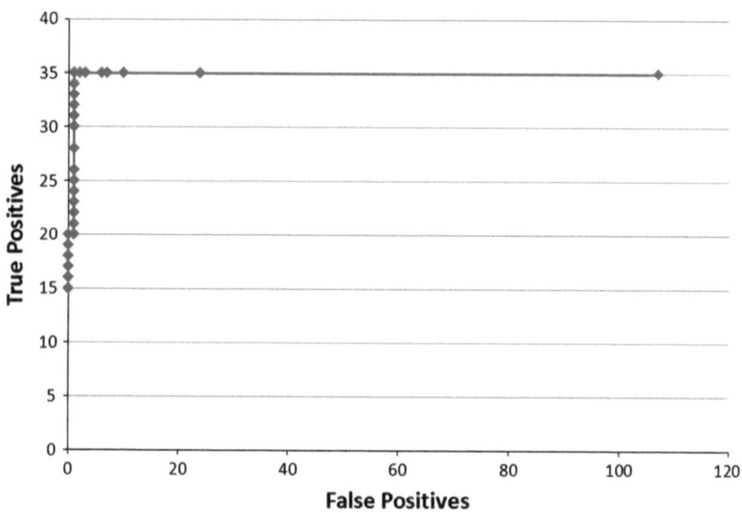

Fig. 11 The ROC graph for unit 4 based on 2,000 threshold values

age (30–115 years) and yield density (0.66–1.02). For each forestry unit we generated a ground truth map (Fig. 10). This map contains multiple kinds of information:

- All trees in the unit are marked with a red pixel.
- Everything that does not belong to the test unit is marked black.
- Every other pixel has to be marked in another color than black and red.

This representation of ground truth data is simple to generate: We used the grey scale image of the nDSM (Fig. 10 left). In a first step the tops of the crown of all trees were marked in the map and in a second step all pixels that belong to trees which are located outside of the unit were painted with black color. These

Table 3 Thresholds for the 11 units

Unit	Dominant height (m)	Yield factor	Range of thresholds	Average
1	21.11	0.75	[0.2;1.0]	0.6
2	16.37	0.98	[0.2;1.0]	0.6
3	12	0.87	[0.0;0.0]	0
4	31.8	0.66	[9.2;30.0]	19.6
5	19.86	1.02	[0.2;1.0]	0.6
6	30.89	0.7	[4.2;18.0]	11.1
7	25	0.74	[2.2;4.0]	3.1
8	19.19	0.9	[0.2;1.0]	0.6
9	16.66	0.88	[0.2;1.0]	0.6
10	27.5	0.84	[5.2;23.0]	14.1
11	28	0.86	[1.2;11.0]	6.1

operations can be performed with a simple graphics editor in the field and allow a fast recording of the required ground truth data.

During the calculation this image was first used to filter the relevant parts of the nDSM. Afterwards the positions of the red dots were extracted. We then calculated all tree candidates with the Volumetric Algorithm and performed a mapping between the generated trees and the marked ground truth trees for 2,000 settings of the threshold value. In this dataset we counted the true positives and false positives for each threshold value and generated a ROC graph for the unit. Figure 11 shows the ROC graph for unit 4 as an example.

We chose to accept a trade-off between true positives and false positives, because we wanted to achieve a balanced behavior of the algorithm which is acceptable for all parties in the forestry administration and industry. Therefore we decided to use the third rule from Sect. 4.1 that would look for the contact point of the 45° tangent. This balances the generation of additional TPs against the addition of unwanted FPs. In most graphs this point did not correspond to a single threshold but to a range of threshold values, that produce the same detection result. Table 3 lists the ranges and the average value of the thresholds for the 11 units.

The next step is to find a relation between the attributes of the forestry units and the threshold values. We tried the attributes dominant height and yield factor for this relation, because these attributes can be monitored from remote sensing data (Rossmann et al. 2009). It turned out that there is obviously a connection between the dominant height and the threshold (Fig. 12 top). We observed a correlation factor of 0.854. For the eleven test units we could not see a clear connection between the yield factor and the threshold (Fig. 12 bottom, correlation factor – 0.633). This observation could change for a larger set of test units, because the yield factor is a rather fuzzy attribute and it is common that this attribute varies by 10–20 % when it is determined by different surveyors.

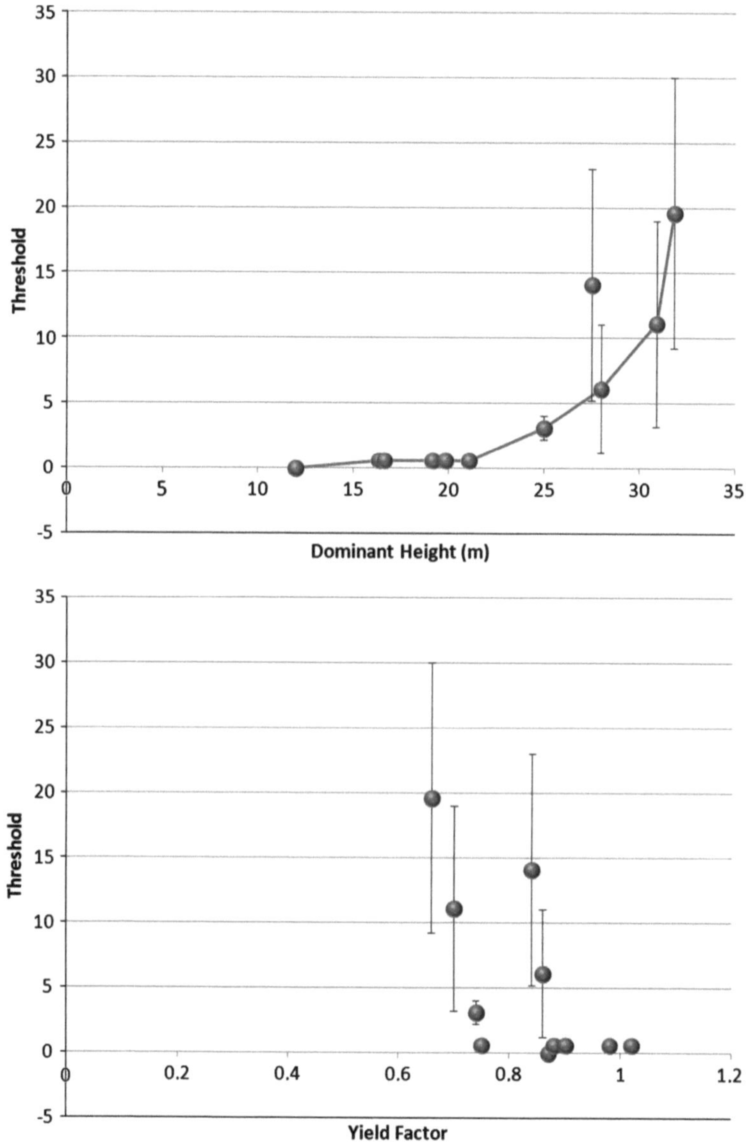

Fig. 12 The relation between the thresholds and the dominant height (*top*), respectively the thresholds and the yield factor (*bottom*)

5 Results

We implemented the sweep-plane variant of the Volumetric Algorithm and added the parameter curve as a heuristic that provides a connection between dominant

Table 4 Comparison of the number of delineated trees for fourteen test units

Region	Number of Trees	Human Operator	Heuristic
1	16	15	15
2	124	115	113
3	122	110	109
4	56	53	52
5	78	75	75
6	147	132	130
7	102	93	93
8	126	121	116
9	139	132	131
10	49	47	47
11	105	92	91
12	78	70	69
13	53	48	45
14	71	63	63
Total	1266	1166	1149
Hit-rate		92.1	90.8

height and threshold. With this implementation we tested both, execution speed and the detection rate of the algorithm.

5.1 Detection Result

Besides the eleven units with 237 trees that were used to determine the shape of the ROC graph, we have got ground truth data for fourteen additional regions which contain another 1,266 trees. These trees were used to test the segmentation results of the new heuristic. We used the sweep-plane implementation of the Volumetric Algorithm to calculate the volumes for all local maxima. Afterwards we determined the dominant height for the regions and calculated the appropriate threshold by means of the Receiver Operator Characteristic as explained in the previous chapter. The same area was then also processed by an experienced human operator with the interactive graphical user interface. Table 4 compares the results of the algorithm and the operator for the ground-truth data.

It turned out that the detection result of the time-consuming interactive segmentation is only slightly better than the fully-automated delineation. Only seventeen trees (which is less than 1.5 % of the total number of trees) were additionally detected in the manual process.

Table 5 Comparison of the performance of the original version and the sweep-plane implementation of the volumetric algorithm on a standard pc (core 2 duo, 2,13 GHz)

Unit	Size (ha)	Original version (min)	Sweep-plane implementation (min)
1	0.4	0:10	0:15
2	0.92	0:11	0:14
3	3.2	2:02	1:26
4	5.37	3:55	2:03
5	7.92	6:23	3:39
6	12.02	13:58	5:08

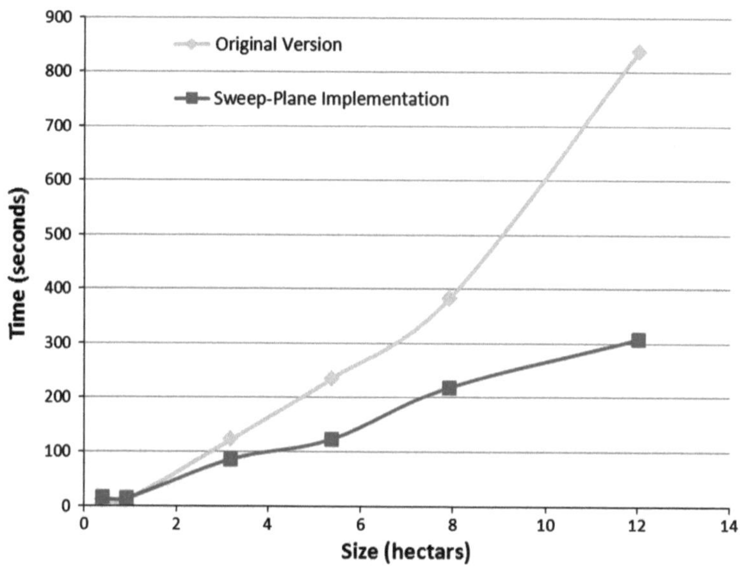

Fig. 13 A plot of the running-times for the test-units

5.2 Processing Speed

In addition to the good detection results it was also essential that the algorithm performs fast enough for large areas. We therefore analyzed the performance of the algorithm for several individual units of different sizes. Table 5 and Fig. 13 show the results of this test. It turns out that the sweep-plane implementation is faster for large areas but a little slower for small units. The time that is used to initialize the data structures effects the total running time as well as the considerably high, constant number of iterations, that the sweep-plane algorithm has to undergo.

After these first tests, we decided to use the whole test-area of Schmallenberg as a benchmark. On a single high-performance personal computer the algorithm took less than 4 h to delineate the trees of all forestry units within the 280 km^2 of this test-area.

If we extrapolate this performance linear from the time that was required to process the 280 km^2 to the time that will be required for the delineation of the area of the whole state of North-Rhine Westphalia (34,088 km^2), it is possible to process this area within a month on a single pc or within days on a small cluster.

6 Conclusions

The new implementation and the added heuristic significantly enhance the usability on the Volumetric Algorithm for single tree delineation. While the 2007 version of the algorithm was suitable to process a small number of individual forestry units, the new version presented in this paper has got the required performance and the level of automation to process areas of the size of a state. Due to the linear complexity it becomes possible to extrapolate the calculation time only based on the total area of the test site.

The quality of the detection-results of the fully-automated process comes close to the results that a human operator can achieve on the same areas with the interactive implementation.

The ROC proved to be a powerful tool that can be used to estimate free parameters objectively. While a parameterization based on a manual process always mirrors personal attitudes of the operator and therefore requires that a greater number of operators to adjust the algorithm for the same samples, the ROC reduces the effort to one calculation pass. With this property, the Receiver Operator Characteristic has shown to be not only valuable for single tree delineation, but also for a number of other algorithms in the context of remote sensing in forestry applications.

Until now, we tested the heuristic only on LIDAR data that was recorded with the same parameters (point-density, beam-diameter, etc.) as the Schmallenberg data-set, because all of our test-areas were recorded with these settings. Currently, we are deriving a toolset to allow us to make "educated guesses" concerning the performance of our approach in areas with different forest types, as well as different data densities and qualities.

The delineated trees were used to initialize a single-tree database, which will provide the foundation for a single-tree forest inventory.

Acknowledgments The project "Virtual Forest" is co-financed by the European Union and North-Rhine-Westphalia—European Regional Development Fund (EFRE)

References

Buecken A, Rossmann J (2007) Using 3D-laser scanners and image-recognition for volume-based single-tree-delineation and -parameterization for 3D-GIS-applications. In: van Oesterom,

Zlatanova, Penninga, Fendel (eds) Advances in 3D geoinformation systems, lecture notes in geoinformation and cartography LNG&C, Springer, Delft

Diedershagen O, Koch B, Weinacker H, Schütt C (2003) Combining LIDAR- and GIS data for the extraction of forest inventory parameters. In: Proceedings scand laser 2003

Erikson M (2003) Structure-preserving segmentation of individual tree crowns by brownian motion. In: Proceedings SCIA 2003, pp 283–289

Fawcett T (2003) ROC graphs: notes and practical considerations for data mining researchers. HP http://www.hpl.hp.com/techreports/2003/HPL-2003-4.pdf. Accessed 10 January 2012

Garcia R, Suárez J, Patenaude G (2007) Delineation of individual tree crowns for LiDAR tree and stand parameter estimation in Scottish woodlands. In: Fabrikant, Wachowski (eds) The European information society—leading the way with geo-information lecture notes in geoinformation and cartography, Springer, Berlin

Gougeon FA (1998) Automatic individual tree crown delineation using a valley-following algorithm and a rule-based system. In: Proceeding of international forum on automated interpretation of high spatial resolution digital imagery for forestry. 11–23

Gougeon FA (2010) Forest remote sensing in canada and the individual tree crown (ITC) approach to forest inventories. J Fac Agric Shinshu Univ 46:85–92

Hyyppä J, Inkinen M (1999) Detecting and estimating attributes for single trees using laser scanner. Photogram J Finland 2:27–42

Obuchowski N (2005) Fundamentals of clinical research for radiologists, ROC analysis. Am J Roentgenol 184:364–372

Persson Å, Holmgren J, Söderman U (2002) Detecting and measuring individual trees using an airborne laser scanner. Photogramm Eng Remote Sens 68:925–932

Popescu S, Wynne R (2004) Seeing the trees in the forest: using lidar and multispectral data fusion with local filtering and variable window size for estimation tree height. Photogramm Eng Remote Sens 70:589–604

Rossmann J, Krahwinkler PM (2009) Tree species classification and forest stand delineation based on remote sensing data – large scale monitoring of biodiversity in the forest. In: Proceedings of the ISRsE33

Rossmann J, Schluse M, Buecken A, Hoppen M (2009) Advances in forestry geo-information systems enabling new approaches in the bioenergy sector. In: Proceedings of the bioenergy 2009 conference

A Service-Based Concept for Camera Control in 3D Geovirtual Environments

Jan Klimke, Benjamin Hagedorn and Jürgen Döllner

Abstract 3D geovirtual environments (3D GeoVEs) such as virtual 3D city models serve as integration platforms for complex geospatial information and facilitate effective use and communication of that information. Recent developments towards standards and service-based, interactive 3D geovisualization systems enable the large-scale distribution of 3D GeoVEs also by thin client applications that work on mobile devices or in web browsers. To construct such systems, 3D portrayal services can be used as building blocks for service-based rendering. Service-based approaches for 3D user interaction, however, have not been formalized and specified to a similar degree. In this paper, we present a concept for service-based 3D camera control as a key element of 3D user interaction used to explore and manipulate 3D GeoVEs and their objects. It is based on the decomposition of 3D user interaction functionality into a set of services that can be flexibly combined to build automated, assisting, and application-specific 3D user interaction tools, which fit into service-oriented architectures of GIS and SDI-based IT solutions. We discuss 3D camera techniques as well as categories of 3D camera tasks and derive a collection of general-purpose 3D interaction services. We also explain how to efficiently compose these services and discuss their impact on the architecture of service-based visualization systems. Furthermore, we outline an example of a distributed 3D

J. Klimke (✉) · B. Hagedorn · J. Döllner
Hasso-Plattner-Institut, University of Potsdam,
Prof.-Dr.-Helmert-Str. 2-3,
14482 Potsdam, Germany
e-mail: jan.klimke@hpi.uni-potsdam.de

B. Hagedorn
e-mail: benjamin.hagedorn@hpi.uni-potsdam.de

J. Döllner
e-mail: doellner@hpi.uni-potsdam.de

J. Pouliot et al. (eds), *Progress and New Trends in 3D Geoinformation Sciences*,
Lecture Notes in Geoinformation and Cartography, DOI: 10.1007/978-3-642-29793-9_6,

geovisualization system that shows how the concept can be applied to applications based on virtual 3D city models.

Keywords 3D camera control · Service-oriented architecture · 3D portrayal · Virtual 3D city models · 3D thin client

1 Introduction

3D geovirtual environments (3D GeoVEs) such as virtual 3D city models and 3D landscape models serve as integration platforms for complex 2D and 3D geospatial information. They provide a conceptual and technical framework to integrate, manage, edit, analyze, and visualize that information, facilitate use and communication of geospatial information, and represent a key functionality for IT solutions based on 3D GeoVEs. There is a growing number of application fields for 3D GeoVEs, in particular in those fields that require a true three-dimensional representation as in the case of virtual 3D city models. Because 3D GeoVEs commonly rely on massive, heterogeneous, and complex structured 3D geodata, high-quality, interactive 3D geovisualization systems usually demand for high processing power, large memory, and hardware-accelerated 3D graphics. These demands, however, make the development of robust, efficient, and compatible client applications a challenging task.

Service-oriented architectures (*SOA*) (Papazoglou et al. 2007), as a paradigm for design and development of distributed information systems, represent a common approach to address these challenges. 3D geovisualization systems, based on the SOA paradigm, encapsulate resource intensive tasks, such as management, processing, transmission, and rendering of massive 2D and 3D geodata as services that can be reused by various client applications. While 2D geovisualization systems can rely on standardized and robust services such as the Web Map Service (WMS), specified by the Open Geospatial Consortium (OGC), only first approaches and service implementations for 3D geovisualization, namely the *Web 3D Service* (W3DS) and the *Web View Service* (WVS), have been suggested (Schilling and Kolbe 2010; Hagedorn 2010). Recent developments in service-oriented architectures for interactive 3D portrayal aim at making 3D geodata, geodata management functionalities, and geospatial knowledge available even through *thin clients*, i.e., applications with low requirements concerning processing power and 3D graphics capabilities designed for lightweight platforms such as mobile phones or web browsers (Hildebrandt et al. 2011).

Besides capabilities for the presentation of 3D geodata, client applications need to offer tools to interact with 3D GeoVEs. 3D camera control, as the major 3D interaction type, enables users to explore and use a 3D GeoVE; it is crucial for its usability, as "a 3D world is only as useful as the user's ability to get around and interact with the information within it" (Tan et al. 2001). Existing 3D portrayal services provide only rudimentary support for user interaction or camera control. Service-based approaches for 3D interaction have not been specified and formalized so far. Thus, 3D camera control functionality still needs to be designed and implemented separately for each client application.

In this paper, we present a concept for service-based 3D camera control as a key element of 3D user interaction used to explore 3D GeoVEs and their objects. It is based on the decomposition of 3D user interaction functionality into a set of services that can be flexibly combined to build automated, assisting, and application specific 3D user-interaction tools, which fit into service-oriented architectures of GIS and IT solutions based on spatial data infrastructures (*SDI*).

We discuss 3D camera control, categories of 3D camera tasks, and derive a collection of general-purpose 3D interaction services. We also explain how to efficiently compose these services and discuss their impact on the architecture of service-based visualization systems. Furthermore, we show by example how to apply this concept and to decompose a specific camera control technique into a set of services.

The remainder of this paper is organized as follows: Section 2 provides an introduction to service-based 3D geovisualization and 3D camera control in 3D GeoVEs as well as related work. Section 3 presents our concept for the decomposition of 3D camera control functionalities and their provisioning as services. Section 4 gives an example of a distributed camera control system that is based on the described services. Section 5 provides a discussion of the properties of such systems. Section 6 gives conclusions and an outlook.

2 Basics and Related Work

In this paper we build onto research in the area of service-based geovisualization and user interaction in virtual environments. Service-based geodata provisioning, processing and visualization have been standardized in recent years and systems implementing this paradigm are continuously evolving. In the following we provide an introduction to service-based 3D geovisualization and provide related work in the area of 3D camera control.

2.1 Service-Based 3D Geovisualization

The interoperability of systems and applications dealing with geodata is a central issue to build systems out of interoperable software components for geodata access, processing, and visualization. Beside a common understanding on information models (Bishr 1998), definitions of service interfaces are necessary. The Open Geospatial Consortium (OGC) defines a set of standardized services, models, and formats for geodata encoding and processing. For example, a Web Feature

Service (WFS) (Panagiotis and Vretanos 2010) can provide geodata, encoded in the Geography Markup Language (GML) (Portele 2007) or City Geography Markup Language (CityGML) (Gröger et al. 2008), and processed by a Web Processing Service (WPS) (Schut 2007).

For geovisualization processes a general portrayal model is provided by the OGC that describes three principle approaches for distributing the tasks of the general visualization pipeline between portrayal services and consuming applications (Altmaier and Kolbe 2003; Haber and McNabb 1990). While the OGC Web Map Service (WMS), providing map-like representations of 2D geodata, is widely adapted and used, 3D geovisualization services have not been elaborated to a similar degree. Several approaches for 3D portrayal have been presented (Basanow et al. 2008) and are currently discussed as standard proposal in the context of the OGC (Schilling and Kolbe 2010; Hagedorn 2010). These approaches differ in the type of data that is exchanged between client and service: Either filtered feature data, graphical representations (display elements), or rendered images are transmitted. Each type of data is generated by one specific OGC service:

- A WFS provides *feature* data, encoded in standardized formats, to a service consumer. This data can be processed at the client side; for visualization a *thick client* has to derive graphical representations and to perform the rendering.
- A Web 3D Service (W3DS) provides *display elements* to a service consumer (e.g. X3D or KML, organized as 3D scene graph) (Schilling and TH 2010; Altmaier and Kolbe 2003). This representation includes, e.g. geometry information and texture data. For visualization, a *medium client* needs to be able to process and render this graphics data.
- A Web View Service (WVS) provides *images* of a 3D scene to a potentially *thin client* (Hagedorn et al. 2009). In the simplest case, a client displays finally rendered images to a user. More advanced clients may also allow for more interactive visualizations using a WVS.

These segmentations lead to different requirements regarding 3D rendering capabilities of the portrayal services and corresponding client applications and to different types of interaction techniques that can be implemented within client applications.

2.2 Camera Control for 3D GeoVEs

On a technical level, the 3D camera control process generally includes (a) recognizing navigation intentions, (b) deriving path information, and (c) adjusting the visualization. Users of a 3D GeoVE express their navigation intentions by inputs provided to a client application, such as pressing UI controls, selecting objects in the scene, or sketching paths or gestures (Hagedorn et al. 2009). User input is evaluated and camera animations, i.e., camera positions alongside with camera orientations, are derived.

2.2.1 Assisting Camera Control Techniques

3D camera control in a virtual environment is a complex task, especially for non-expert users. Therefore, techniques for camera control in virtual environments were presented that assist users to explore 3D space by avoiding confusing or disorienting viewing situations (Buchholz et al. 2005). Task-oriented camera techniques generate camera paths with respect to a high-level navigation intention, such as "go to the closest landmark".

A virtual camera's behavior can depend on the semantics of the underlying model data of the 3D GeoVE (Döllner et al. 2005). Such semantics-based camera interaction techniques need client-side data and computing capabilities to perform camera path computations. Due to network and computational limitations, it is hard to make such capabilities available on thin clients or for large datasets.

2.2.2 Camera Control in Service-Based 3D GeoVEs

For distributed applications using thick and medium clients (as described) the rendering stage of the visualization pipeline is implemented on client-side, so the necessary data, such as model geometry or points of interest, is generally available. Therefore various camera-control techniques can be implemented within such applications, while thin clients need service-side support for reaching corresponding results since there is usually only a very limited set of client-side information about the geometry or topology of the 3D environment.

For example, a W3DS client, running on a machine with high processing capabilities and high speed connection to the W3DS server, could provide highly interactive real-time visualization and camera control, based on the retrieved 3D graphics data.

In contrast, a WVS provides multi-layer images, including not only color images but also, e.g. depth information per pixel, which allows for implementing (a) clients that only display images and provide only a step-by-step navigation as well as (b) more complex clients that reconstruct the virtual environment from information contained in such images and could even provide real-time navigation.

The complexity of camera-control techniques achievable for these client classes differs: Thick clients can easily consider semantic information from underlying geodata, which is not per se available through a W3DS or WVS. However, each of these 3D portrayal approaches could benefit from providing camera-control capabilities as distributed, reusable resources.

2.3 Challenges for Camera Control in 3D GeoVEs

This paper is motivated by the goal to implement interactive 3D GeoVEs on thin clients. Compared to desktop-based, thick client 3D GeoVEs, thin client 3D GeoVEs face several challenges regarding network capabilities as well as device constraints such as computing capacity, presentation, and interaction issues.

For distributed systems implementing camera-control functionality, various requirements need to be considered for achieving effective 3D camera control for 3D GeoVEs:

- Visualization and interaction should be decoupled to facilitate reuse of camera control functionalities as independent building blocks of 3D geovisualization systems.
- The separation of camera-control functionalities should support the implementation as services, but also as integrated part of a client application. So the decision of the location of network boundaries in concrete system architectures can be made per client application.
- Feedback (about available and pending camera movements, as well as system state information) should be provided by a distributed system for 3D camera control.
- A distributed system for 3D camera control should be designed to deal with limitations of wireless communication networks in connection with mobile clients (e.g. connection loss, latency times, available bandwidth etc.).
- The system architecture should not be restricted to a special type of input. Especially mobile devices provide more than one sensor that can serve as user input device. For example, device location, orientation, speed or other data delivered by device sensors could influence the way camera control has to be performed, e.g. to support building a relation between a user's actual position and the position and orientation inside the 3D GeoVE.

2.4 Further Work

Döllner et al. (2005) present an approach for interactive visualization of 3D GeoVEs on mobile devices using server-generated video streams. A user expresses his/her navigation intention by sketches instead of specifying the parameters for and steering the virtual camera explicitly. Sketch data is transmitted to a server, which interprets the input data, depending on the semantics of underlying objects, and computes a resulting camera path. A camera path animation is rendered as video and streamed to the requesting client. This way, only minimal demands are put to the mobile device. Our approach for interactive camera control introduces a more general and more flexible model of integration of 3D camera control into service-based 3D geovisualization systems. This allows for a larger set of input and output methods and decoupling camera control functionality from visualization functionality where possible.

Nurminen et al. (Nurminen and Helin 2005; Nurminen 2008) introduce a mobile application, which heavily uses rendering and transmission optimizations for 3D city models. They provide an interactive 3D GeoVE that integrates dynamic data, provided by remote servers, into the visualization. Geodata is preprocessed at server side optimized and transmitted for rendering. Camera control in the virtual environment is implemented completely client-side on the mobile device. However, for devices with limited input capabilities a higher level camera control could improve the usability of such 3D map applications. Due to the iterative transmission of model data to the mobile client depending on the camera parameters, camera navigation techniques that use model semantics cannot operate in many situations due to the lack of data available on the user's mobile device. Efficient 3D camera control could be integrated more easily in such an environment using a service-based approach for the separation and distribution of camera control functionality.

Chen and Bowman (2009) advocate that design for 3D interaction techniques should be application domain specific. They propose to decompose the interaction tasks into subtasks that consist of universal interaction tasks (e.g. navigation, selection or manipulation). The subtasks are implemented by concrete interaction techniques. Here, Chen focuses more on the question how to design domain-specific interaction techniques. In contrast, the focus of this paper is more on system engineering. We describe how such techniques could be designed as components of a service-oriented system, which facilitates reuse of specifically designed camera interaction techniques wherever the specific application domains come into play.

3 Concept for a Service-Based 3D Camera Control System

Since network bandwidth and end-user hardware and software is very heterogeneous, the development of robust, compatible, and efficient applications that provide interactive access to 3D GeoVEs represents a complex software architecture problem. 3D geovisualization systems using thin clients can bypass such limitations by designing a software architecture that can cope with hardware and software limitations of end-user devices and platforms. With thin clients, only small parts of the overall geodata are available on client side and could therefore be considered for camera path computation. Thus, we propose to move major parts of functionality for 3D camera control away from client applications to services. This loosens the dependency of camera interaction techniques from specific client implementations. Service components can be run in a scalable, controlled server environment and can, therefore, be maintained and optimized more efficiently.

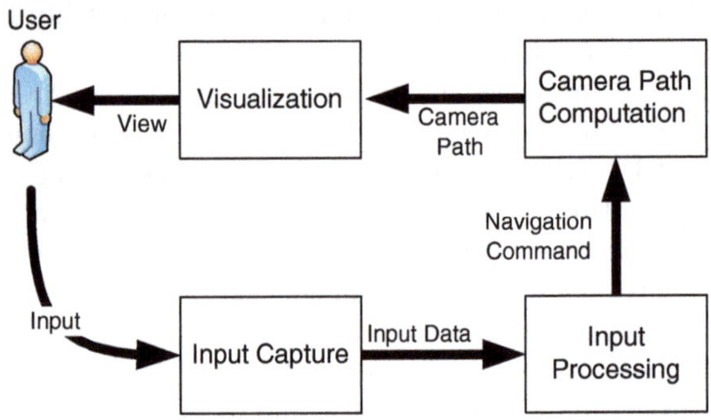

Fig. 1 Conceptional tasks of a 3D camera control process

Server-side access to geodata is usually more efficient due to lower network latencies and better performing hardware. Each of such service components is required to expose its capabilities, e.g. their operations, parameters, and effects, as well as their technical requirements. Capability information should also contain quality of service information, e.g. expected operation times such as minimum, maximum and average processing time for requests to allow raw latency estimations.

To structure the interaction cycle of service-based 3D visualization systems, we divide the process for 3D camera control into four core tasks (Fig. 1):

- *Input Capture*: Input provided by a user has to be captured and encoded in a way that allows for efficient evaluation.
- *Input Processing*: User input is preprocessed, e.g. converted, transformed, smoothed, or patterns are recognized and a navigation command is derived from the resulting data. This command is used to select the 3D camera service for camera path computation.
- *Camera Path Computation*: Camera positions and orientations, and transitions between them are computed. Specifications for camera paths are the result of this stage.
- *Visualization*: The computed camera specifications have to be applied for the client-side visualization of the 3D GeoVE. Visual or non visual (e.g. audible) feedback has to be generated and integrated in order to complete a 3D camera-control cycle.

While input capture has to be implemented by a client application, input processing, camera path computation and visualization can be implemented by one or multiple services. Figure 2 illustrates our concept for decomposing the core tasks for 3D camera control into functional independent *3D interaction services* and depicts their collaboration and the types of data exchanged between them. In the following, we present these major 3D interaction services: input preprocessing services, command recognition services, 3D camera services and composition services.

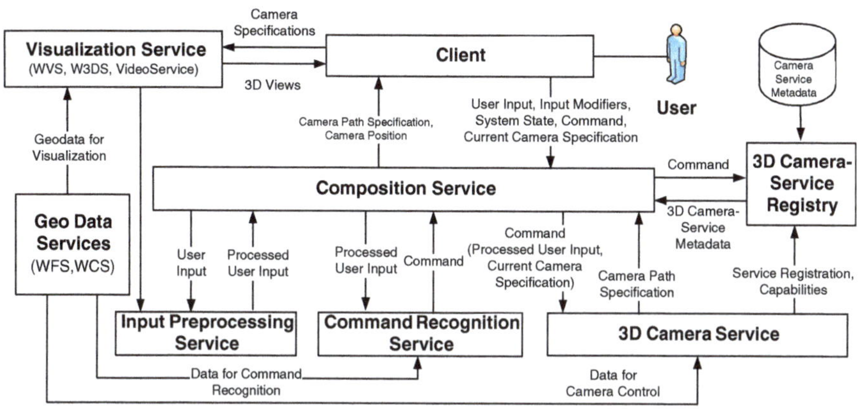

Fig. 2 Abstract component architecture and data flow of a geovisualization system using service-based 3D camera control

3.1 Input Capture

An end-user client-application must provide a description of user inputs. These could be, e.g. a higher-level navigation command (e.g. "look into a certain direction"), button events, or captured mouse-cursor respectively finger positions. Thus, a specification of user inputs is required that supports a variety of user inputs. Each data sample is annotated with timestamps to allow, e.g. for segmenting user input in time and space (e.g. series of sketches or device positions) and computing velocities of movements.

Additional information that is not directly originating from user actions could be required for service-based camera interaction. Client state information, such as input modifiers (e.g. pressed keys) or previous navigation commands, may affect the mapping from input parameters to navigation commands to be executed. Further, a user's current view (including, e.g. camera specification and visible objects) specifies the geospatial context of the user input, which may affect the evaluation of a user input.

3.2 Input Processing

Input processing is divided into two steps: input preprocessing and command recognition.

Depending on the type of input captured by the client, an *input preprocessing step* may be necessary (a) to improve the quality of the input, e.g. by filtering or smoothing and (b) to convert the input data to an analytic representation, e.g. recognizing geometry from a series of 2D input samples.

In a *command recognition step*, navigation commands are derived from the preprocessed input. These represent a more abstract description of a user's intention and include all the parameters required for their execution. We define three categories of navigation commands in respect of the camera-control task they describe (Hagedorn and Döllner 2008):

- *Direct Camera Manipulation*: A command directly influences the values of camera parameters, such as position or orientation vectors, which specify the current view. Commands like 'turn by 30°' or 'move 100 m into camera direction' are examples for such direct camera manipulation commands.
- *Path Oriented Navigation Command*: A command includes a path description that has to be followed by a camera path. The desired path has been computed in the input processing step or has been specified by the client directly (explicit or implicitly, e.g. by providing a target name).
- *Task Oriented Navigation Command*: A command contains a description of a task, which has to be fulfilled by a 3D camera service. Commands like "go to the next feature of class X" or "inspect feature X" belong to this command category.

To describe a command and to support command recognition, a generic, structured command schema is required that specifies, e.g. command parameters (types and possible values). The command recognition step can involve retrieval of additional geoinformation, e.g. from geodata or geovisualization services such as WFS or WVS.

Input preprocessing as well as command recognition are optional steps in the 3D camera control process. Simple camera-control tasks can be transmitted by a client as navigation command, e.g. "move one meter to north" for a stepwise camera control. Such commands may be handled directly by an appropriate 3D camera service.

The functionalities of the input preprocessing and command recognition steps are encapsulated by respective service types, *input preprocessing services* and *command recognition services*.

3.3 Camera Path Computation

Camera path computation is the core task for camera-control in 3D GeoVEs. *3D camera services* compute camera paths from navigation commands and their parameters. Conceptually, one 3D camera service implements one technique for camera path computation. This includes the generation of camera path components, e.g. camera positions, orientations, or other information that can be associated with a camera transition, e.g. textual annotations. Each of those can be computed by distinct functional components that apply specific algorithms and navigation constraints per path component. For example, a specific position component could determine camera positions only along a street network, while a specific orientation component could aim to keep nearby landmarks visible.

A 3D camera service may request additional geodata, e.g. from a WFS, W3DS or WVS to provide, e.g. a higher-level, semantics-based camera control or to fulfill constraints for camera parameters. To ensure a consistent behavior, these additional services have to be based on the same geodata as the visualization services themselves.

Based on a navigation command, a 3D camera service is selected using a 3D camera-service registry, which holds information about the available 3D camera service instances, the navigation commands they support, and additional metadata.

3.3.1 3D Camera Service

In order to be managed in a service registry and allowing consumers to bind correctly to their operations, 3D camera services are required to express general service information as well as functional and additional non-functional metadata (ISO 2003, 2005). 3D camera service capabilities should include metadata regarding the following aspects:

- *Service identification*: Type and version of the service, service description.
- *Camera control metadata*: Available path description formats, covered geospatial region, supported spatial reference systems, available navigation commands including command parameters.
- *Quality of service metadata*: Information such as expected computation time, result accuracies, available collision avoidance (e.g. guaranteed, best-effort, or no avoidance).
- *Application context*: Information regarding, e.g. user information, usage information, network conditions, and device properties.

3.3.2 Camera Path Specification

Camera paths, generated by 3D camera services, represent transitions from one set of camera parameters to another. Parameters required for the definition of a view of a 3D virtual environment are the camera position, its orientation in 3D space and its projection parameters. Additionally, a camera specification can provide annotations that can be used to enrich the 3D GeoVE with thematic information (e.g. distances or relevant objects) using overlays generated by specialized visualization services.

We distinguish two types of representations of camera paths (see Fig. 3): sampled and analytical representations:

Sampled representations of camera paths include of a series of camera-specification samples. Those can be created either using fixed or variable sample times. An adaptive sampling rate of camera specifications allows for more efficient representations of camera paths. This means more dense sampling for time periods where camera parameters change more rapidly, e.g. because of increased movement speed or sharp camera turns.

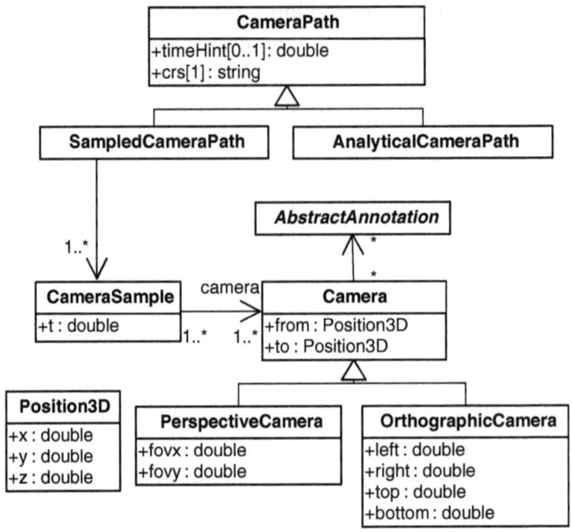

Fig. 3 Camera path specification. A `Camera Path` can be described either by samples of camera parameters or analytically using functional representations. Camera definitions can be associated with annotations containing additional information for user feedback

Analytical representations provide a separate function definition for each camera parameter to compute their values for an arbitrary point in time during a camera transition (Fig. 3). Each of those functions can be expressed as piecewise function, which eases the definition of camera paths by different kinds of functions per time slice, like Bézier curves, splines, linear or even constant functions. Furthermore, this enables the definition of story-board-like camera transitions. The overall time for the complete camera path animation is normalized. A camera path specification contains a recommended overall animation time, which would produce a comfortable camera motion.

A client may specify which type of path representation it requests. Analytical descriptions are more favorable for clients that are capable of interactive 3D rendering themselves, instead of using service-based image synthesis. In contrast, image-based clients may prefer sampled representations of a camera path, as it allows them to request images from portrayal services with a minimum of implementation effort and computational requirements.

A set of utility services can provide functionalities that can be used by several services of the distributed 3D camera control system. Functionality implemented by utility services may include, e.g. path manipulation (conversion, transformation, smoothing, composition), camera orientation computation, sketch recognition, and overlay creation (creation of image overlays as additional user feedback). Further, existing standards-based implementations of services, e.g. for geocoding of locations or conventional 2D routing using street networks may be used for camera path generation.

3.4 Visualization

The final task of the distributed, service-based 3D camera control process is to adjust the visualization of the 3D GeoVE according to a generated camera path, and to display the result to a user. Depending on the type and capabilities of a client application, several processing and visualization services could be involved for this:

- The retrieval of camera specifications from a camera path (e.g. by interpolating camera samples) could be implemented by a client application itself or could be provided by additional utility services.
- A client application that implements the 3D rendering itself needs to process the camera path specification, adjust the visualization accordingly, and generate new visual representations of the 3D GeoVE; utility services could support this process. Graphics data (e.g. X3D scene-graph) could be requested, e.g. from a W3DS.
- A client that is not capable of high-quality 3D rendering would incorporate a visualization service (e.g. a WVS) for creating visual representations of the 3D GeoVE, which could be served as image, set of images, or video to a client application.

Besides generating views of a 3D GeoVE, there are several possibilities for providing additional feedback about the camera control process. For example, textual or graphical annotations can be included in a camera path specification (Fig. 3).

3.5 Service Composition

Deploying the functionalities of the 3D camera control process as independent services enables assembling user input and camera control functionalities aligned to the requirements and capabilities of a specific client application as well as client device and network.

For a 3D camera control system, service composition includes (a) the composition of relevant input processing, command recognition, and 3D camera services and (b) the combination of multiple 3D camera services for reaching a specific camera dramaturgy.

3D camera services themselves could compose other services for implementing higher-level camera-control functionalities, e.g. task-oriented camera control.

4 Example

In this section we provide an example of a service-based 3D geovisualization system that supports 3D camera control for a mobile client applications. Images of a virtual 3D city model, are generated by 3D portrayal service (WVS). Figure 4

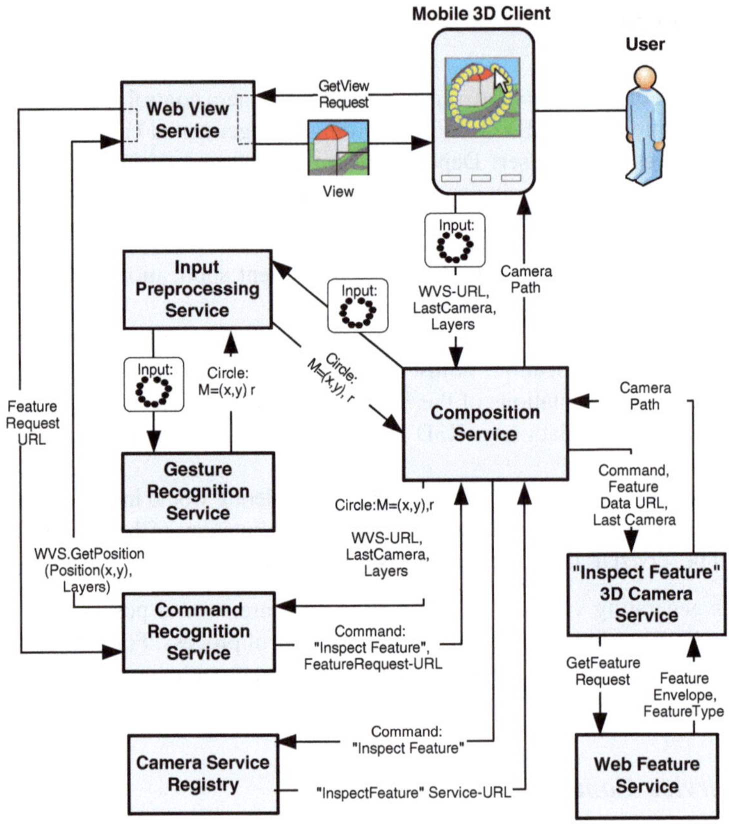

Fig. 4 Example of a distributed, service-based camera control system. The call sequence and concrete input and output data per service are depicted for one type of sketch-based navigation command

illustrates the services involved, the sequence of service calls and the data exchanged between them. As one purpose, the application allows users to inspect features of the 3D GeoVE. A user specifies a desired camera control task by performing gestures on a tangible display, which are captured as series of 2D positions. The gesture data, together with the client's current camera specification, a description of currently visible features (layers), and information about the visualization service used for image generation, is passed to the composition service, which manages the workflow for camera path generation.

Input data is handed over to an input preprocessing service, which performs gesture recognition. The gesture recognition results in geometric primitives that serve as basis for the command recognition. For example, performing a circle gesture around an object on the display could mean to inspect that specific feature. In this case, the gesture recognition service detects a circle primitive from the input positions.

The command recognition service matches the recognized geometry (circle) to the corresponding "inspect feature"-command. The object to be inspected is determined by requesting the object identifier for the center pixel of the circle from a WVS using its "GetFeatureInfo" operation and used as command parameter.

The 3D camera service registry is used to identify the 3D camera service that is able to process the "inspect feature" command. The service is invoked with this command, the current camera parameters, and the reference to the feature data. The service resolves the reference and retrieves the feature data from a WFS. Depending, e.g. on the type or the size of an feature an appropriate technique for camera positioning and alignment can be chosen by the 3D camera service. For example, the HoverCam (Khan et al. 2005) might be better applicable for the exploration of buildings, while area-like features, such as green spaces, may be explored more efficiently through a flyover from a bird's eye perspective.

The 3D camera service delivers the generated camera path specification to the calling client. To adjust its view of the 3D GeoVE, the client interprets the camera path, adjusts the virtual camera parameters accordingly, and requests new image data from the 3D portrayal service.

5 Discussion

The decomposition of interaction functionality into independent 3D interaction services requires to decide which kind of functionality to implement at the client side and which functionality to provide by services. On the one hand, a client-side implementation allows for application, user, and task specific interaction techniques, which are typically developed in detailed knowledge of a concrete use case of an application. On the other hand, reusable services permit to provide uniform interaction mechanisms for a number of different client applications and configurations. For example, the externalization of techniques for camera path computation can enable interaction techniques that need a global view of model data even for thin clients that are unable to deal with the amount and the complexity of massive 3D models.

The system provided as example in the previous section represents a quite complete implementation of the camera interaction functionality using external services. As central element for camera interaction, camera path computation, which includes determination of good views regarding various criteria, is most likely to be usefully implemented as external service. Input preprocessing services and command recognition services on the other hand are mostly useful for ultra thin, purely image-based clients (e.g. browser-based clients that displays rendered images, received from WVS instances). Such clients pass user input to an interaction service chain and receive a new camera specification to request new views from the portrayal service. This relieves clients of all the complexity of 3D computations. Though, such a client is easy to implement and has low hardware and software requirements.

Compared to thick client applications implementing user interaction functionality completely on client-side, the use of interaction services for camera control raises challenges concerning response times and bandwidth as well as their effect on the perceived responsiveness and interactivity of a consuming client application. However, these effects get attenuated by the reduction of the amount of data that has to be transferred over the network due to possible server-side preprocessing and aggregation of complex data needed for the computation of camera paths. Furthermore, 3D camera services can use precomputed camera positions or paths as well as caching strategies to reduce server-side processing, communication effort, and response times.

Dynamic geodata, relevant for the computation of camera paths, can be integrated efficiently by 3D interaction services, since the service encapsulates retrieval and evaluation of data at a single point in the system. Again, clients are relieved from data access and processing. For example, a 3D camera service could integrate sensor or other live data, e.g. extreme temperatures inside a building indicating a possible fire, to derive situation dependent camera paths. In general, the decoupling of the complexity of model geodata, model management, access and usage from camera computations enables to develop general purpose client applications. These implement only a set of core visualization and interaction functionalities that are relatively domain independent. This reduces the complexity of applications that are deployed on end-user devices and platforms and, therefore, raises the compatibility of such application.

6 Conclusions and Outlook

In this paper, we presented a concept for a distributed 3D camera control system for 3D GeoVEs, which integrates into existing service-based geovisualization approaches. This concept provides categories of services and their interconnection so that functionalities for 3D camera control can be decomposed into and deployed as independent services. Principal service classes for input processing, camera path computation, and 3D visualization are introduced; relevant data such as 3D camera paths and service metadata is modeled. By an example, we demonstrate how to compose these services to support the exploration of a mobile 3D GeoVE.

With our concept, 3D camera control functionalities become available through the Internet as distributed resources rather than being hard-coded in specific client applications. Services for input processing and 3D camera control form major building blocks for the implementation of interactive service-based 3D GeoVEs. Using these services, client applications can be relieved from complex computation tasks (e.g. for generating high-quality 3D camera positions and orientations) or implementing adapters to services providing geodata relevant for computations (e.g. WCS or WFS); they only need to interact with the distributed camera control system. Thus, 3D camera control functionalities become available even for thin clients, e.g. running on mobile phones or in web-browser environments.

Our approach enables to flexibly select from alternative 3D camera services for adapting the interaction process to specific application requirements (e.g. user tasks), device properties (e.g. input devices), and network capabilities (e.g. latencies or network bandwidth). Additionally, the proposed interaction services can be reused for achieving a consistent system behavior of the same quality across various applications. Service reuse also could enable a faster development of service-based 3D geovisualization systems and client applications. The approach allows for systematically accessing additional geoinformation provided as distributed services and using this information for 3D camera control (e.g. dynamic sensor data for camera path computation). Thin client applications for 3D geovisualization will become more popular in future due to increasing computing and 3D rendering capabilities of mobile devices (flanked by developments towards browser-based 3D rendering) and their increasing distribution. Such applications would directly benefit from a distributed, service-based 3D camera control system.

References

Altmaier A, Kolbe TH (2003) Applications and solutions for interoperable 3d geo-visualization. In: Fritsch D (ed) Proceedings of the photogrammetric week 2003. Wichmann, Stuttgart, pp 251–267

Basanow J, Neis P, Neubauer S, Schilling A, Zipf A (2008) Towards 3D spatial data infrastructures (3D-SDI) based on open standards–experiences, results and future issues. In: Advances in 3D geoinformation systems, Springer, Berlin, Lecture notes in geoinformation and cartography, pp 65–86. http://www.springerlink.com/content/u45062mr4hh54547

Bishr YA (1998) Overcoming the semantic and other barriers to gis interoperability. Int J Geograph Inf Sci 12(4):299–314

Buchholz H, Bohnet J, Döllner J (2005) Smart and physically-based navigation in 3D geovirtual environments. In: 9th international conference on information visualisation (IV'05), IEEE, pp 629–635.

Chen J, Bowman D (2009) Domain-specific design of 3D interaction techniques: an approach for designing useful virtual environment applications. Presence: Teleoper Virtual Environ 18(5):370–386.

Döllner J, Hagedorn B, Schmidt S (2005) An approach towards semantics-based navigation in 3D city models on mobile devices. In: Proceedings of the 3rd symposium on LBS& teleCartography, Springer, Vienna.

Gröger G, Kolbe T, Nagel C, Häfele K (2008) OpenGIS city Geography Markup Language (CityGML) encoding standard version 1.0.0. http://www.opengeospatial.org/standards/citygml

Haber RB, McNabb DA (1990) Visualization in scientific computing, IEEE Computer Society Press, chap visualization idioms: a conceptual model for scientific visualization systems, pp 74–93

Hagedorn B (2010) OGC web view service. In: OGC discussion paper

Hagedorn B, Döllner J (2008) Sketch-based navigation in 3D virtual environments. In: Proceedings of the 9th international symposium on smart graphics. Lecture notes in computer science, vol 5166. Springer, Berlin, pp 239–246.

Hagedorn B, Hildebrandt D, Döllner J (2009) Towards advanced and interactive web perspective view services. In: Developments in 3D geo-information sciences, Springer, Berlin, pp 33–51

Hildebrandt D, Klimke J, Hagedorn B, Döllner J (2011) Service-oriented interactive 3d visualization of massive 3d city models on thin clients. In: 2nd international conference on computing for geospatial research and application cOMGeo 2011.

ISO (2003) ISO 19115. Geographic information–metadata, international standard.

ISO (2005) ISO 19119. Geographic information–services, international standard.

Khan A, Komalo B, Stam J, Fitzmaurice G, Kurtenbach G (2005) HoverCam: interactive 3D navigation for proximal object inspection. In: Proceedings of the 2005 symposium on interactive 3D graphics and games, ACM, vol 1, pp 73–80

Nurminen A (2008) Mobile 3D city maps. IEEE Comp Graph Appl 28(4):20–31

Nurminen A, Helin V (2005) Technical challenges in mobile real-time 3D city maps with dynamic content. In: Kokol P (ed) Software engineering

Panagiotis P, Vretanos A (2010) OGC web feature service implementation specification. http://www.opengeospatial.org/standards/wfs

Papazoglou MP, Traverso P, Dustdar S, Leymann F (2007) Service-oriented computing: state of the art and research challenges. Computer 40(11):38–45. doi:10.1109/MC.2007.400. http://ieeexplore.ieee.org/lpdocs/epic03/wrapper.htm?arnumber=4385255

Portele C (2007) OpenGIS geography markup language (GML) encoding standard. http://www.opengeospatial.org/standards/gml

Schilling A, Kolbe TH (2010) Draft for candidate openGIS web 3D service interface standard. http://portal.opengeospatial.org/files/?artifact_id=36390

Schut P (2007) OGC Web Processing Service. http://www.opengeospatial.org/standards/wps

Tan D, Robertson G, Czerwinski M (2001) Exploring 3D navigation: combining speed-coupled flying with orbiting. In: Proceedings of the SIGCHI conference on human factors in computing systems. ACM New York, pp 418–425

Representing Three-Dimensional Topography in a DBMS with a Star-Based Data Structure

Hugo Ledoux and Martijn Meijers

Abstract For storing and modelling three-dimensional topographic objects (e.g. buildings, roads, dykes and the terrain), tetrahedralisations have been proposed as an alternative to boundary representations. While in theory they have several advantages, current implementations are either not space efficient or do not store topological relationships (which makes spatial analysis and updating slow, or require the use of a costly 3D spatial index). We discuss in this paper an alternative data structure for storing tetrahedralisations in a DBMS. It is based on the idea of storing only the vertices and *stars* of edges; triangles and tetrahedra are represented implicitly. It has been used previously in main memory, but not in a DBMS—we describe how to modify it to obtain an efficient implementation in a DBMS. As we demonstrate with one real-world example, the structure is around 20 % compacter than implemented alternatives, it permits us to store attributes for any primitives, and has the added benefit of being topological. The structure can be easily implemented in most DBMS (we describe our implementation in PostgreSQL) and we present some of the engineering choices we made for the implementation.

1 Introduction

Several data models to represent 3D topographic objects (e.g. buildings, roads, the terrain and dikes) and to store them in a database management system (DBMS) have been proposed. Some of them store only the geometry of single objects while

H. Ledoux (✉) · M. Meijers
Delft University of Technology, Delft, The Netherlands
e-mail: h.ledoux@tudelft.nl

M. Meijers
e-mail: b.m.meijers@tudelft.nl

J. Pouliot et al. (eds), *Progress and New Trends in 3D Geoinformation Sciences*, Lecture Notes in Geoinformation and Cartography, DOI: 10.1007/978-3-642-29793-9_7,

others permit us to explicitly store the topological relationships between objects (and also those between the lower-dimensionality primitives of the representation of an object). Geometry models usually store objects with a boundary representation, called *b-rep*, and one popular data structure used in practice is GML (OGC 2007). It can be seen in the overview of Zlatanova et al. (2004) that many topological models are variations of the *formal data structure* (FDS) (Molenaar 1990; Molenaar 1998), which, unlike its name implies, is a conceptual model—with rules and constraints to preserve the validity—that can be implemented in a DBMS. FDS is a b-rep model in which four primitives are kept (nodes, arc, edge and face; bodies are implicit), and where the topological relationships between them are stored.

An alternative to b-rep models is to use *tetrahedralisations*, where 3D objects are decomposed into tetrahedra and where empty space (e.g. between buildings) is also decomposed into tetrahedra and integrated in the model.[1] Carlson (1987) and Pilouk (1996), among others, argue that tetrahedralisations have several advantages to represent 3D objects, in the same way that triangulations have advantages in 2D (Frank and Kuhn 1986). The advantages the most often cited are: storage is simplified as only one convex primitive is needed (Penninga 2005), spatial analysis operations can perform more efficiently (Ledoux and Gold 2008), and the overall implementation is simpler, thus more robust. However, tetrahedralisations also have theoretical drawbacks as Zlatanova et al. (2004) state in their comparison of the different topological models: "An additional disadvantage of TEN is its much larger database size compared with other representations."

As Penninga (2008) states, the storage penalty is true only if all the primitives of the tetrahedra are explicitly represented (nodes, edges and faces, as with the FDS). He proposes a data structure where only vertices and tetrahedra are represented, and the other primitives are extracted on-the-fly. His data structure can be seen as a variation of *Simple Features* (OGC 2006): vertices are stored in one table and have an ID, and one tetrahedron is formed by the concatenation of 4 vertex IDs into one series of bits. He ensures that all tetrahedra are correctly *oriented* (i.e. the ordering of the vertices is the same for all tetrahedra), which speeds up spatial analysis and the incremental update of a model.

While Penninga (2008)'s data structure is compact—it takes only about 20 % more space than Oracle Spatial's polyhedra for a few tested real-world datasets—the topological relationships between the tetrahedra are not explicitly stored (neither adjacency between tetrahedra nor incidence from one tetrahedron to primitives in other tetrahedra). That means that spatial analysis operations will perform slowly on big datasets, and so will incremental updates to a model. Penninga (2008) advocates using an auxiliary spatial index for each tetrahedron, such as an R-tree (Guttman 1984), although that has not been implemented nor

[1] Notice that this model is often called "TEN", which stands for either TEtrahedral Network, TEtrahedral irregular Network, TEtrahedronised irregular Network or TEtrahedron Network, depending on the authors. For the purpose of this paper, they are all equivalent and a TEN is a tetrahedralisation, as defined in Sect. 2.

tested. The main problems with an auxiliary index are: (1) storage space, the bounding box of a tetrahedron (required by the R-tree) has only one point less than the tetrahedron itself; (2) a R-tree is a rather complex structure and updating incrementally a large tree can be slow. Triangulations in 2D often do not index at the triangle level for the same reasons, see Finnegan and Smith (2010).

We discuss in this paper an alternative data structure for storing tetrahedralisations in a DBMS. Instead of storing explicitly tetrahedra, we store only two lower-dimensionality primitives (vertices and edges), and the *stars* of edges. It has been used previously in 3D in main memory (Blandford et al. 2005), but to our knowledge no attempts has been made to implement it in a DBMS. We define in Sect. 2 the concept of star, which is related to that of a tetrahedralisation. One important advantage of our star-based structure is that it permits us to avoid the use of an auxiliary 3D spatial index, instead the topological relationships between the tetrahedra are exploited (only a standard B-tree is needed to index the tetrahedra). As explained in Sects. 3 and 4, the structure is very compact (around 20 % compacter than that of Penninga (2008)'s) and can be implemented easily in a DBMS. We have implemented it in PostgreSQL, and tested it with one real-world 3D city model. We are currently working on the development of the structure, and in Sect. 5 we discuss some of the engineering choices we made so far, and our future challenges.

2 Constrained Tetrahedralisation and Stars

2.1 Tetrahedralisation

Given a set S of points in 3D space, a tetrahedralisation decomposes the convex hull of S into non-overlapping tetrahedra. It is possible to construct the *Delaunay* tetrahedralisation (DT) of S, i.e. a tetrahedralisation where every tetrahedron has an empty circumsphere; the DT has desirable properties that make them popular in several domains.

2.2 Constrained Delaunay Tetrahedralisation

If the input set contains also edges and/or surfaces (boundaries that have to be respected in the tetrahedralisation; think of a 3D city model, the walls of the buildings have to be present in the resulting tetrahedralisation), then the problem is more complex.

In 2D, a constrained Delaunay triangulation can be used, Fig. 1a shows an example. However, as Shewchuk (2002) explains, in 3D there are two approaches. The first one is the *conforming* DT, where additional vertices (known as *Steiner*

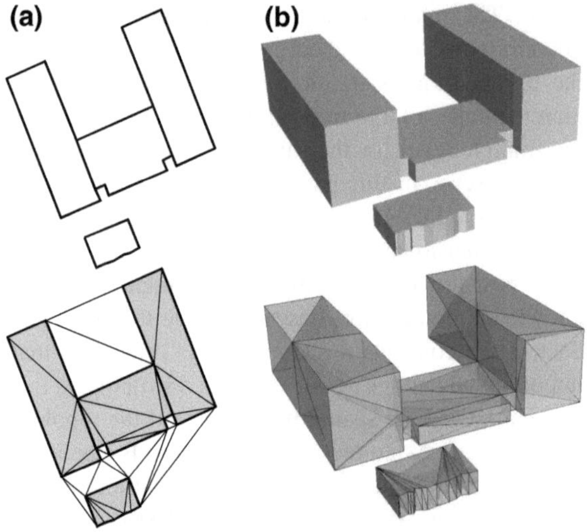

Fig. 1 **a** Two-dimensional polygons representing buildings footprints, and their constrained Delaunay triangulation. **b** Three-dimensional representation of the same buildings [polyhedra in this case, obtained by extruding the footprints in (**a**)] and their constrained Delaunay tetrahedralisation (for clarity only the tetrahedra inside the polyhedra are shown here)

vertices) are inserted to ensure that edges/surfaces are recovered (these extra vertices do not modify the shape of the surfaces/polyhedra). The main problem for a data structure is that *several* new vertices can be required (Cohen-Steiner et al. 2004), and that would require more space.

The second solution, the one we use in this paper, is the *constrained* Delaunay tetrahedralisation (CDT) (Si 2008), as Fig. 1b shows. The tetrahedra created are not fully Delaunay, but this is not a problem and the number of newly inserted vertices is minimised in comparison to the previous two approaches. Another advantage is that robust and efficient implementations of CDTs exist, as they are extensively used in engineering.

2.3 Stars and Links

Let v be a vertex in a d-dimensional triangulation. Referring to Fig. 2, the star of v, denoted star(v), consists of all the simplices that contain v; it forms a star-shaped polytope. For example, in 2D, all the triangles and edges incident to v form star(v), but notice that the edges and vertices disjoint from v—but still part of the triangles incident to v—are not contained in star(v). Also, observe that the vertex v itself is part of star(v), and that a simplex can be part of a star(v), but not some of its facets.

The set of simplices incident to the simplices forming star(v), but 'left out' by star(v), form the *link* of v, denoted link(v), which is a $(d - 1)$ triangulation. For example, if v is a vertex in a tetrahedralisation, then link(v) is a two-dimensional triangulation formed by the vertices, edges and triangular faces that are contained by the tetrahedra of star(v), but are disjoint from v.

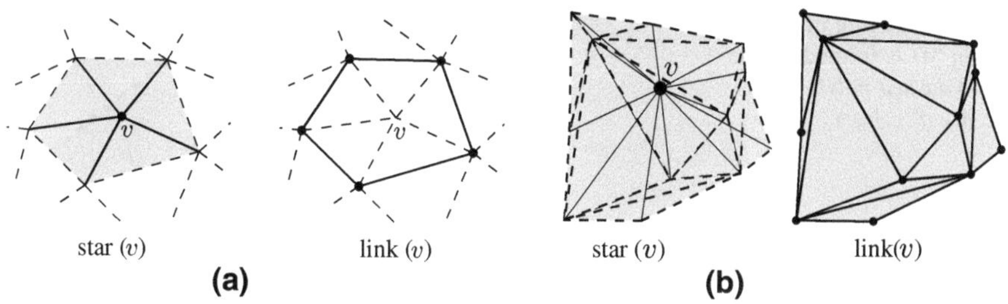

Fig. 2 The *star* and the *link* of a vertex v in **a** 2D and **b** 3D

The closely related concepts of star and link also apply to edges in 3D: in Fig. 3 star(ab) is formed by all the incident simplices (6 in this case), and link(ab) is a 1-dimensional triangulation (a polyline).

3 A Star-Based Data Structure

To our knowledge, Cline and Renka (1984) are the first to design a data structure where stars of vertices are used to represent a triangulation, albeit in 2D. Their structure is compact, but does not allow incremental updates (which is arguably important for a GIS data model). Shewchuk (2005) uses a star-based data structure to manipulate 3D triangulations, but his structure is very verbose since the star of every vertex is stored as a 2D triangulation, itself stored with stars. Blandford et al. (2005) fix both issues with their structure, which is valid for 2D and 3D triangulations. Their representation indeed uses about a factor 3 less memory than traditional representations in 3D and at the same time can be queried and dynamically modified. To achieve this compression, they designed a pointer-less structure where each vertex is assigned a label (an integer) and they compress based on labels: if possible they store the integers using 4 bits (they use differences between the labels to achieve that) and use different optimisations to keep the memory footprint low.

While it would theoretically be possible to implement the compression in a DBMS, we are interested in their basic idea of using labels for vertices and store the stars of edges in 3D. In a nutshell, a star-based data structure for a tetrahedralisation in a DBMS is as follows. For an edge ab of the tetrahedralisation, we store its link as an ordered list of labels. The orientation is consistent with the *right-hand rule*: if the thumb points from a to b, the vertices of link(ab) are ordered in the direction of the curled fingers of the right hand. In Fig. 3, link(ab) is the ordered list is $[c, d, e, f, g, h, c]$; notice that this is a circular list and that the starting vertex could be any vertex. The link list is of variable length: its minimum is 2 (one tetrahedron) and its theoretical maximum is the number of vertices in the tetrahedralisation minus 2. A tetrahedron is formed by ab and 2 consecutive vertices in the list; a, b, c, d and a, b, h, c are two examples of tetrahedra implicitly

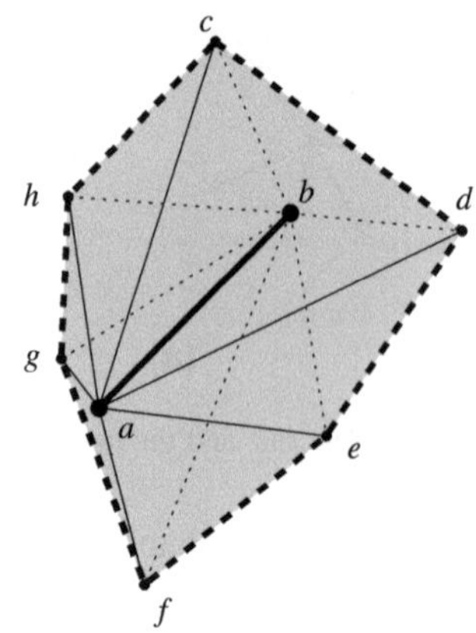

Fig. 3 The link of the edge *ab* in 3D is formed by the *bold dashed polyline* (c, d, e, f, g, h, c)

represented in link(ab). Notice that the length of the list gives the number of incident tetrahedra to ab. Also, lower-dimensionality simplices (triangles and edges) are present in links: for instance, referring to Fig. 3, the triangle abc is present in the links of its 3 edges. Since the link is an ordered list, the simplices implicitly represented are also ordered.

The key idea behind the structure is that if we represent the link of each edge, then we obtain a data structure where relationships such as incidence and adjacency between tetrahedra are present. It is the overlap between the (ordered) links that permits us to represent explicitly that information. Observe that each tetrahedron is represented in the link of 6 edges, and that since these are ordered, we can easily navigate from tetrahedron to tetrahedron. We show in Sect. 4.2 a few examples of queries.

3.1 Representative Edges

Storing the link for each edge of a tetrahedralisation yields a powerful and topological data structure, but also one that is not space efficient. Indeed, if the CDT of a set S of n points contains t tetrahedra, then the number e of edges is significantly higher: Blandford et al. (2005) estimate it at $(7/6)t$ for a CDT where the points are uniformly distributed in space.

To reduce the number of edges whose star is stored, we store only the *representative edges* (RE), as Blandford et al. (2005) suggest. If the label given to each vertex is an integer, a RE is one where its 2 vertex labels are either odd or even. If we randomly label the vertices, that should reduce by a factor of about 2 the number of edges to be stored and still permits us to represent at least once each triangle and each tetrahedron (which is fundamental to ensure that all topological relationships are present). Indeed, it ensures that each triangle has at least one RE:

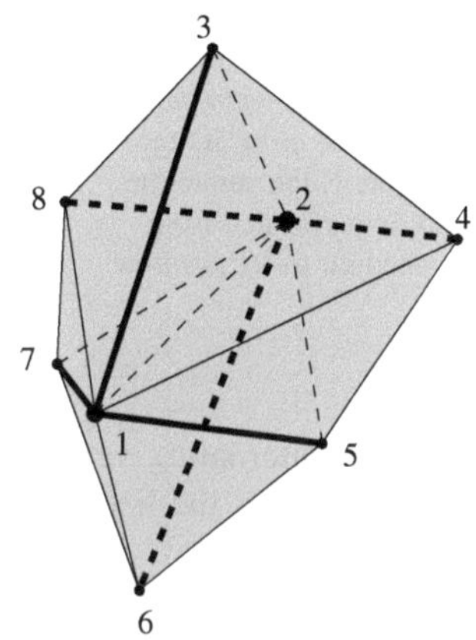

Fig. 4 A set of 8 vertices yields a tetrahedralisation with 8 tetrahedra. Out of the total 19 edges the 6 representative edges (REs) are stored and shown in *bold*: $\langle 1,3\rangle$, $\langle 1,5\rangle$, $\langle 1,7\rangle$, $\langle 2,4\rangle$, $\langle 2,6\rangle$, $\langle 2,8\rangle$

a triangle has either 3 REs (3 odd or 3 even labels) or one RE (1 odd/2 even; 2 odd/1 even).

Figure 4 shows the same 6 tetrahedra as Fig. 3 where the REs are highlighted (in bold). It can be seen that out of the 19 edges, 6 are representative, and that each triangle contains at least one RE.

3.2 Storage Space

Evaluating the theoretical storage space is difficult since the number of tetrahedra in a CDT depends on the locations of the points and the constraints. However, to obtain an order of magnitude, we can state that we need on average 3 labels per tetrahedron. Indeed, each tetrahedron has 4 triangles, which are shared by 2 tetrahedra (if we ignore those on the convex hull); thus 2 triangles per tetrahedron. A triangle has 3 labels, but since it appears in 3 stars and that only half of the edges are represented, we obtain $1\frac{1}{2}$. Thus: $2 \times 1\frac{1}{2} = 3$ labels per tetrahedron. Our experiment with a real-world dataset corroborates that, see Sect. 4.3.

3.3 Attributes

Attaching attributes to the tetrahedra is possible, although one must be careful since tetrahedra are present in multiple stars. We exploit the fact that each vertex has a unique label (which can be ordered) and attach the attributes to the star of the REs whose origin is the lowest; in case there are more than one, the one having the lowest destination is chosen. Given a tetrahedron *abcd*, we can find out in constant time which RE stores its attributes. The attributes can be stored either in the same

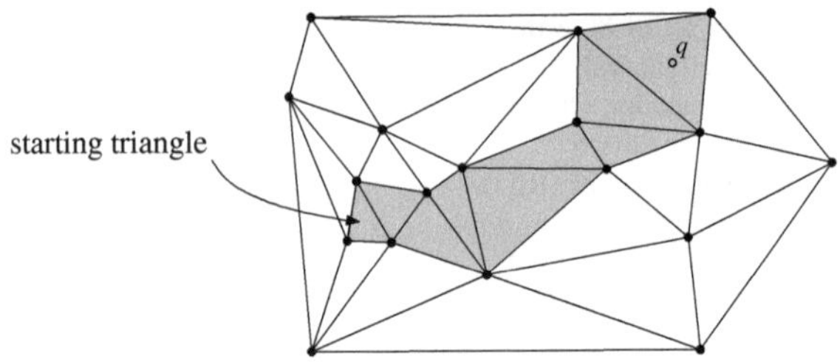

Fig. 5 Walking in a 2D triangulation, starting from a given *starting triangle* to the query point q. In 3D the principle is the same: the walk is performed from tetrahedron to tetrahedron

link list (alternating vertex labels with attributes), or in another list (having the same length as the list of the star).

3.4 Spatial Indexing: The Tetrahedralisation Itself

An advantage of a star-based structure—or of any structure in which adjacency and incidence relationships are stored—is that a spatial index, such as an R-tree (Guttman 1984), is not necessary to access efficiently the tetrahedra. Instead, the tetrahedralisation itself can be used to determine which tetrahedron contains a query point q: the adjacency relationships between the tetrahedra are used to navigate in the tetrahedralisation. The latter can be implemented with the *walking* algorithm as described in Mücke et al. (1999). It is a sub-optimal algorithm that is favoured by practitioners since it does not require an auxiliary data structure and yields fast practical performances (Mücke et al. 1999; Devillers et al. 2002).

The idea is as follows: starting from a given tetrahedron σ, we move to one of the neighbours of σ (we choose one neighbour such that the query point q and σ are on each side of the triangular face shared by σ and its neighbour) until there is no such neighbour, then the tetrahedron containing q is σ. In Fig. 5, only the grey triangles are visited during the walk.

To minimise the number of triangles visited, the starting triangle should be close to q. Mücke et al. (1999) investigated a 'bucketing' approach where a certain number of triangles are randomly selected, and each walk starts from the closest one (selected by a simple Euclidean distance test); it is called the *jump-and-walk* method. The result of a query can be either a tetrahedron, or one representative edge in that tetrahedron. Modifying this algorithm to start from a representative edge is trivial.

4 Implementation in a DBMS and Experiments

This section describes a prototype implementation of the star-based data structure in a specific, object-relational DBMS (PostgreSQL), but since the data structure is based solely on lists of labels, implementing it in another DBMS should be

```
-- Vertex table
CREATE TABLE pgtet_vertex (
    gid bigint,
    x numeric,
    y numeric,
    z numeric
);
ALTER TABLE pgtet_vertex ADD PRIMARY KEY (gid);

-- Edge table
CREATE TABLE pgtet_edge (
    start bigint,
    end bigint,
    link bigint[] -- array of integers
);
ALTER TABLE pgtet_edge ADD PRIMARY KEY (from_gid, to_gid);
```

Fig. 6 Schema definition of the star-based data structure in PostgreSQL

straightforward. Several engineering decisions had to be taken when implementing the structure in PostgreSQL, and we report here on the main ones.

4.1 PostgreSQL Tables

Figure 6 shows that the schema definition of the data structure is straightforward if the DBMS supports an array type of variable length: two tables are created, one table for vertices (points) and one for representative edges. A unique ID is assigned to each vertex and is stored together with the ordinates of each point. For the IDs of the points, the type `bigint` (64-bit integers) is used since 32-bit integers would limit the size of the datasets that could be stored. We define the column 'gid' as a primary key, which creates a binary-tree index on the column (B-tree). This ensures efficient access and enforces uniqueness.

An edge is stored as a reference to its start and end vertices (the primary key of the table is composed of the concatenation of both IDs), and its link is stored as an array of type `bigint` (which refer to the gid in the vertex table). At a later stage we will aim at compressing the array stored for the link (by using differences in vertex labels), so that less storage space is needed; this then will have a direct impact on how much data needs to be read from disk by the DBMS. Also, to be able to represent triangles and tetrahedra two custom types are defined. Both types are a sequence of vertex IDs, where triangles are represented by 3 vertex IDs and tetrahedra by 4. Based on these custom types a view can be defined that 'glues' the geometry of the vertices and edges together to triangles and tetrahedra (performed by a DBMS join).

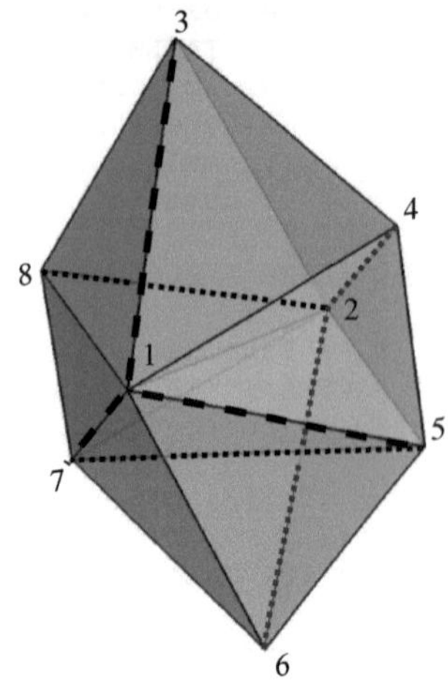

Fig. 7 Small example dataset stored in PostgreSQL, see Table 1 for the information that is stored

4.2 Examples of Topological Queries

We describe how a few typical queries could be performed with a star-based structure. All the queries refer to the example in Fig. 7, and the resulting tables in PostgreSQL (Table 1). Since triangles and tetrahedra can be present multiple times in the structure, querying the structure has to be done with care. For example, tetrahedron $\langle 5, 7, 1, 2 \rangle$ is present in the edge table in the links of all its representative edges: $\langle 1, 5 \rangle$, $\langle 1, 7 \rangle$ and $\langle 5, 7 \rangle$.

Is tetrahedron $\langle 5, 7, 1, 2 \rangle$ *present*? First, one RE has to be found: $\langle 1, 5 \rangle$ is one of them. Observe that a RE is found in constant time only by finding locally 2 odd or even IDs in the tetrahedron. Second, the IDs 2 and 7 have to appear consecutively in the link of that edge (which is the case, therefore the tetrahedron is present).
What tetrahedra are adjacent to $\langle 5, 7, 1, 2 \rangle$? First, find one RE as above ($\langle 1, 5 \rangle$) and find the position of vertices 2 and 7 in the link. The vertices before and after the tuple give 2 adjacent tetrahedra: $\langle 1, 5, 6, 7 \rangle$ and $\langle 1, 5, 2, 4 \rangle$. Second, find another RE of $\langle 5, 7, 1, 2 \rangle$ and repeat the same operations in its link. Since we know that each triangle is represented in at least one RE, this operation will always return the 4 tetrahedra.
Total number of tetrahedra? Here we apply the same criteria as for storing attributes in Sect. 3.3: the lowest concatenation of the IDs is the one representing the tetrahedron. For instance, tetrahedron $\langle 1, 3, 4, 2 \rangle$ is conceptually stored in edge $\langle 1, 3 \rangle$ and not $\langle 2, 4 \rangle$. Thus, it suffices to scan the edge table and take a local decision to extract tetrahedra.

We are currently investigating which custom functions are necessary for modelling 3D topographic datasets. Other examples than the ones already mentioned above are insertion of a new point or a constraint, point location, attaching attributes to a

Table 1 Storing the tetrahedra from the example dataset of Fig. 7. In the link column, ∅ means that the link of the edge does not form a cycle, i.e. the edge is on the convex hull of the dataset

Vertex table

id	x	y	z
1	5.0	2.5	6.0
2	5.0	11.5	6.0
3	5.0	6.0	12.0
4	9.0	6.0	8.0
5	9.0	6.0	4.0
6	5.0	6.0	0.0
7	1.0	6.0	4.0
8	1.0	6.0	8.0

Edge table

Start	End	Link[]
2	4	{∅, 3, 1, 5}
5	7	{6, 1, 2}
1	7	{∅, 8, 2, 5, 6}
1	5	{∅, 6, 7, 2, 4}
2	6	{∅, 5, 7}
2	8	{∅, 7, 1, 3}
1	3	{∅, 4, 2, 8}

specific tetrahedron, etc. These functions will be programmed in PL/pgSQL (the procedural language that PostgreSQL offers) or C.

4.3 Experiments With Real-World Data

To test the star-based data structure in PostgreSQL, we have made an experiment with one real-world dataset. It is the 3D city model of our university campus obtained by extrusion; the process used to construct it is described in Ledoux and Meijers (2011). The original dataset covers an area of 2.3 km^2 and has 370 buildings. We have created the CDT of the model with TetGen[2] (Si 2008). Table 2 gives the details of the 3D extruded dataset, and the results of the construction of the CDT.

Figure 8 shows a part of the extruded TU Delft campus, once tetrahedralised. As can be seen, each polyhedron is decomposed into a set of tetrahedra. Notice also that while the tetrahedra representing the 'air' are not shown, they are still stored. In addition, we have constructed six extra planes forming the bounding box of the dataset and added them to bound the area.

[2] www.tetgen.org

Table 2 Details concerning the datasets used for the experiments

Input 3D model		CDT				Star
Vertices	Constraints	Vertices	Edges	Triangles	Tetrahedra	Representative edge
5,978	3,982	6,938	56,291	95,420	47,707	25,697

The CDT has added around 1,000 vertices to the original model (the Steiner points), and the total number of tetrahedra is 47,707, for an input of only 370 polyhedra. However, it should be noticed that while the total number of edges in the CDT is 56,291, less than half of these are REs and thus the edge table in the DBMS is only about 25,000 rows. If the data structure of Penninga (2008) was used—which is, to the best of our knowledge, the most compact structure for storing tetrahedralisations in a DBMS—the tetrahedra table would have 47,707 rows, and each row would have exactly 4 IDs. The total number of IDs required for this dataset would thus be 190,828, if we omit the vertex table.

With our structure, the vertex table is exactly the same as Penninga's. The edge table has 25,697 rows with 2 IDs (start and end vertices), plus a total of 101,694 IDs in all the links (this number was obtained by querying the DBMS; the average length of a link is 4.93, the minimum is 3 and the maximum is 28). Thus, the total is 153,078 IDs, which makes it around 20 % compacter for this real-world dataset.

For populating the DBMS, we have created a program that takes as input the result of the tetrahedralisation (a list of vertices and tetrahedra) and outputs a list of REs and their links. Currently, only bulk loading of data is supported in our prototype.

5 Discussion and Future Work

We have shown that a star-based data structure implemented in a DBMS can be *both* compact and topological at the same time, two criteria that are usually contradictory. Our structure uses in theory only 3 IDs per tetrahedron, which is an improvement of 33 % over the most compact structure implemented in a DBMS so far (that of Penninga (2008)). We should add that these results ignore the fact that with a non-topological structure an auxiliary spatial index must be used, which increases greatly the storage space (3 extra vertices per tetrahedron are needed, plus the size of the tree) and is complex to maintain when objects are deleted. With a star-based structure, only a standard B-tree is needed, and furthermore less rows need to be indexed (in our real-world dataset, we had around twice as many tetrahedra as representative edges). Another strong point of the star-based structure is that it can easily be implemented in any DBMS with two simple tables.

While a star-based structure seems more cumbersome to maintain when the data are updated, the users need not be aware that this is the structure used. We plan to add functionalities to the DBMS so that views can be created over the edges and stars so that only features (e.g. buildings) are shown to the user, which is

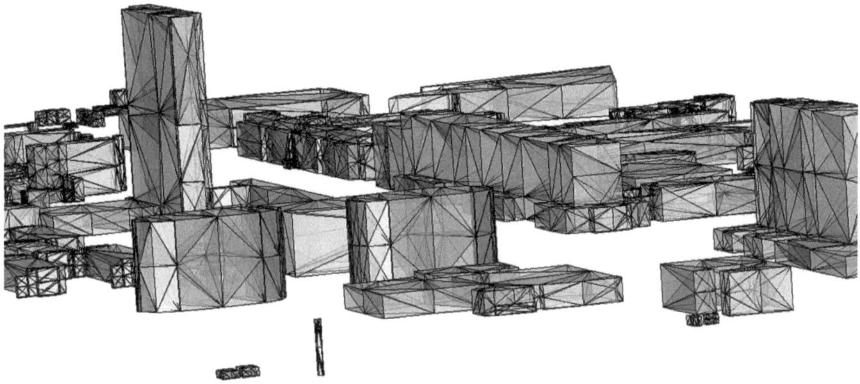

Fig. 8 Part of the tetrahedralised 3D model of our campus, which was obtained by extrusion

what van Oosterom et al. (2002) advocate for storing GIS datasets in a DBMS. We have plans to make the data structure fully dynamic (i.e. supporting insertions and deletes of tetrahedra and of features, updating the already stored tetrahedra in the database).

Object relational DBMS systems keep evolving and new powerful features have been added to mainstream systems. One example is Common Table Expressions (SQL-99) which paves the way to deal with more complex data structures like trees and graph storage inside such systems natively. We intend to see how far these features are useful during implementation of the query part of our data structure and how much custom procedural functionality still needs to be build.

Apart from 3D topography, 3D models can also be useful for modelling space and map scale in one integrated 3D data structure, e.g. van Oosterom and Meijers (2011). Hence, one operation we want to investigate in more detail is to create a cross section through the stored 3D model: selecting intersecting tetrahedra, performing intersection and then create a topologically clean output of the intersected elements to see whether this tetrahedra based model can be a useful underlying technology for producing vario-scale data.

Acknowledgments This research is supported by the Dutch Technology Foundation STW, which is part of the Netherlands Organisation for Scientific Research (NWO) and partly funded by the Dutch Ministry of Economic Affairs, Agriculture and Innovation (project codes: 11300 and 11185).

References

Blandford DK, Blelloch GE, Cardoze DE, Kadow C (2005) Compact representations of simplicial meshes in two and three dimensions. Int J Comput Geom Appl 15(1):3–24

Carlson E (1987) Three-dimensional conceptual modeling of subsurfaces structures. In: Proceedings 8th international symposium on computer-assisted cartography (Auto-Carto 8), Falls Church, VA, pp 336–345.

Cline AK, Renka RJ (1984) A storage-efficient method for construction of a Thiessen triangulation. Rocky Mountain J Math 14:119–139

Cohen-Steiner D, Colin de Verdire E, Yvinec M (2004) Conforming delaunay triangulations in 3D. Comput Geom Theor Appl 28:217–233

Devillers O, Pion S, Teillaud M (2002) Walking in a triangulation. Int J Found Comp Sci 13(2):181–199

Finnegan DC, Smith M (2010) Managing LiDAR topography using Oracle and open source geospatial software. In: Proceedings GeoWeb 2010, Vancouver, Canada.

Frank A, Kuhn W (1986) Cell graphs: a provable correct method for the storage of geometry. In: Proceedings 2nd international symposium on spatial data handling, Seattle, USA.

Guttman A (1984) R-trees: a dynamic index structure for spatial searching. In: Proceedings 1984 ACM SIGMOD international conference on management of data, ACM Press, pp 47–57.

Ledoux H, Gold CM (2008) Modelling three-dimensional geoscientific fields with the Voronoi diagram and its dual. Int J Geograph Inf Sci 22(5):547–574

Ledoux H, Meijers M (2011) Topologically consistent 3D city models obtained by extrusion. Int J Geograph Inf Sci 25(4):557–574

Molenaar M (1990) A formal data structure for three dimensional vector maps. In: Proceedings 4th international symposium on spatial data handling, Zurich, Switzerland, pp 830–843.

Molenaar M (1998) An introduction to the theory of spatial object modelling for GIS. Taylor& Francis, London

Mücke EP, Saias I, Zhu B (1999) Fast randomized point location without preprocessing in two- and three-dimensional Delaunay triangulations. Comput Geom Theor Appl 12:63–83

OGC (2006) OpenGIS implementation specification for geographic information-simple feature access. Open Geospatial Consortium inc., document 06–103r3.

OGC (2007) Geography markup language (GML) encoding standard. Open Geospatial Consortium inc., document 07–036, version 3.2.1.

Penninga F (2005) 3D topographic data modelling: Why rigidity is preferable to pragmatism. In: Cohn AG, Mark DM (eds) COSIT-Proceedings international conference on spatial information theory, Lecture Notes in Computer Science, vol 3693. Springer, pp 409–425.

Penninga F (2008) 3D topography: a simplicial complex-based solution in a spatial DBMS. PhD thesis, Delft University of Technology, Delft, The Netherlands.

Pilouk M (1996) Integrated modelling for 3D GIS. PhD thesis, ITC, The Netherlands.

Shewchuk JR (2002) Constrained Delaunay tetrahedralization and provably good boundary recovery. In: Proceedings 11th international meshing roundtable, Ithaca, New York, pp 193–204.

Shewchuk JR (2005) Star splaying: an algorithm for repairing Delaunay triangulations and convex hulls. In: Proceedings 21st annual symposium on computational geometry, ACM Press, Pisa, pp 237–246.

Si H (2008) Three dimensional boundary conforming Delaunay mesh generation. PhD thesis, Berlin Institute of Technology, Berlin.

van Oosterom P, Stoter J, Quak W, Zlatanova S (2002) The balance between geometry and topology. In: Richardson D, van Oosterom P (eds) Advances in Spatial Data Handling-10th International Symposium on Spatial Data Handling, Springer, pp 209–224.

van Oosterom P, Meijers M (2011) Towards a true vario-scale structure supporting smooth-zoom. In: Proceedings of 14th ICA/ISPRS workshop on generalisation and multiple representation, Paris, pp 1–19.

Zlatanova S, Abdul Rahman A, Shi W (2004) Topological models and frameworks for 3D spatial objects. Comp Geosci 30(4):419–428

Can Topological Pre-Culling of Faces Improve Rendering Performance of City Models in Google Earth?

Claire Ellul

Abstract 3D City Models are becoming more prevalent, and have many applications including city walk-throughs or fly-throughs to show what a new building would look like in situ, or whether a view or light will be blocked by a new structure, flood modeling, satellite and signal modeling. Often, these models are created using a process of extrusion of 2D topographic mapping, resulting in Level of Detail 1 buildings with flat roofs. The models can contain many thousands of polyhedra, which in turn results in performance issues when attempting to visualize such models in virtual earth applications such as Google Earth. This paper presents the results of a series of tests to determine whether using a topological approach to pre-cull hidden Faces from the model can bring about performance improvements. Such an approach could also be said to be one step towards the generalization of such models to support multiple levels of detail.

Keywords 3D city models · Topology · Rendering · Performance · Intersection

1 Introduction

Three-dimensional (3D) City Models are becoming more prevalent and have applications including utility infrastructure validation ("call-before-you-dig"), comparing existing buildings with building and planning regulations, engaging the public in planning issues (Batty et al. 2001; Coors et al. 2009; Isikdag and Zlatanova 2010), real estate sales, noise studies (Stoter et al. 2008a; Stoter et al. 2008b),

C. Ellul (✉)
Department of Civil, Environmental and Geomatic Engineering, University College London, Gower Street, London, WC1E 6BT, UK
e-mail: c.ellul@ucl.ac.uk

J. Pouliot et al. (eds), *Progress and New Trends in 3D Geoinformation Sciences*, Lecture Notes in Geoinformation and Cartography, DOI: 10.1007/978-3-642-29793-9_8,

cadastral systems (for example Stoter and Salzmann 2003) and the work undertaken by the many participants in the recent 3D cadastral conference (such as Pouliot and Vasseur 2011; Khoo 2011; Aien et al. 2011; Stoter et al. 2011), augmented reality (Coors 2004), personalized tourist information (Blechschmied et al. 2006), Schulte and Coors, 2008 in Boguslawski et al. 2011), line of sight analysis (Fredericque and Lapierre 2009, vehicle positioning in situations where satellite out-takes occur (Lowner et al. 2010), energy consumption in buildings (e.g. comparing surface area to volume ratios, Carrión et al. 2010), shadow effects on photo-voltaic cells (Alam 2011), 3D navigation on mobile phones (Basanow et al. 2008) and advanced driver assistance systems (Van Essen 2008). Such models also facilitate integration of heterogeneous 2D and 3D data (Glander and Dollner 2008)—for example Richmond and Romano (2008) construct a 3D City Model and then integrate it with geo-demographic data to automatically identify 'residential' neighborhoods in the city.

A number of methods can be used to generate data for City Models, including the manual creation of detailed 3D buildings in applications such as Sketch-Up (2011), or more automated processes such as extrusion, where a 2D footprint of a building is 'grown' to a given height (Evans et al. 2007; Ledoux and Meijers 2011; Richmond and Romano 2008). Models can also be created from LiDAR and Laser Scanning datasets (Richmond and Romano 2008; Wang and Sohn 2011). Increasingly, applications such as *neo-photogrammetry* (Heipke 2010; PhotoSynth 2012) are also providing sources of detailed 3D data.

The process of extrusion is most efficient when a larger area is to be covered (for example an entire city), and where high levels of detail (e.g. sloping roofs) are not required, and has the advantage of integrating 3D buildings with a 2D footprint (Kada 2009), resulting in Level of Detail 1 (LoD1) buildings (Kolbe et al. 2005). However, a common problem resulting from this process of model creation is the complexity of the outcome with respect to individual components, their computer graphics and the rendering resources (Glander and Dollner 2008). Indeed, the resulting 3D data is generally quite large in volume, and thus potentially difficult to visualize in its entirety utilizing 3D packages such as ArcGIS 3D Analyst (ESRI 2011) or Google Earth (2012).

Two aspects can be considered. Firstly, the cognitive load of the resulting model—does the result provide sufficient information for the applications described above? Will end users suffer from information overload? Will the resulting clutter lead to degradation in user performance (Baudisch and Rosenholtz 2003 in Kazar et al. 2008)? Secondly, what is the performance of such systems when confronted with large datasets? Efficient and scalable techniques for storing, querying and visualizing such datasets are fundamental for City Modeling (Kazar et al. 2008) and the large volume of data presents one of the challenges when serving City Models over the web, particularly in real time (Sester 2007; Curtis 2008).

This paper focusses on this second issue, and examines the use of topological concepts to underpin the identification of shared and hidden Faces a 3D City Model. Can performance gains be made by a process of pre-culling to remove these Faces from the model before displaying the results in Google Earth?

The remainder of the paper is structured as follows: Sect. 2 presents background information into the generation of datasets for 3D City Modeling, options for structuring such data and issues involved in visualization. Section 3 describes the preparation of two tests datasets—one urban, one suburban—for two London areas, including the creation of the 3D model by extrusion and the identification and removal of shared Faces. Section 4 presents the results of the tests carried out using the Google Earth Application Programmers Interface and Web Plug-In, and Sect. 5 discusses the potential of the methods developed and identifies areas for further work.

2 Background

This section reviews the process of generating a 3D City Model from a number of sources, and includes information on the data storage methods used within a spatial database for the resulting data. An overview of the general process used to render such models is given and existing approaches to improving the performance of this rendering process outlined.

2.1 Methods Used to Generate 3D Datasets and Resulting Levels of Detail

In 2001, Batty et al. (2001) published an initial list of methods for generating City Models, which ranged from digital ortho-photos (having very low geometric content), 2.5D image draping and extrusion (block modeling, with a medium level of geometric content) and highly detailed, fully volumetric buildings derived from Computer Aided Design (CAD) models. In addition to this, both Light Detection and Ranging data (LiDAR, airborne laser scanning generating highly detailed point clouds from which models can be extracted), vertical photogrammetry (either professional or through applications such as PhotoSynth, 2011) and terrestrial Laser Scanning can be used to generate 3D models of buildings to varying levels of detail (LoD). In a 3D context, the different levels of detail have been defined by Kolbe et al. (2005)—where LoD0 corresponds to a digital terrain model, LoD1 is a block model without any roof structures, and moving up to LoD4, which includes roofs, and also the interior structures of the buildings.

While it may be possible to derive a 3D model from a single source—as exemplified by Tse et al. (2008) who attempt to use LiDAR to generate a 3D City Model including roof structures—in practice, 3D models are generated using a combination the above methods and data sources. For example, Van Essen (2008) describes the production of 3D city maps starting from the 2D base and adding height, roof representations and textures—in this case deriving building height from photogrammetric elevation models, subtracting terrain height from normal digital elevation models and façade textures from TeleAtlas' mobile mapping images. As the basic method only applies to standard buildings with planar facades, a separate set of 3D landmarks has been created and added into the map. Pu (2008) starts by extracting important building features—walls, windows, roofs, doors, from a point cloud captured via terrestrial laser scanning. Then visible building geometries are determined by direct fitting of polygons to the feature segments. Geometric assumptions are made for the occluded parts. These elements are then combined to generate the solid building models. Richmond and Romano (2008) describe constructing a City Model through extrusion using a combination of LiDAR and floor height estimates. More recently, Wang and Sohn (2011) propose fusing terrestrial and airborne laser scanning with architectural drawings to generate a seamless 3D building model.

Given the varying input sources, the output of the varying methods can differ widely in terms of the level of detail about individual buildings. In general, however, methods such as extrusion (combining 2D topographic mapping with height information derived from LiDAR data) provide a rapid mechanism of generating an entire City Model to LoD1, where as detailed terrestrial laser scans or CAD drawings may be suitable to obtain greater detail (e.g. LoD4) but for fewer buildings within the city. Extruded models also provide separate 'building' entities which cannot easily be identified from LiDAR or terrestrial scanning, where blocks of buildings form one point cloud. While CAD models do provide separate buildings, they are not generally available in large quantities, making extrusion the most appropriate mechanism to generate the data required for testing performance of Google Earth.

2.2 *Spaghetti and Topology*

Geographical Information Systems (GIS) distinguish between two approaches to modeling lines, polygons or polyhedra. Firstly, a simple features approach, where each object is stored individually and data is represented as a planar configuration of points, arcs and areas with no explicit representation of the topological inter-relationships of the configuration, such as the adjacency relationships between constituent areas or volumes (Longley et al. 2011; Worboys and Duckham 2004). The structure of such simple feature polylines or polygons is sometimes called 'spaghetti' because, like a plate of cooked spaghetti lines (strands of spaghetti) and polygons (spaghetti hoops) can overlap and there are no relationships between individual features (Longley et al. 2011). Polygons are represented as independent cycles of coordinates and in a 2D context, the 'spaghetti' approach is also useful to rapidly render data onto screen (Worboys and Duckham 2004) or to color individual features according to their attributes (Rigaux et al. 2000). However, given that each feature is stored as a separate entity, this model requires duplication of

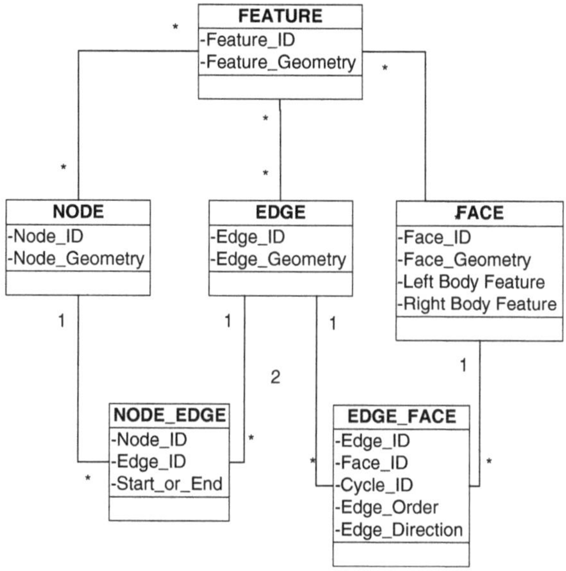

Fig. 1 Basic topological data structure, showing node, edge and face primitives and the links between them

information—for example, the shared wall between two adjacent buildings will be represented twice. This increases the amount of data that is held within the model.

A second approach to data storage is to create a topological model, where topological features are simple features (Nodes, Edges, and Faces) structured using topological rules (Longley et al. 2011). There is no redundancy in the model as the shared boundaries between objects are stored once and then combined to make up the specific object. Thus topological queries (adjacency, intersection) can be efficiently computed, and there is also update consistency—when you move the boundary between two buildings, the same boundary object is moved for both buildings (there is no redundancy in the shared boundaries, Rigaux et al. 2000). As shared boundaries are only represented once, data structured topologically requires less storage than that structured as spaghetti—and the lack of data duplication means that any edits to the data are automatically propagated to related objects (e.g. adjacent buildings). A topological structure pre-calculates the topological relationship between features and stores it in a data structure optimized for quick retrieval.

Figure 1[1] shows a simplified topological data structure for 3D data. The basic components of the model consist of a series of 0-dimension primitives (Nodes)—which in turn contain X, Y and Z coordinates. These Nodes are joined together to construct 1-dimensional primitives (Edges) which in turn are ordered to form

[1] Note that for simplicity the volumetric object, which would permit the identification of the components of a building and of which buildings share a Face, has been omitted as the work described here focusses on visualisation and on reduction of Face primitives. In general such an object should be added to have a complete 3D model.

2-dimensional primitives (Faces). The Faces are linked together to form 3D polyhedra, which represent the buildings in the City Model. For convenience, the Face primitives have also been retained as geometry objects in the model—these are used to detect intersecting Faces (see Sect. 3.3 below).

The process of extrusion described in Sect. 2.1 results in a spaghetti structure, where each Face between buildings is represented twice. Details of how this data can be transformed into the topological structure shown above are given in Sects. 3.2 and 3.3.

2.3 Displaying 3D City Models in Google Earth

Google Earth is a commonly used virtual globe which allows the user to explore the world and shows satellite imagery, maps, terrain and 3D buildings. Users can zoom in and out, tilt and rotate the map and navigate to places of interest. The software is based on software created by Keyhole Corporation, which Google acquired in 2004, is the first version of Google Earth was launched in June 2005 (Google 2012a) and Google estimate that there have been over 1 billion downloads of the software (IT World 2012).

Users can also add their own data to Google Earth, making use of the Keyhole Markup Language (Google 2012b) which allows the specification of 3-dimensional points, lines and polygons. Having built a model (which can show the detail of a building or can encompass an entire city) users can publish this to the Google Earth environment for other users to view.

The graphics engine inside Google Earth is proprietary. However, users can run Google in two Graphics optimization modes—DirectX or OpenGL, depending on their hardware.

Figure 2 shows part of the graphics rendering process used by OpenGL.[2] The full rendering process to display data on screen requires a transformation from the 3D dataset, having real world coordinates, to a set of 2D pixels corresponding to the layout of the screen. For each object to be displayed the system must 'scan' the object by generating a series of scan lines (rays) and calculate where these intersect the object. This computational geometry process is coupled with a process of clipping (using computational geometry to determine which objects are actually visible in the view on screen) and anti-aliasing (to improve rendering of curved objects). Hidden surface removal algorithms are used to determine which objects (and which parts of each object) are visible to the user, and which are hidden by other objects (and hence should not be drawn) (Chen and Chen 2008). Rendering algorithms are run for every object to be displayed. Therefore, to

[2] Open GL is a framework developed over 20 years ago by the Silicon Graphics Lab, and is now maintained by the Khronos Group [0]. It is a standard for writing computer graphics based programs, making use of the computer's graphical processing unit (GPU) which is specialized hardware designed to optimize image display on computer screens.

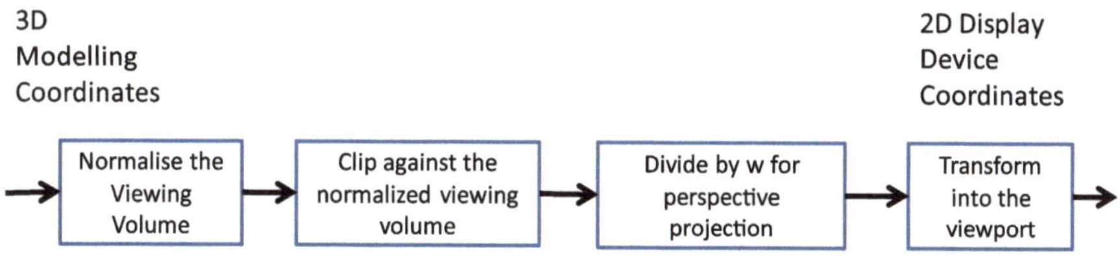

Fig. 2 Overview the graphics rendering pipeline (from Chen and Chen 2008)

minimize the number of times such operations must be repeated, a process of data reduction is required, to ensure that the data volume 3D City Model can be rendered by the graphics engine in a timely manner.

2.4 Current Approaches to Data Reduction

A number of approaches to the task of reducing the volume of data of a 3D City Model can be identified

Firstly, data compression has been identified by Van Essen (2008) as a partial solution to reducing the size of their City Model, which was stored in the Virtual Reality Mark-up Language. However, this approach, which reducing the time to transmit the model to a client in a web-based environment, does not reduce the number of Faces that are required to be rendered by the graphics pipeline. Similarly, a process of mesh simplification is often employed in Computer Graphics to remove extraneous detail from 3D triangular meshes. As Sester (2007) notes, these methods are more suitable for large meshes representing a single object and having many redundant points, rather than the multiple individual features held in 3D City model, where removing one point may cause a problem with an individual building.

Secondly, the concept of *Levels of Detail*, (Kolbe et al. 2005) where different representations of the data is used at different scales, is used to limit the size of datasets—the effective visualization of complex 3D City Models requires an abstraction of City Model components (Glander and Dollner 2008). Whilst in 2D Geographic Information Systems (GIS), the equivalent concept (known as generalization) is well understood, this is less so in 3D. A number of approaches to 3D generalization can be identified. For example, Glander and Dollner (2008) investigated methods including cell-based generalization, convex hull creation (using average heights of buildings in a block) and a voxel based approach, assigning objects to specific cells on a grid. Generalization using a process of subdividing building parts into half-planes is suggested by Kada (2007) and Guercke et al. (2009) describes approaches that take into account the importance of semantics to the end result, including the identification and emphasis of key features in a city, proposing a parametric approach towards the problem. The authors in Guercke et al. (2009) also cite a number of other approaches such as that by Lal 2005 (in Guercke et al. 2009) who makes a distinction between micro, meso and macro models for generalisation and Dollner and Buchholz (2005), in (Guercke et al. 2009) who introduce the concept of continuous level of quality buildings that allows the user to model buildings with a custom level of granularity according to the task at hand. An additional review of approaches to and techniques for generalisation can be found in Fan et al. (2009), who themselves investigate generalisation at block level, using an approach to extract the exterior shell of building models that contain interior and exterior surfaces for walls and roofs, and examining the issue of generalisation of windows.

A third approach to data reduction is that of defining scales and zoom levels at which certain features within the model become visible. For example, when opening a City Model, it may be possible to see only key landmark features in 3D model. As the user zooms in on a particular area additional features and details become available.

Although both the second and third approaches above show promise, the former restricts the user to specific view scales and the latter to pre-defined levels of detail. Section 3 proposes a complementary method to reduce the data volume of a 3D City Model, which could in turn increase the flexibility and range of both these approaches.

3 Methodology

The aim of the experiments described in the remainder of this paper is to determine whether transforming a 3D City Model from a spaghetti to a topological data structure (as described in Sect. 2.2), and hence reducing the data volume of the model, results in a performance improvement when rendering the data.

The dataset used for the experiments is a detailed topographic mapping dataset for London, UK Map, provided by GeoInformation Group, and includes height information for each polygon. Two tests datasets, one covering a small portion of central London (the area surrounding University College London, consisting mainly of high rise buildings) and the other covering a more suburban area (Petts Wood, to the South East of the city, consisting of low-rise buildings and residential housing) were chosen to provide insight into the extent to which topologically structuring the data could reduce the complexity of the dataset in both urban and suburban situations (Fig. 3). Table 1 summarizes the datasets.

The data was uploaded to an Oracle Spatial database using Snowflake's GO Loader software (Snowflake 2012). Oracle Spatial was selected for its ability to model and index 3D spatial datasets, and in particular for the inbuilt *extrude* function which allows the automatic generation of 3D polyhedra from the 2D polygons. Given the focus on buildings, tables containing the subset of buildings data were created for further work.

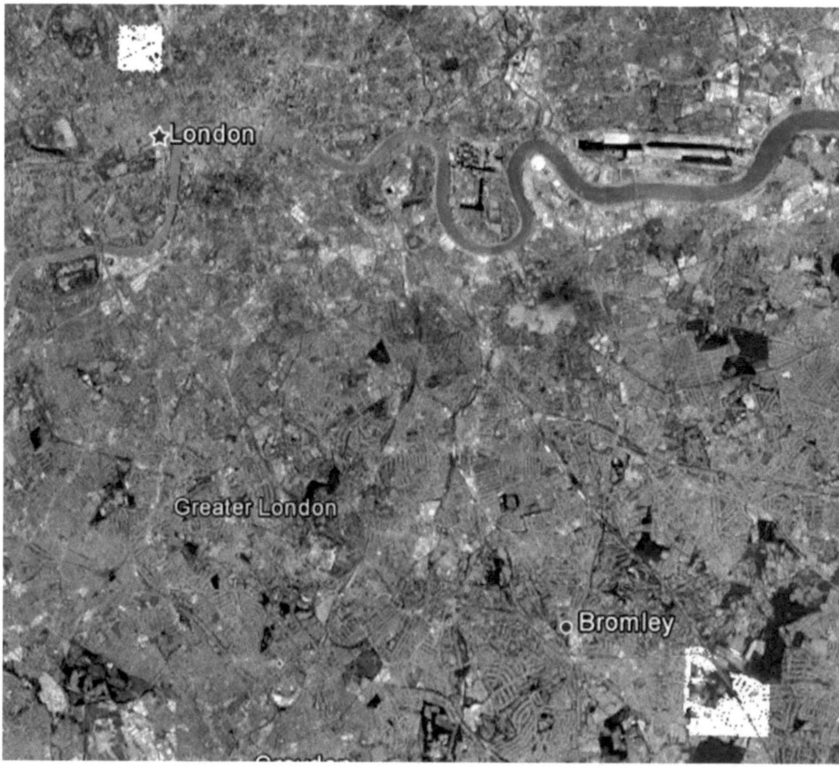

Fig. 3 Overview map of London in Google Earth showing UCL test data (*top left*) and Petts Wood test data (*bottom right*)

3.1 Extruding the 3D Dataset

Oracle Spatial provides built-in functionality to extrude 3D spatial data (Oracle 2011). However, preliminary tests revealed that Oracle's *SDO_UTIL.EXTRUDE* function is not able to handle polygons with internal holes. A Java program has been written to iterate through each building, and identify and extract the shell and Oracle's *SDO_UTIL.EXTRUDE* utility is then applied to this polygon. The internal details of any building, where present, are discarded.

3.2 Topologically Structuring the Data: Nodes, Edges and Faces

The result of the extrusion process in Oracle Spatial is a set of 3D features stored in a spaghetti structure (see Sect. 2.2). Each feature is made up of a single polyhedron, which is in turn constructed from a series of individual Faces held in an

Table 1 Topographic mapping polygon count by theme

Theme	Number of polygons—UCL	% Area, UCL	Number of polygons—Petts Wood	%Area, Petts Wood
Buildings	3147	40.46	12757	8.23
Man-made surfaced areas	2778	49.35	4106	17.35
Man-made structures (not buildings)	26	0.40	35	0.07
Vegetated/Natural areas –grass	486	9.52	6356	45.83
Water	0	0.00	20	0.03
Natural areas—scattered trees	0	0.00	9	0.79
Natural areas—mostly trees	1	0.26	75	27.70
Total features	6438	100	23358	100

Oracle SDO_GEOMETRY. The next step in the data preparation is to identify the topological relationships (in particular adjacency) between the constituent Faces forming each polyhedron, and to convert the data from spaghetti to a topological structure.

Each 3D polyhedron created by the extrusion process was split into constituent Nodes, Edges and Faces. Oracle Spatial stores 3D objects internally as lists of Faces, which in turn are made up of ordered lists of 3D coordinates. The coordinates were first extracted and Node objects created for each triplet (with checks for duplicates). The Node order was then used to construct Edges (formed of two Nodes) and individual Face polygons created for each of the Faces in the 3D Object. The results were stored in the topological data structure described in Sect. 2.2 above. For each new Node and Edge component, the corresponding table was first queried to determine whether it had already been created. Given the absence of a 3D *EQUALs* in Oracle Spatial operation, this was done by comparing the coordinate values at *Node* level, and then the linked *Nodes* at *Edge* level. The following section describes the process used to identify shared and intersecting Faces.

3.3 Topologically Structuring the Data: Identifying Shared Faces

As can be seen in Fig. 1 above, the final topological structure permits a direct query on the resulting Face objects to determine whether the Face has both a *left* and *right* feature. If this is the case, then it can be concluded that the Face is shared between two buildings and is an internal wall. Therefore, the second part of the topological structuring process involved the identification of adjacent Face objects and hence shared internal walls.

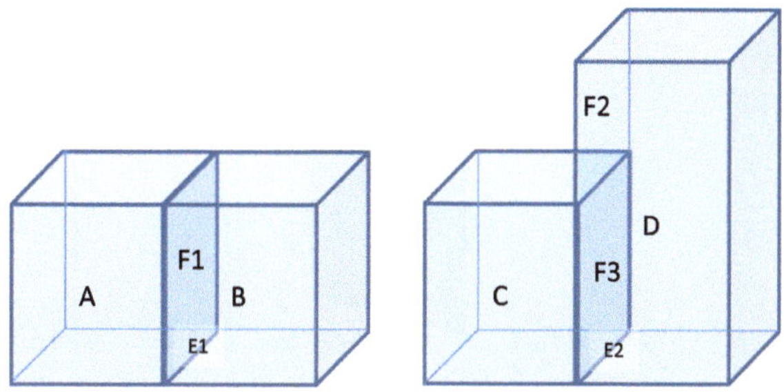

Fig. 4 Identifying shared faces between adjacent 3D buildings—step 1

To date (late 2011), Oracle Spatial does not provide a full 3D intersection query would easily identify shared Faces.[3] Therefore an alternative option was identified to determine whether two buildings were adjacent in some manner and hence shared a Face, and if so to create this 'shared' Face geometry. Two situations were considered—buildings that were identical in height and width, and buildings that were identical in width but not in height, as shown in Fig. 4, where Buildings A and B share identical Face F1 and Buildings C and D share identical Face F3 but Building D also includes another Face F2 on the same plane as F3.

The shared Faces were identified by locating common shared base edges—such as E1 and E2 in Fig. 4. The identified Faces were then projected into 2D space, by extracting the X and Z coordinates from each Node, as shown in Figs. 5 and 6.

Once the data was projected, 2D intersection and difference queries were used in Oracle Spatial, on each pair of Faces that shared an edge in 3D, to determine the intersection geometry and also any difference geometry. The results of this were then projected back into 3D space through a reversal of the process shown in Figs. 5 and 6, ensuring that the links between the original Face (e.g. F2 and its replacement Faces (F3 and F5) were maintained.

3.4 Creating Subsets of Faces for Testing

Once shared Faces were identified, and split topologically where necessary, the following datasets were created to be utilized performance testing:

- All Faces—these are the Faces derived from the spaghetti structure in the original dataset, and this dataset includes any duplicate Faces or overlapping Faces, which are represented twice.
- Single Faces—in this dataset, any overlapping Faces identified by the process in Sect. 3.3 were replaced by a single Face which formed the internal wall between two buildings, and by any split external Faces where building heights differed.

[3] Oracle Spatial at Version 11g supports the following queries in full 3D mode: SDO_ANYINTERACT, SDO_FILTER, SDO_INSIDE (for solid geometries only), SDO_NN and SDO_WITHIN_DISTANCE—http://docs.oracle.com/cd/B28359_01/appdev.111/b28400/sdo_intro.htm#BABIDJJB Accessed 12th January 2012.

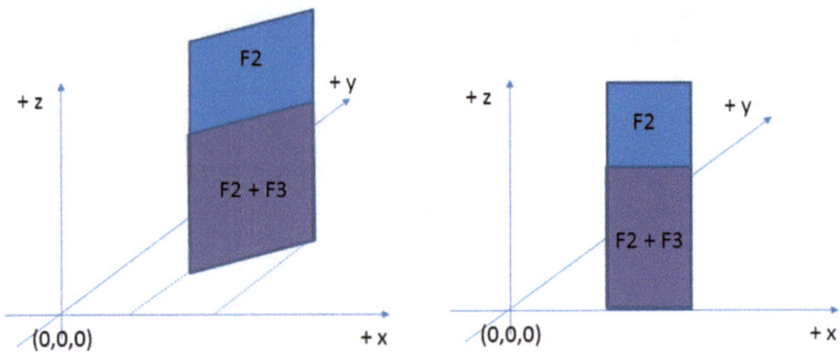

Fig. 5 Identifying intersections between shared faces—step 2—project the faces into the X/Z plane

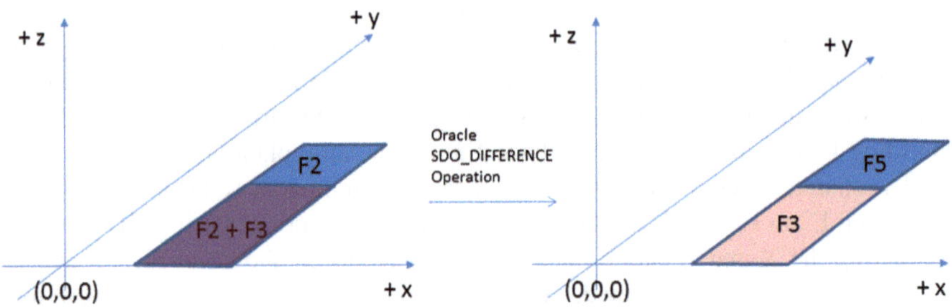

Fig. 6 Identifying intersections between shared faces—step 3—project the faces into the X/Y plane and then use Oracle's intersection queries to determine intersection and difference polygons

- No Internal Faces—in this dataset, the shared internal walls were removed entirely, and only the full shell of the block of buildings was represented.
- No Floors—in this dataset, the Faces making the floors were identified (as having all height values 0) and removed.

3.5 Performance Testing Using Google Earth

As described in Sect. 2.3, Google Earth is a desktop product designed to read files storing spatial data in the Keyhole Markup Language. For each of the test datasets listed above, KML files were generated by iterating through each Face and adding it as a 3D polygon to the KML file. As the original dataset was stored in the British National Grid projection, and Google Earth requires geographic coordinates (latitude/longitude) the data was transformed as part of this process, using Oracle's SDO_CS.TRANSFORM functionality.

KML files were generated for each of the four options described above (All Faces, Single Faces, No Internal Faces and No Floors) for both the Petts Wood and UCL

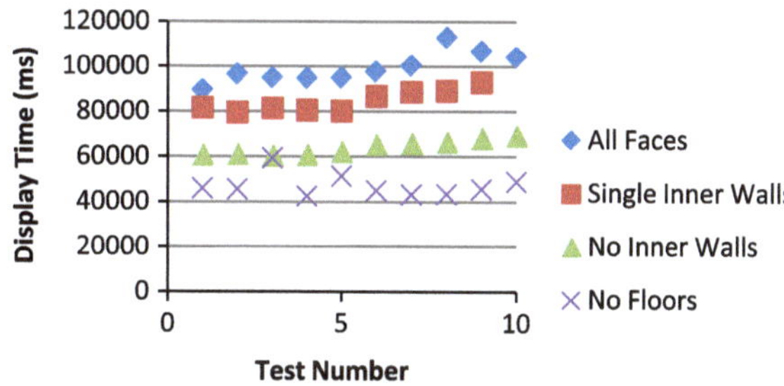

Fig. 7 Results for 10 valid rendering tests for the four face datasets—Petts Wood

datasets (8 files in total). The files were generated via a Java export routine, which iterated through the lists of Faces and created each Face as an individual polygon in the KML. These were then opened in the desktop version of the Google Earth package, to validate the KML. The results obtained are shown in Fig. 3, which shows the UCL data at the top left of the screen, and the Petts Wood data at the bottom right (the map area covers the greater London area).

The desktop version of the Google Earth package does not provide opportunities for customization to perform timed display testing. Therefore, the Google Earth Application Programmers Interface was used, along with JavaScript and HTML, to undertake the necessary tests. A test script was written that tracked two specific events in the Google Earth display process—firstly, the loading of the dataset was timed. Loading takes place via an AJAX (Asynchronous Java and XML) request, and time was measured from request to response. Secondly, the time to display the datasets was measured by trapping the Google Earth 'frameend' event, which is triggered when a drawing event is completed. As multiple 'frameend' events are triggered during a draw process, the test script was designed to log each of these and allow the user to determine when the final frame was drawn (and hence measure the time between the start of the draw process and the final 'frameend' event).

Google Earth displays a series of 'base' images underpinning any other data. As these images are retrieved over the internet, each KML file in each dataset was set to fly to an identical location on startup, thus causing identical base map tiles to load. GeoChalkboard (2012) explain how to use the concept of 'regions' to improve performance in Google Earth—this involves sub-dividing the data into a series of more and more detailed 3D volumes, which 'switch on' when the user has zoomed into a particular scale. However, this concept was not applied in this case as the aim of the test was to compare performance when displaying a full dataset.

All tests were carried out in the Google Chrome browser, Version 16.0.912.75, on a desktop machine having a basic Intel HD Graphics Renderer Video Card, with 3844 MB of Adapter RAM, on an Intel Core i7 CPU with a processing speed of 2.80 GHz, 8 GB RAM and running Windows 7 64-bit. It was planned to load each KML dataset ten times.

4 Results

4.1 Topological Data Storage

Storing the data in topological format reduced the number of coordinate triplets from 151595 to 32889 for UCL and from 508449 to 114040 for Petts Wood. A similar reduction in Edges can also be observed—from 124680 to 56267 for UCL and from 413961 to 190423 for Petts Wood.

4.2 Issues With the Google Earth Plug-In

During preliminary tests, a number of issues were noted with the Google Earth plug-in in the context of the larger Petts Wood dataset. Three issues were encountered, as follows:

- Excessively long time required loading and display data, where results obtained for individual tests were significantly higher than those previously obtained for the same dataset. For example, a load time of 17492 ms and a display time of 288591 ms for the 'Single Inner Walls' dataset for Petts Wood, and load time of 6740 ms and display time of 174212 ms for the 'No Floors' dataset. It was hypothesized that this could be due to internet connection responsiveness and given the overall consistency of the readings obtained (see Fig. 7) these values were considered outliers and removed.
- Excessively rapid time required to load and display data, again with results that can be considered outliers—for example a display time of 19048 ms for the 'Single Inner Walls' dataset for Petts Wood. The issue could be detected by a change in behavior in the plug in—instead of zooming directly to the location of interest, it first zoomed to the North of Scotland. These results appeared to be due to data caching, and were again excluded from the measurements. To overcome this issue, the Google Chrome browser was closed and re-opened as necessary during the tests.
- Frequent crashes of the Google Earth plug-in, despite re-installation. This was observed to happen in particular when the test dataset was changed (e.g. when testing firstly on 'All Faces' and then on 'Single Internal Walls').

Given the above issues, tests were run for the Petts Wood data until ten valid results were obtained for each test (requiring approximately 5 extra tests per dataset). Figures 7 and 8 below show the results obtained. As can be seen, once the outliers are eliminated the results obtained are consistent across all datasets. The smaller size of the UCL dataset meant that the above issues were not encountered.

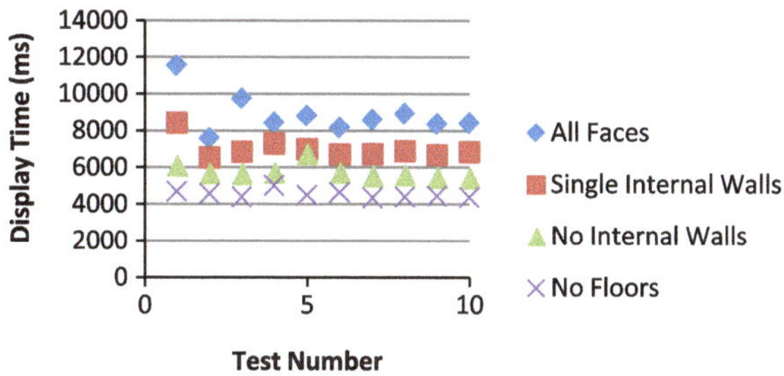

Fig. 8 Results for 10 valid rendering tests for the four face datasets—UCL

Fig. 9 Google Earth display for Petts Wood data—fixed display for all tests

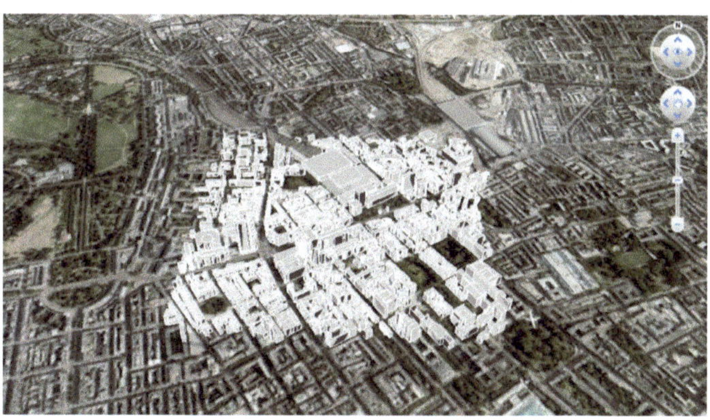

Fig. 10 Google Earth display for UCL data—fixed display for all tests

Table 2 Petts Wood results—load and display time, number of faces and KML file size

Petts Wood	All faces	Single inner walls	No inner walls	No floors
Average load time (ms)	2628.30	2422.90	2439.20	3424.80
Average display time (ms)	99287.30	84850.30	64096.70	47167.30
Number of faces	94488.00	88246.00	76622.00	63868.00
KML file size	41.39	38.76	33.84	27.73

Table 3 UCL results—load and display time, number of faces and KML file size

UCL	All faces	Single inner walls	No inner walls	No floors
Average load time (ms)	1337.90	1140.20	862.30	711.00
Average display time (ms)	8866.90	7010.40	5777.00	4558.20
Number of faces	26915.00	24138.00	18612.00	15477.00
KML file size	12.11	10.93	8.59	6.91

Table 4 Decreasing numbers of faces as hidden faces removed

	All faces	Single inner walls	No inner walls	No floors
UCL	100	89.68	69.15	57.50
Petts Wood	100	93.39	81.09	67.59

4.3 Summarizing the Results

Figures 9 and 10 show the resulting Web pages with the embedded Google Earth plug-in, set to the fixed view in each case to ensure that the base map tiles loaded were identical and thus did not influence the comparative results.

Tables 2 and 3 summarize the results obtained for UCL and Petts Wood respectively. As expected, the size of the KML file decreases with the number of Faces in the file, dropping from 41.39 MB to 27.73 MB for Petts Wood and from 12.11 MB to 6.91 MB for UCL. The ratio between the number of Faces and number of buildings reflects the urban and suburban nature of the locations—with a total of 94488 Faces for 12757 buildings in Petts Wood (7.4 Faces per building), and 26915 Faces for 3147 buildings (8.5 Faces per building) perhaps reflecting the more 'complex' building footprints in the UCL area.

The above tables highlight the differences between the UCL and Petts Wood datasets, both in terms of the overall number of Faces (UCL has 26915 versus 94488 for Petts Wood) but also the impact of removing shared internal Faces, all internal Faces and the floors. Table 4 summarizes the resulting number of Faces, expressed as a percentage of the total.

As can be seen, the percentage of Faces identified as shared is higher for the UCL area, which may be due to the more central location and hence more densely packed, terraced (buildings on both sides) buildings. It could be assumed that in a suburban area, houses may be detached (not touching other buildings) or semi-detached (only one other building to one side). The urban versus suburban nature of these neighborhoods is also reflected in the impact of removing the Faces

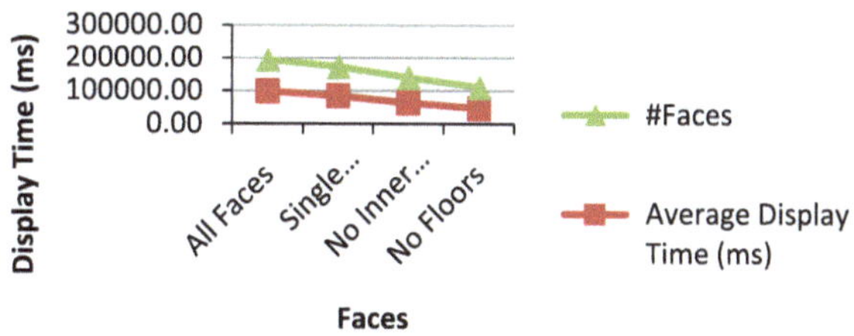

Fig. 11 Comparing the average number of faces and average display time—Petts Wood

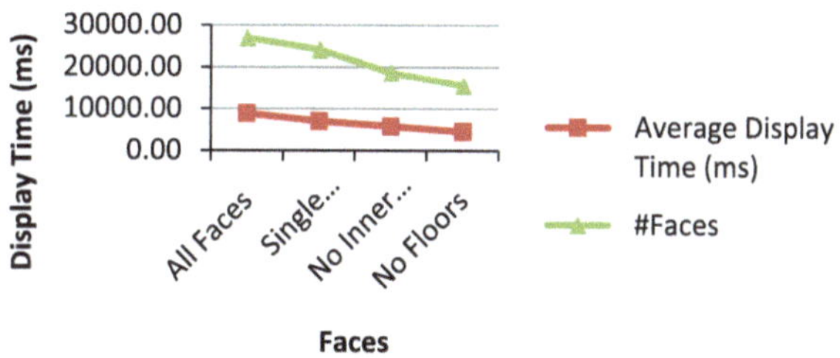

Fig. 12 Comparing the average number of faces and average display time—UCL

corresponding to the floor of each building—in the case of UCL, this removed 11.64 % of the Faces, whereas for Petts Wood, the total percentage of Faces removed was 13.50 %.

Figures 11 and 12 graph the correspondence between the number of Faces and the time taken to display the dataset. In both the UCL and the Petts Wood case, these graphs show an almost linear drop in display time as the number of Faces decrease. In the case of UCL, going from All Faces to No Floors results in an overall rendering time nearly half that of the original (51 %), with the display time for Petts Wood taking less than half the original (47 %).

4.4 Visualizing the Results

Figure 13 shows a screen shot of the Google Earth plug in and the UK Map data around the 'Cruciform' building which forms part of UCL. The KML dataset shown on the left contains All Faces, whereas the KML dataset on the right is the 'No Floors' dataset. As can be seen, the resulting performance gain can be obtained with no loss of external detail of the buildings.

5 Discussion and Further Work

The results obtained above show that there are significant performance gains to be obtained by making use of topological data structures to store 3D City Models generated by extrusion, pre-culling some Faces and hence generating KML files including only the Faces required for display of a building's outer shell.

Fig. 13 UCL cruciform building and surroundings—all faces (*left*) and UCL cruciform building and surroundings—No Floors (*right*)

Removing internal and floor information does not impact the overall quality of the resulting City Model. Additionally, the methods developed take full advantage of Oracle Spatial's inbuilt spatial data management functionality (e.g. the extrude option) and do not require additional algorithm development. In general, the selected tools (Oracle Spatial, Java and Google Earth's web plug-in) also proved adequate. Although not applied in this case, where a simple test was carried out, the methods described above to identify shared 3D faces (project to the X/Z plane, project into 2D, intersect, re-project upwards and into X/Y/Z) are directly applicable to situations where the shared Edge between two faces is not identical—i.e. where one building is offset from the either, either forwards, backwards or in both directions. It is anticipated that extending the method will result in a higher number of shared faces. However, in the offset situation the overall number of faces will not be reduced—two overlapping offset Faces, when split, will result in three Faces in total, one of which will represent the shared area (hidden Face) and will hence be removed.

Given the high quality dataset used for this project, there was no requirement to define a tolerance setting to identify shared Nodes –this could be added for use with poorer quality data or with data sources such as CAD, where dividing walls may be represented by two separate Faces that define the physical width of the wall itself. Additional work is also required to improve the Face matching methods for more complex building types—can it be adapted to LoD2, 3 and 4? The findings of Ledoux and Meijers (2011) are particularly relevant—they note that there are "no guarantees that a set of footprints will yield a topologically consistent city model" and present methods to overcome issues of topological inconsistency.

As a desktop application, Google Earth is well equipped to handle large datasets, and additionally provides 'region' functionality to improve performance. However, as described in Sect. 4.2 a number of issues were encountered with the Web-Based Google Earth plug-in when handling the larger dataset. Increasingly City Models are being used in a web-based context—thus the data reduction approach described above may become even more significant with the increase in uptake in mobile devices and tablets—currently a Google Earth App is available for both the Android and iPhone Platforms.

The techniques described above do not, by themselves, solve the performance problems encountered in Virtual Globes. Given the limitations of generalization and fixed-zoom viewing described in Sect. 2.4, it is suggested that the techniques described above can be used in combination with these approaches to gain an overall performance improvement for rendering while offering the possibility of viewing a larger area of the city in more detail, enhancing the user experience. Such combinations of methods can then be tested with larger datasets, at city level. The results obtained could also be compared to the rendering capacity of the Google 3D City layer, and results for other 3D web mapping tools tested.

The use of an Oracle Spatial database as a back-end to the project opens up the possibility to create a more dynamic system where the KML file could be automatically regenerated when the underlying data changes. Indeed, as KML provides a very useful data transfer format but is not optimized for editing, additions or changes to the data could be made using tools similar to those described in Ellul et al. (2009), with results stored directly in the Oracle Spatial database and shared through regenerated KML files, moving towards a more fully functional 3D toolbox. Such an approach could also take advantage of the inbuilt 3D queries within the Oracle database to provide additional functionality such as area and volume measurement.

A second benefit of tighter integration with the database would be the option to take advantage of Oracle's three-dimensional R-Tree query and again 'pre-cull' the number of Faces to be rendered by only retrieving data for the current area of interest. This approach would be similar to the oct-tree described in Fabritius et al. (2007) but would generate data subsets dynamically as required. Combining such approaches with the outcome of work carried out on 3D generalization may also permit such a tool to retrieve multiple levels of detail in one request—highly detailed data for the objects closest to the viewer, and more generalized data for objects further away, perhaps selecting the data using a 3D distance query inside the database. Given the time required to query and generate such datasets within the database, a comparison with current methods (as described in the graphics chain) would be required to determine whether the trade off in terms of extended SQL query time brings an overall benefit to the display process.

Beyond performance improvements, topology in GIS is well known as a tool for improving 2D data quality—the process of generating a topological data structure (identifying shared Nodes, Edges and Faces) can eliminate inconsistencies in geometry such as slivers. Research into data quality in 3D—and in particular the definition of valid geometries—is ongoing (Kazar et al. 2008; Ledoux and Meijers 2011). It is envisaged that the methodology presented here could be extended to assist in the resolution of such inconsistencies, resulting in the improved geometry quality required to apply 3D City Models to applications beyond visualization.

Finally, it should be noted that the experiments described in this paper addressed only one of the two issues faced when using 3D models—the volume of data. It is hoped that the improvement in rendering performance gained by using the methods described above will open up opportunities for further experimentation on the second issue—that of providing 3D City Models with appropriate content for the task at hand. The importance of gaining an understanding of both what is technically possible (in terms of performance) and cognitively useful (in terms of detail) within a City Model cannot be underestimated.

Acknowledgments The author would like to thank the GeoInformation Group for the provision of the UK Map dataset, and in particular Alun Jones for his initial encouragement. Thanks also go to the team at Snowflake Ltd. for the GO Loader software used to load the map data into Oracle Spatial, and to Richard Goodman and Intergraph Corporation for the use of Geomedia Professional to visualize the datasets stored in Oracle.

References

Aien A, Ali A, Kalantari M, Rajabifard A, Williamson I (2011) Advanced principles of 3D cadastral data modelling. In: Proceedings of the 2nd international workshop on 3D cadastres, organized by FIG, EuroSDR and TU Delft, Delft, The Netherlands, November 2011, 271–290 [online] Available from: http://3dcadastres2011.nl/programme/ Accessed 3rd Jan 2012, 377–396

Alam M (2011) GISt Report No. 5 shadow effect on 3D city modelling for photovoltaic cells, ISBN: 978-90-77029-27-5 ISSN: 1569-0245

Basanow J, Neis P, Neubauer S, Schilling A, Zipf A (2008) Towards 3D spatial data infrastructures (3D SDI) based on open standards—experiences, results and future issues. In: Van Oosterom P, Zlatanova S, Penninga F, Fendel E (eds) Advances in 3D GeoInformation systems, Springer, Chapter 2, pp 19–46

Batty M, Chapman D, Evans S, Haklay M, Keupers S, Shiode N, Hudson Smith A, Torrens P (2001) Visualising the city: communicating urban design to planners and decision-makers. In: Brail R, Klosterman R (eds) Planning support systems, models and visualisation tools. ESRI Press and Center Urban Policy Research, Rutgers University, Redland, pp 405–443

Blechschmied H, Coors V, Etz M (2006) Augmented reality and location-based services projects. In: Zlatanova S, Prosperi D (eds) Large-scale 3D data integration: challenges and opportunities, Taylor and Francis

Boguslawski P, Gold C, Ledoux H (2011) Modelling and analysing 3D buildings with a primal/dual data structure. ISPRS J Photogram Rem Sens 66:188–197

Chen J, Chen C (2008) Foundations of 3D graphics programming: using JOGL and Java3D, 2nd edn. Springer publishing company, ISBN: 9781848002838

Coors V, Hunlich K, On G (2009) Constraint-based generation and visualization of 3D city models. In: Lee J, Zlatanova S (eds) 3D geoinformation sciences. Springer, Berlin

Curtis E (2008) Serving CityGML via web feature services in the OGC web services –Phase 4 Testbed, Models, In: Van Oosterom P, Zlatanova S, Penninga F, Fendel E (eds) Advances in 3D geoinformation systems, Springer

Carrión D, Lorenz A, Kolbe T (2010) Estimation of the energetic rehabilitation state of buildings for the city of berlin using a 3D City model represented in CityGML. International archives of the photogrammetry, remote sensing and spatial information sciences, vol XXXVIII-4/W15

Ellul C, Haklay M, Francis L, Rahemtulla H (2009) A mechanism to create community maps for non-technical users, GEOWS '09 Proceedings of the 2009 international conference on advanced geographic information systems and web services, IEEE computer society

ESRI (2011) ArcGIS 3D analyst [online] Available from: http://www.esri.com/software/arcgis/extensions/3danalyst/index.html Accessed 12th Jan 2012

Evans S, Hudson-Smith A, Batty M (2007) 3-D GIS: Virtual London and beyond an exploration of the 3-D GIS experience involved in the creation of virtual London, CyberGeo—Eur J Geogr

Fabritius G, Kranigg J, Krecklau L, Manthei C, Hornung A, Habbecke H, Kobbelt L (2007) City virtualization—coursework [online] Available from http://openmesh.org/uploads/media/vrar_01.pdf, Accessed 12th Jan 2012

Fan H, Meng L, Jahnke M (2009) Generalization of 3D buildings modelled by CityGML. In: Cartwright W, Gartner G, Meng L, Peterson M (eds) Lecture notes in geoinformation and cartography. Springer, Berlin

Fredericque B, Lapierre A (2009) 3D City GIS—A major step towards sustainable infrastructure—a Bentley white paper [online] Available from, 3D City GIS—A major step towards sustainable infrastructure Accessed 20th Jan 2012

GeoChalkboard (2012) Using KML regions to display large datasets in Google Earth, [online]. Available from: http://geochalkboard.wordpress.com/2008/01/14/using-kml-regions-to-display-large-gis-datasets-in-google-earth-part-1/ Accessed 13th Jan 2012

Glander T, Dollner J (2008) Techniques for generalizing building geometry of complex virtual 3D city models. In: Van Oosterom P, Zlatanova S, Penninga F, Fendel E (eds) Advances in 3D geoinformation systems, Springer

Google (2012) Google company history [online] Available from : http://www.google.com/about/corporate/company/history.html Accessed 21st Jan 2012

Google (2012b) Keyhole markup language reference [online] Available from http://code.google.com/apis/kml/documentation/ Accessed 21st Jan 2012

Google Earth (2012)—3D Buildings showcase [online] Available from: http://www.google.co.uk/intl/en_uk/earth/explore/showcase/3dbuildings.html Accessed 12th Jan 2012

Guercke R, Brenner C, Sester M (2009) Generalization of semantically enhanced 3d city models. Proceedings of the GeoWeb 2009 conference, Vancouver, Canada

Heipke C (2010) Crowdsourcing geospatial data. ISPRS J Photogram Rem Sens 65(6):550–557

Isikdag U, Zlatanova S (2010) Interactive modelling of buildings in Google Earth: a 3D tool for urban planning. In: Neutens T, De Maeyer P (eds.) Developments in 3D geo-information sciences, Springer, 52–70

IT World (2012) Google Earth Announces 1 Billion Downloads [online] Available from: http://www.itworld.com/cloud-computing/210825/google-earth-announces-1-billion-downloads Accessed 21st Jan 2012

Kada M (2007) A contribution to 3D generalisation, Photogrammetric week, 41–51

Kada M (2009) The 3D Berlin project, Photogrammetric week, 331–340

Kazar B, Kothuri R, van Oosterom P, Ravada S (2008) On valid and invalid three-dimensional geometries, in models. In: Van Oosterom P, Zlatanova S, Penninga F, Fendel E (eds) Advances in 3D geoinformation systems, Springer

Khoo V (2011) 3D cadastre in Singapore. In: proceedings of the 2nd international workshop on 3d cadastres, organized by FIG, EuroSDR and TU Delft, Delft, The Netherlands, November 2011, 271–290 [online] Available from: http://3dcadastres2011.nl/programme/ Accessed 3rd Jan 2012

Kolbe T, Groger G, Plumer L (2005) CityGML—Interoperable access to 3D city models. In: van Oosterom P, Fendel E, Zlatanova S (eds) Proceedings of the international symposium on geo-information for disaster management, Delft, Springer Verlag

Ledoux H, Meijers M (2011) Topologically consistent 3D city models obtained by extrusion. Int J Geogr Inf Sci 25(4):557–574

Longley P, Goodchild M, Maguire D, Rhind D (2011) Geographical information systems and science, 3rd edn. Wiley, Hoboken

Lowner M, Sasse A, Hecker P (2010) Needs and potential of 3D city information and sensor fusion technologies for vehicle positioning in urban environments. In: Neutens T and De Maeyer P (eds) Developments in 3D geoinformation sciences, Springer

Oracle (2011)—Oracle Spatial SDO_UTIL.EXTRUDE [online] Available from: http://docs.oracle.com/cd/B28359_01/appdev.111/b28400/sdo_util.htm#BJECJIIE Accessed 12th Jan 2012

PhotoSynth (2012) About Photosynth [online] Available from: http://photosynth.net/about.aspx Accessed 12th Jan 2012

Pouliot J, Vasseur M (2011) Spatial representation of condominium/co-ownership: comparison of Quebec and French cadastral system based on LADM specifications. In: Proceedings of the 2nd international workshop on 3D cadastres, organized by FIG, EuroSDR and TU Delft, Delft, The Netherlands, November 2011, 271–290 [online] Available from: http://3dcadastres2011.nl/programme/ Accessed 3rd Jan 2012

Pu S (2008) Automatic building modelling from terrestrial laser scanning. In: Van Oosterom P, Zlatanova S, Penninga F, Fendel E (eds) Advances in 3D geoinformation systems, Springer

Richmond P, Romano D (2008) Automatic generation Of residential areas using geodemographics. In: Van Oosterom P, Zlatanova S, Penninga F, Fendel E (eds) Advances in 3D geoinformation systems, Springer

Rigaux P, Scholl M, Voisard A (2000) Introduction to spatial databases: applications to GIS Morgan Kaufmann

Sester M (2007) 3D Visualization and generalization. 51st Photogrammetric week, Stuttgart Germany, 285–295

Sketch-Up (2011) 3D modelling for everyone [online] Available from http://sketchup.google.com/ Accessed 17th Jan 2012

Snowflake (2012) GO Loader—Load GML into your database of choice [online] Available from http://www.snowflakesoftware.com/products/goloader/ Accessed 21st Jan 2012

Stoter J, de Kluijver H, Kurakula V (2008) Towards 3D environmental impact studies—example of noise. In: Van Oosterom P, Zlatanova S, Penninga F, Fendel E (eds) Advances in 3D geoinformation systems, Springer, 2008, Chapter 2, 19-46

Stoter J, de Kluijver H, Kurakula V (2008b) 3D noise mapping in urban areas. Int J Geogr Inf Sci 22(8):907–924

Stoter J, Salzmann M (2003) Where Do Cadastral Needs and Technical Possibilities Meet? In: van Oosterom P, Lemmen C (eds.) Computers, environment and urban systems, 27:4, 395-410

Stoter J, Hendrik P, Louwman W, van Oosterom P, Wünsch B (2011) Registration of 3D situations in land administration in the Netherlands. In: Proceedings of the 2nd international workshop on 3D cadastres, organized by FIG, EuroSDR and TU Delft, Delft, The Netherlands, November 2011, 27–290 [online] Available from: http://3dcadastres2011.nl/programme/ Accessed 3rd Jan 2012, 377-396

Tse R, Gold D, Kidner D (2008) 3D City Modelling from LiDAR Data. In: Models van Oosterom P, Zlatanova S, Penninga F, Fendel E (eds) Advances in 3D geoinformation systems, Springer

Coors V (2004) 3D modelling and visualisation. Comput Graphics 28(4):519–526

van Erp J, Cremers A, Kessens J (2011) Challenges in 3D geoinformation and participatory design and descision. In: Kolbe T, Konig G, Nagel C (eds) Advances in 3D geoinformation sciences, Springer

Van Essen R (2008) Maps get real: digital maps evolving from mathematical line graphs to virtual reality models. In: van Oosterom P, Zlatanova S, Penninga F, Fendel E (eds) Advances in 3D geoinformation systems, Springer

Wang L, Sohn G (2011) An integrated framework for reconstructing full 3d building models. In: Kolbe T, Konig G, Nagel C (eds), Advances in 3D geoinformation sciences, Springer

Worboys M, Duckham M (2004) GIS: a computing perspective, 2nd edn. CRC Press

Zabiki M (2011) OpenCL/OpenGL approach for studying active Brownian motion

On Problems and Benefits of 3D Topology on Under-Specified Geometries in Geomorphology

Marc-O. Löwner

Abstract The science of geomorphology is working on natural 3D landforms. This includes the change of landforms as well as the processes causing these changes. The main concepts of geomorphology, i.e. the sediment budget and the sediment cascade approach can definitely be enhanced by introducing 3D geometrical and topological specifications of the Open Geospatial Consortium. The ISO 19107, Spatial Schema, implements OGC's Abstract Specification. It enables the modelling of real world 3D phenomena to represent them as formal information models. Unfortunately, OGC's concepts are not widely applied in the science of geomorphology. In this article we are going to show the explicit benefit of 3D topology for the science of geomorphology. Analysing topological relationships of landforms can be directly related to geomorphic insights. This includes firstly, the process-related accessibility of landforms and therefore material properties, and secondly, the chronological order of landform creation. Further, a simple approach is proposed to use the benefits of the abstract specification 3D topologic model, when only *under-specified geometries* are available. Often, no sufficient data is available on natural landforms to model valid 3D solids. Following clearly defined geometric conditions the introduced class `_UG_Solid` mediates between primitives of lower dimension and a `GM_Solid`. The latter is the *realisation* of a `_UG_Solid` that definitely holds the 3D geometry we need to associate with the 3D topological concepts.

M.-O. Löwner (✉)
Institute for Geodesy and Photogrammetry, Technische Universität Braunschweig, Pockelsstraße 3, 30106 Braunschweig, Germany
e-mail: m-o.loewner@tu-bs.de

J. Pouliot et al. (eds), *Progress and New Trends in 3D Geoinformation Sciences*, Lecture Notes in Geoinformation and Cartography, DOI: 10.1007/978-3-642-29793-9_9,

1 Introduction and Problem Statement

The Open Geospatial Consortium's Abstract Specifications (OGC 2012) enable the modelling of real world phenomenon to represent them as formal information models (Kottman and Reed 2009). These information models may include geometry, attributes and topological relationships of real world objects. The main advantage of international accepted standards like OGC' Abstract Specification is interoperability. This means the seamless exchange of data and a simplified application of analysis concepts. The main document presenting the Abstract Specification is the ISO 19107 'Spatial Schema' (Herring 2001) defining geometric primitives and complexes from 0D to 3D according to the boundary representation (Foley et al. 1995). Next to other concepts, Spatial Schema is implemented in the Geography Markup Language (GML) (Lake et al. 2004). The release of GML led to a number of application schemas e.g. City Geography Markup Language (CityGML) (Gröger et al. 2012). However, CityGML mainly represents models on manmade environments.

Spatial Schema also provides a topology package mainly to convert computational geometry algorithms into combinatorial ones (Herring 2001, p. 104). Topological primitives (i.e. node, edges, faces and solids) need realizations in the form of geometric primitives with the same dimension. Thus, if no valid 3D geometry is provided for features that are known to be 3 dimensional, no 3D topology can be applied.

In the science of the land's surface, geomorphology, objects under examination are definitely volumetric. Built of sediment that is allocated by mainly externally driven processes geometry concepts of the Spatial Schema would be helpful to resent such sediment storages. Topological concepts may support the analysis of geomorphic systems in two aspects. Firstly, identifying neighbouring features and features connected via material transporting processes and, secondly, supporting analysis of landform's chronological order within a geomorphic system.

However, OGC's 3D concepts are not widely accepted in geomorphology. This is different with the simple feature concept implemented by main GIS companies. The main reason is that 3D data is difficult to collect due to complex phenomena and limited prospecting methods. Thus, especially 3D topology is not applicable to the science of geomorphology, since the topology package of Spatial Schema needs to refer to a valid geometry representation.

In this article a new class for 3D objects with under-specified geometry is proposed. _UG_Solid mediates between Spatial Schema's geometric primitives with a dimension less than 3 on the one side and a GM_Solid on the other. Constraints to aggregate a _UG_Solid are defined. The introduction of _UG_Solid enables the application of 3D topological concepts to geometric objects that are known to be volumetric but have to be constructed from sparse data.

In the next section the nature and main concepts of geomorphology will be outlined. A special focus is put on the topological aspects of landforms. Special cases of topological relationships between 3D solids will directly be related to geomorphic insights (Sect. 2.2). In Sect. 3 an application model on geomorphic objects and processes will be reviewed. Data acquisition and modelling problems have been identified as the main problems for the acceptance of 3D concepts (Sect. 3.3). Section 4 focuses on utilization the 3D topological concepts. Constraints for building an under-specified 3D geometry will be defined and proven for geomorphology. Section 5 follows up with a discussion.

2 Geomorphology: The Science of Natural 3D Landforms—Geometrical and Topological Considerations

Geomorphology defines itself as the science of natural landforms (Chorley et al. 1984; Hugget 2003). This does not only include geometric aspects of a 2.5D land surface which are covered by the science of geomorphometry in detail (Evans 1972; Rasemann 2004), but the change of landforms and the processes causing these changes. In general, landforms can be described as units of material, the sediment, which was accumulated under specific conditions and is reworked due to shape, material properties and external forces. The outcrops of these landforms compose the 2.5D boundary surface between solid earth and the atmosphere and the hydrosphere.

Without doubt, climate and gravity are the main external forces of such material transport processes (e.g. soil erosion and the corresponding accumulation or mass movements). In the first place running water erodes and transports material from one landform and accumulates it on the top of another one. Next to climate conditions the eroding power of flowing water is determined by the surface (e.g. slope) and the material's resistance. The same internal properties (e.g. soil texture or bulk density) determine the effectiveness of gravity causing mass movements like rock fall or debris flows (Summerfield 1997). The latter is definitely a property of a 3-dimensional body holding the sediment of a landform. Therefore, in geomorphology the *surface* under consideration is a three-dimensional body divided into neighbouring landforms. These facts are expressed in the broadly accepted term *georelief* coined by the German scientists Kugler (1974) and Dikau (1996). First, it represents the visible and measurable boundary surface between land surface and atmosphere or hydrosphere and, second, the material this surface is composed of (Young 1978). In geomorphology this recognition results in a triad of process, material and form. These three variables are characterised by strong feedback which has to be resolved when describing a geomorphic system.

It is obvious that a 2.5D concept is not sufficient to represent the 3D georelief under investigation. While geomorphology investigates the history of landforms, next to the visible surface, the subsurface is of interest as well. This subsurface, the paleo-surface, forms the starting conditions of the landform under investigation and therefore an important jigsaw piece revealing the landform's history. Anyway, a 2.5D concept is not able to represent more than one surface at a given position.

2.1 On Main Concepts of Geomorphology

Sediment budgets in geomorphology are used to quantify erosion and accumulation processes on a catchment scale. They are expressed by the sediment-delivery ratio. This describes the ratio between sediment eroded and transported in and through a system and material finally pushed out of the system (Cooke and Doornkamp 1990; Reid and Dunne 1996). Thus, sources and sinks of a geomorphic system have to be investigated and quantified. Performed by ground openings, drillings or geophysical exploration, geometry and thus volume of sediment bodies are reconstructed. Internal properties of investigated sediment bodies give hints to the main material transport processes and the periods in which these processes were active. The main material transport process might change in time due to climate variation and others.

The sediment budget concept in geomorphology may definitely be enhanced by the 3D modelling concepts of the Spatial Schema. The approach investigates true 3D geometries to get volumetric information. Application models are needed to store internal properties of the objects under investigation (i.e. soil type, density, chemical composition, etc.). Representing a geomorphic system in an application model would further support the exchange of data within and throughout the community.

The concept of a *sediment cascade* expands the sediment budget approach to the detailed questions concerning residence time of sediment. It is one of the main concepts in geomorphology (Church and Slaymaker 1989; Jordan and Slaymaker 1991). In theory, sediment is captured in storages for a certain length of time. This length depends on eroding processes, their force and, of course, the sediment's internal resistance to these processes. In terms of system theory (Chorley and Kennedy 1971), the output of one storage acts as the input for another one. Regulators like the land surface may divide the eroded material either to stay on the landform or to be transported to one or many others.

One example of a sediment cascade is the interaction of a free face, i.e. a wall, with an underlying talus slope (rf. Fig. 1). As the main storage, the wall feeds the talus slope by stone fall. The talus slope is situated at the walls foot and built up of wasting products of the wall. Nevertheless, the same process could accumulate the material on smaller features sitting on the wall itself, i.e. a band or a cleft. While a band is a step like feature a cleft is a crack present in every wall. Both are able to store material. Due to the formalism used in geomorphology (rf. Schrott et al. 2003), no geometry or time is represented.

While material exchange is more likely to be found between adjacent landforms, topological representation and analysis could definitely support the concept of a sediment cascade in geomorphology. Even the chronological order of landform creation can be analysed using topology (see below). Representing a sediment cascade following the concepts of the Spatial Schema would enable such investigations.

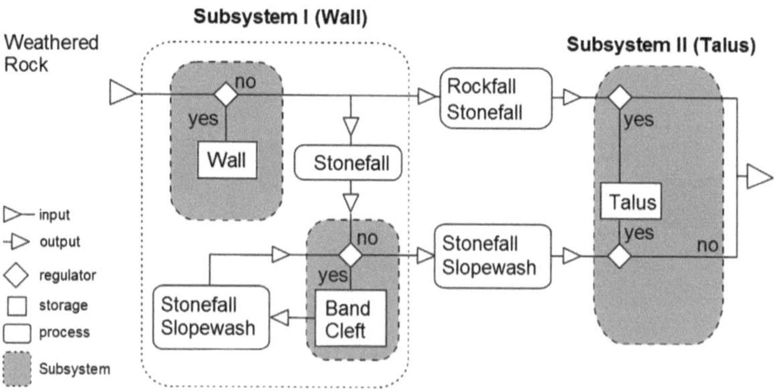

Fig. 1 Sediment cascade of the two subsystems Wall and Talus. Following the notation of system theory, no geometry and time is represented

2.2 Topological Aspects of Landforms

Landforms do not exist in isolation but do interact with others. Their specific association builds up the *georelief* (Kugler 1974; Dikau 1996) and characterizes a specific geomorphic system. While single landforms are scale-dependent, the composition of a geomorphic system follows a spatial hierarchy (Ahnert 1988; Dikau 1989; Brunsden 1996). Smaller landforms are located on the top of larger ones and cover them partly. Therefore, size is a good indicator of a landform's lifetime and age (Ahnert 1996).

Like a single landform, the association of many depends on internal system states, material supply and external driving forces. Following this experience, analysing the composition of landform association allows scientists to understand the main processes reworking a system, the succession of these processes in the past and probable developments in the future.

A lot of the landform's interaction may be analysed using the concept of topology. First, this includes the chance of processes to transport material from one landform to another one. Identifying possible material sources is the first step to expose and explain existing sediment cascades of a geomorphic system. Second, topological investigation may help to identify the chronological order in which landforms were formed. Often, this non-metric or relative dating is an important step towards the understanding of a geomorphic system.

Figure 2 depicts five main relationships of interacting landforms that are characterised using the topological nine-intersection model (Egenhofer and Herring 1990) in Table 1. Here, we follow the notation of Zlatanova (2000) where ∂a is defined as the boundary of a, a^{-} as the interior of a and a° as the exterior of a. Only boundaries and the interior will be considered.

The examples described in Fig. 2 and Table 1 show that geomorphic information can be directly archived from topological analysis. However, this is not

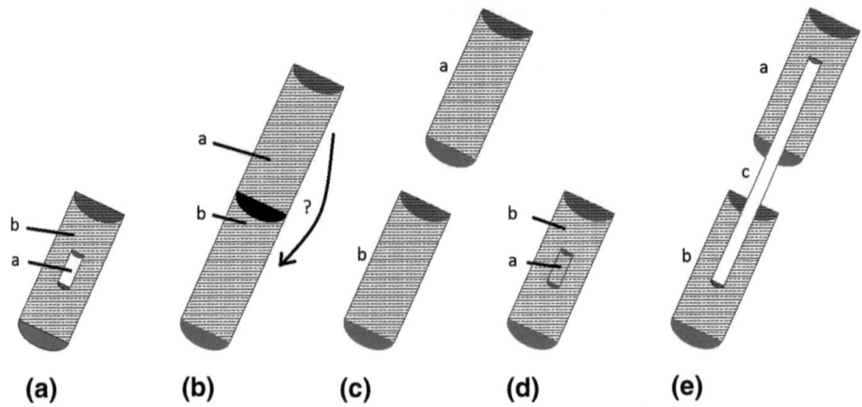

Fig. 2 Topological relationships of landforms within a geomorphic system (rf. Table 1)

Table 1 Geomorphic description of topological relationships of landforms (rf. Fig. 2)

Figure	Topological relationship	Geomorphic description
2.A	$\partial a \cap \partial b = \neg\emptyset$ $a^- \cap b^- = \neg\emptyset$	*a* lies on top of *b* and is definitely younger. *a* is formed by an erosion process on *b* followed by an accumulation. Material of *a* definitely contains material from *b*.
2.B	$\partial a \cap \partial b = \neg\emptyset$ $a^- \cap b^- = \emptyset$	*a* is adjacent to *b*. Chronological order can not be proved directly. Material exchange from the higher situated body to the lower one is most likely (needs further geometric analysis[a]).
2.C	$\partial a \cap \partial b = \emptyset$ $a^- \cap b^- = \emptyset$	*a* and *b* are disjoint. Chronological order can not be proved directly. Material exchange is still possible (needs further geometrical and topological analysis (ref. 2.E)).
2.D	$\partial a \cap \partial b = \emptyset$ $a^- \cap b^- = \neg\emptyset$	*a* lies within the interior of *b*. Genesis of *b* starts before the genesis of *a* and ended later. Implies an interruption of *b*'s building process either by temporarily high accumulation from another source or even temporarily erosion of *b*. Investigators should examine a separation of *b*.[b]
2.E	$\partial a \cap \partial b = \emptyset$ $a^- \cap b^- = \emptyset$ $\partial a \cap \partial c = \neg\emptyset$ $\Lambda\ b^- \cap c^- = \neg\emptyset$	*a* and *b* are disjoined but connected via *c*. Possibly the today filled hollow of *c* acted like a material transport path (formally an open channel?). Finding same material components of *a* in *b* or vice versa is likely (geometric inspection is needed).

[a] Since *above* or *below* are not topological terms, possible pathways have to be analysed applying geometric algorithms using f. i. a DEM

[b] This total inclusion of a smaller landform is definitely a 3D problem. 2.5D concepts are not sufficient

identified in the literature of geomorphology. That definitely shows that topology is not applied in the science of geomorphology.

3 On Problems of Semantical and Geometrical Modelling in Geomorphology

This section exposes significant difficulties in the use of application models within the science of geomorphology. These are different from the three characteristics identified by Kottman and Reed (2009), ignorance, modelling of phenomena not of mutual interest and modelling of phenomena in two different representations. One may argue that the science of geomorphology indeed is a diverse one and therefore, their researchers can be considered as individuals not belonging to the same information community. However, geomorphologists refer to almost the same paradigms and theoretical concepts (ref. Sect. 2). Thus, we argue that the main reason for reluctance to use OGC Abstract Specification based models to represent objects under investigation are firstly acceptance of technical overhead, and, secondly, problems in modelling valid 3D objects from sparse data.

3.1 A Class Model for Objects and Processes

Based on the ISO 19107, Spatial Schema (Herring 2001), Löwner (2010) proposed an application model to represent the aforementioned concepts of geomorphology. The model does not only include an object- oriented view of landforms with a true 3D geometry. It is designed to capture the internal structures and attributes of landforms as well. Both, geometry and internal states of landforms can be represented over different periods of time (ref. Fig. 3).

The abstract class `_Geoobject` represents a solid landform. As a particular spatial unit of the georelief it is a subclass of a `_GeomorphicObject`, which aggregates a `_GeomorphicSystem`. The Class `_Geoobject` has one or more associations to an abstract class `_State`. This is to represent different versions of a `_Geoobject`. Since a landform's characteristics, like geometry, may change from time to time by the impact of processes, its semantical identity remains.

A `_State` of a `_Geoobject` is characterized by its geometry and material. The latter is modelled by an `_AttributeSet`, which is not depicted here as it has no relevance in this context.

A `_Slope` as a synonym for *landform* is a specialisation of the abstract class `_Geoobject`. Referring to Dalrymple's et al. (1968) and Caine's (1974) slope model a `_Slope` may again *contain* `_Slopes`. Thus, the association *contains* represents the nested hierarchy of landforms. A `_Slope` *consists of* one or more abstract class `_Layer`. A `_Layer` may *contain* one or more subLayers. Because `_Layer` is derived from `_Geoobject`, it exhibits the association to a `_State`, too.

The proposed model seems to be a sensible approach to cover the main geomorphic concepts from a semantical viewpoint. It is able to represent the internal structure of landforms, i.e. the slope as the main landform in

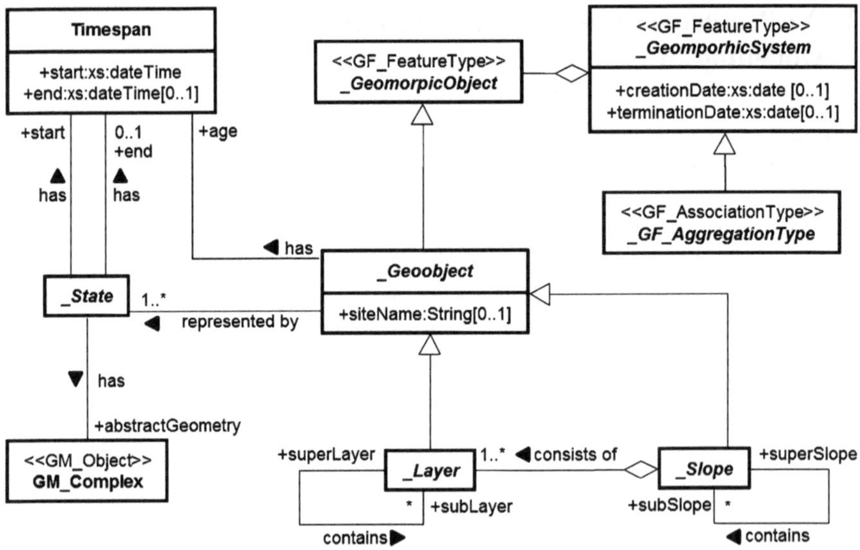

Fig. 3 Application model to represent a landform (`_Geoobject`) and its 3D geometry during different time steps

geomorphology. It consists of volumetric bodies bearing homogeneous material. These bodies themselves may be subdivided, which is an approach that makes sense, since a slope may consist of a soil layer and a regolith lying below it. Furthermore, the soil layer may be structured in different layers of homogeneous material as a result of soil-building processes over time or different sources of accumulated soil material.

In this formal representation of land surface's features the modelling of a class `_Geoprocess` serves three goals: First to store the interconnection of two or more `_Geoobjects` as a process-related accessibility; second to represent the main process that built up the landform an third to store information about the genesis of a `_Geoobject` (Fig. 4).

A `_Geoprocess` has two associations to a `_Geoobject`. It alters one or more `_Geoobjects` while a `_Geoobject` enables one or more `_Geoprocesses`. It is driven by a `_Processforce`, which might be specialised. The association of a `ComplexGeoprocess` is meant to store the genesis of a `_Geoobject`. Thus, the `_Geoobject` can be viewed as an integral of all processes over a given time span.

The formally modelled interrelationship between landform and process enables a Graph like representation of a sediment cascade (ref. Löwner and Otto 2008). The landform acting as a sediment source may then be interpreted as the "from-node" and the sink as the "to-node". Nevertheless, representation of geometry remains the greatest obstacle when applying the proposed model to geomorphology.

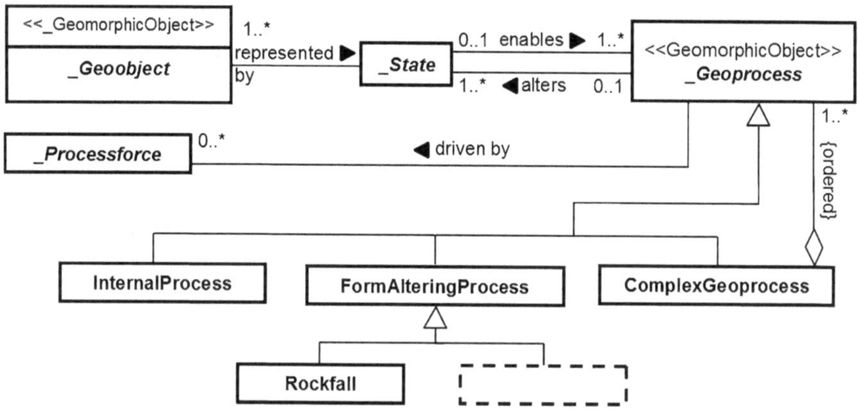

Fig. 4 Class model representing the relationship between a `_Geoobject` (landform) and a `_Geoprocess`

3.2 Acceptance of Overhead

The reviewed application models of geoobjects and geoprocesses represent geomorphology's concepts of the sediment budget and the sediment cascade. It is able to map a landform's 3D shape. Even different states of a landform and process-related accessibility can be stored. Nevertheless, geometric representation following the ISO 19107 seams to be the main problem to apply it to the science of geomorphology. Implementation int a DBMS will produce reasonable overhead and dissuades a geoscientist from leaving known but only 2.5D GIS. Realising the `_GMComplex` representing a `_State's` geometry (Fig. 3) needs to regard another 20 geometrical classes from spatial schema. Although the main literature on implementing the Spatial Schema (Lake et al. 2004) using the Geography Mark-Up Language is well known, only few geomorphologists really work with these techniques.

Describing a geomorphic system 2D or 2.5D maps are widely used (ref. Otto and Dikau 2004). Landforms are mapped using polygons. Depth information is given by semantic attributes. Actually, this is supported by the application model described above (but not depicted here). Löwner (2010) proposes an optional `FieldRepresentation`. It holds an association to a `RectifiedGridCoverage`, which is a common raster dataset. Additionally, it has a planar `LinearRing` to map the feature's 2D shape, e.g. when creating a digital geomorphological map. Unfortunately, this representation does not exploit the main advantages of real 3D modelling in terms of geometry and topology, respectively.

3.3 Data Acquisition and Modelling Problems

The science of geomorphology is working on natural 3D landforms consisting of sediment transported either by water or other driving forces. Although much is known about processes and their interaction with the land surface, no construction plans of landforms are available. In addition, subsurface boundaries of landforms are developed under different and partly unknown (i.e. regional climate) conditions. Therefore, it is not predictable that they vary in the same manner as the land surface today. On the contrary, the reconstruction of the paleo-surface representing a state of a geomorphic system at a specific time is one important research goal in geomorphology and neighbouring disciplines.

Compared to landforms, features of our manmade environment can be modelled much more easily. Take CityGML's Level of Detail 1 building model as an example (Gröger et al. 2012). Only a few points are needed to reconstruct a LOD1 building representation in terms of geometry. Usually a polygon, representing the ground surface and the height of the building is used to create a valid 3D solid representing a building's geometry. Even a LOD2 model is generated by adding a few more planar surfaces to represent the building's roof structure. Today, at least 2D digital information about infrastructure is easily available. Additional 3D datasets are available by LIDAR technologies (Zheng and Schenk 2000; Kada and McKinley 2009).

Landforms cannot be simplified this way. Extruding a planar 2D-polygon representing a landform's boundary by measured feature depth would neglect the vital role of the land surface. On the other hand, simply copying the land's surface digital elevation field to the measured depth would also be inadmissible. Since landforms are the results of partly unknown material transport processes acting over a long time on not exactly known boundary conditions, it is not possible to derive subsurface boundaries from today's land surface with levity.

Data capture may be identified to explain the significant differences between modelling approaches representing manmade structures and landforms, respectively. It is obvious that capturing a buried feature is much more difficult than measuring a construction above the surface. While the surface information on a landform is available by remote sensing, LIDAR or surveying, reconstruction of the subsurface is more difficult.

Normally, subsurface information is gathered by drillings. Boundary surfaces are identified via abrupt change in sediment properties. These are f. i. grain size or distribution, density, colour or biochemical indicators. The parameters used to determine a boundary surface depends firstly on the scientific problem, and, secondly, the theoretically background of the scientist. In most cases, changes in the environmental boundary conditions lead to changing material transport processes and therefore to different properties of the material accumulated.

In more clastic environments like alpine systems, geophysics is often applied to get subsurface information (Schrott and Hoffmann 2008; Schrott and Sass 2008). Depending on the method used, geophysical devices reveal changes in density, electronic conductivity and others. These changes are to be interpreted as boundary surfaces of the landform.

Figure 5 depicts a typically data situation on 3D landforms. In practise, a well known 2.5D surface in combination with a few points (Fig. 5a) or line information from 2D geophysical prospection method (Fig. 5b) are given to model a 3D solid. Even so-called 3D geophysical devices deliver 2D lines, albeit more than one.

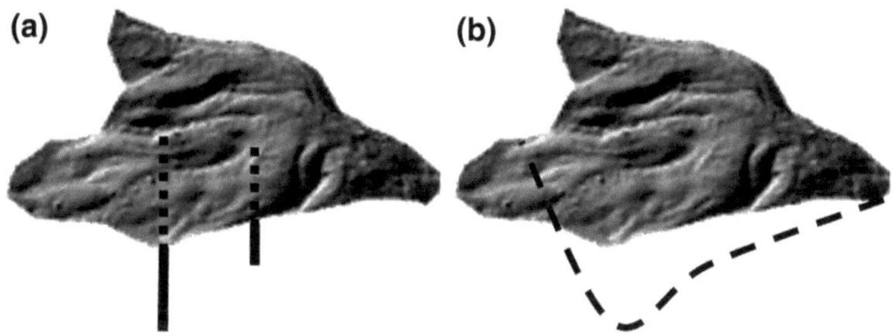

Fig. 5 Typically data situation on 3D landforms. While surface information is available, only few points (**a**) or not very reliable line information (**b**) has to be used to reconstruct a 3D solid

In geomorphology there are two ways to overcome the problems of valid geometric modelling and storing. First, this data is stored using the layer principle. 2D information is represented as a polygon in a GIS. 2.5D data is overlaid, if available. Depth information of a landform resulting from drilling data or geophysical prospection is stored as semantic information (cf. Otto and Dikau 2004; Otto et al. 2009). Second, the data is just represented in form of text, graphics and (not database) tables. As a result, research on landforms may not profit from further developments of the GI community in terms data exchange or geometrically and topologically representation and analysis.

Unfortunately, ISO 19107 Spatial Schema does not offer a valid representation of 3D geometry, neither by aggregating a surface and one or few points nor by aggregating a surface and a line. Consequently, even if the nature of a feature is proved to be three dimensional, Spatial Schema seems to be inadequate for representing under-specified 3D geometries. This directly affects the possibility to apply topological representation.

4 Linking 3D Topology to Under-Specified Geometries

We have outlined that geomorphology is a science on 3 dimensional phenomena that are changing in time (Sect. 2). Many concepts of geomorphology describe three dimensional phenomena. Therefore, a 3D application model for the representation of geometry, semantic and topology is highly desirable. Additionally, topological relationships of landforms directly reflect geomorphic principles like process-related accessibility, possible mass exchange and chronological order. However, the discussed application model in Sect. 3 seems not to be sufficient in supporting geoscientists. This was explained by data acquisition problems and complexity of the object under investigation. Since in Spatial Schema 3D topology is directly linked to a proper 3D geometric model, it may not be applicable to geoscientists.

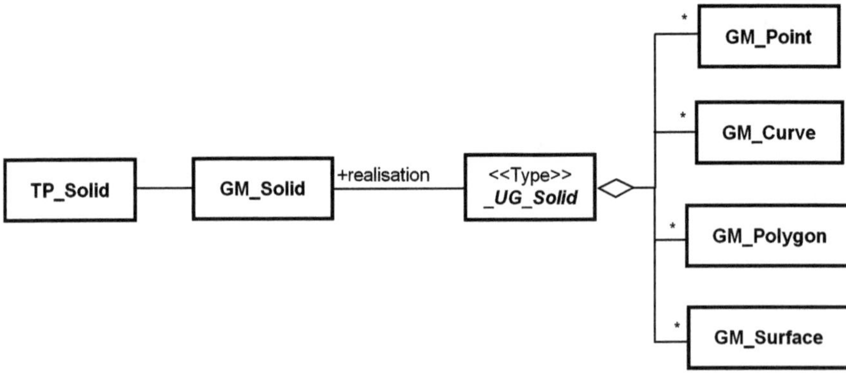

Fig. 6 Aggregation of different `GM_Primitives` to a `_UG_Solid`

Here, we present a simple approach to use the benefits of valid 3D topology even if only sparse data on geometry is available. Therefore, an abstract class `_UG_Solid` is introduced to represent an under-specified 3D geometry. Every object should be modelled as an under-specified 3D geometry that is a 3D object for knowledge reasons but not well defined for geometrical reasons. That means that no real 3D boundary representation is available, but data that enables us to reconstruct such a boundary surface. However, this boundary surface does not mean to represent the real geometry of the object. Following clearly defined conditions, an `_UG_Solid` mediates between the `GM_Primitives` `GM_Point`, `GM_Curve`, `GM_Surface`, the type `GM_Polygon` and a `GM_Solid`. Then, this `GM_Solid` is the *realisation* of `_UG_Solid` that definitely holds the 3D geometry we need to associate with the `TP_Solid` (Fig. 6).

While the realisation of a `_UG_Solid`, the `GM_Solid` class needs to be a real 3D geometry, constraints have to be defined for aggregating an _UG_Solid.

In topology the geometric characteristics are not important, except dimension. Thus, constraints on the association multiplicity only need to make sure that a real solid *could* be modelled from existing data. Of course, this solid must be closed.

Assuming a 2 or 2.5 dimensional (main) surface with an additional geometry representing the depth of a geomorphic feature this constraint can be formulated as (1):

$$(1 + \mathrm{dimMS}) * \mathrm{numMS} + (1 + \mathrm{dimDI}) * \mathrm{numDI} \geq 4$$

with : dim_{MS} = dimension of the geometry representing the main surface (e.g. 0 for a point)num_{MS} = minimum number of geometries to represent the main surface (e.g. 3 points)dim_{DI} = dimension of geometry representing the depth information (e.g. 1 for a line)num_{DI} = minimum number of geometries to represent the depth information (e.g. 1 point)

Table 2 Possible combinations of simple geometries to form an _UG_Solid

Geometries available	Comments
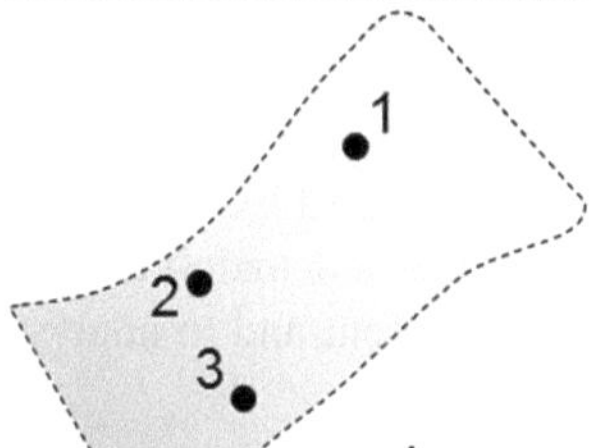	The main surface (e.g. the land surface) is represented by 3 GM_Points. This would result in a triangle representing the surface performing a *realisation*. Depth information is just given by 1 GM_Point (reflecting the normal situation when a drilling is performed). Formula (1) = 4
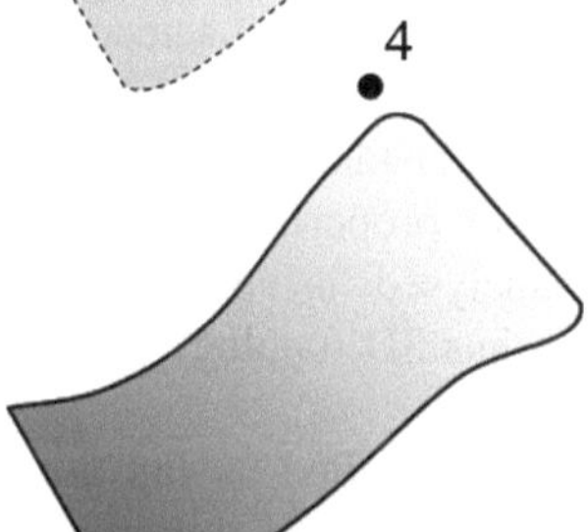	The main surface is represented by a GM_Surface (e.g. a DEM) or a GM_Polygon. Depth information is just given by 1 GM_Point. Formula (1) = 4
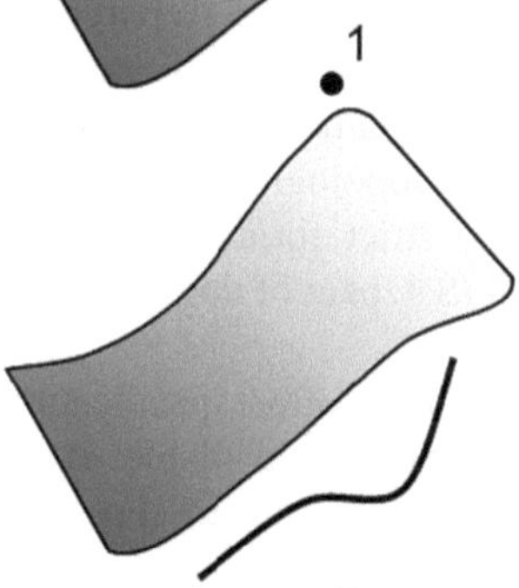	The main surface is represented by a GM_Surface (e.g. a DEM) or a GM_Polygon. Depth information is given by 1 GM_Curve (reflecting normal situation when geophysics are performed). Formula (1) = 5
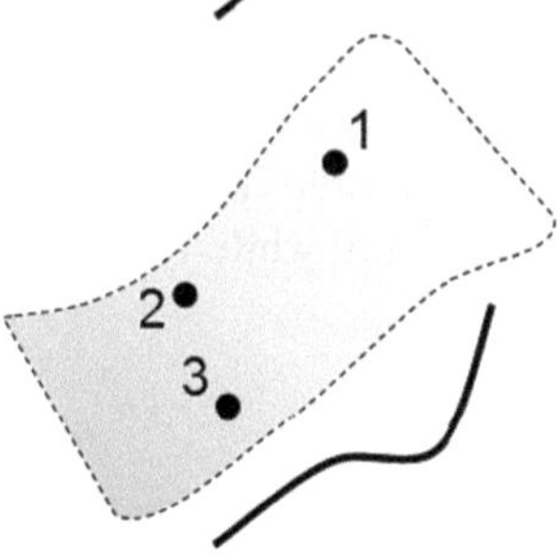	The main surface is represented by 3 GM_Points. Depth information is given by 1 GM_Curve. Formula (1) = 5

Table 2 gives examples of typical data available in through the field work of a geomorphologist. It can be shown that the given data is sufficient to realise a 3D geometry needed to apply 3D topology.

Testing the aggregating geometries against (1) enables a valid realisation that is needed to build a GM_Solid. The realisation itself is a matter of implementation and needs further discussion, elsewhere.

5 Discussion

It was demonstrated that geomorphology is a science investigating natural 3D objects. These objects change their 3D shape in time due to material transporting processes. On the one hand landforms influence these processes and on the other hand they are their product. As a result, the 3D georelief aggregated by landforms is a complex system of neighbouring objects of different age and material. It is argued that OGC's spatial schema concepts are useful to represent and to analyse such geomorphic systems in principle.

Here, for the first time the explicit benefit of 3D topology for the science of geomorphology was brought out by a collection of clear examples. Analysing topological relationships of landforms directly gains our understanding in process-related accessibility of landforms and therefore material properties. Even the chronological order of landform creation can be analysed using topology. Unfortunately, 3D concepts representing geometry and thus 3D topology are not very common in the community of geoscientists.

Data acquisition and modelling problems have been identified as the main reason for the rejection of 3D spatial concepts in geomorphology. Apparently, there is no need to apply the overhead of a 3D concept, when data is sparely available. Moreover, 2D and 2.5D concepts seem to be sufficient to geomorphologist. This, as can be shown with the example of topological analysis is definitely not the case. However, overhead and a very strict formulism hinder geomorphologists to model their perception of do a real (Satzbau: of do) 3D world with 3D concepts.

Here, a simple approach is proposed to use the benefits of the abstract specification 3D topologic model, when only under-specified geometries are available. If no sufficient data is available for a clear 3D object, this approach helps to apply 3D topology on it. It was proven on examples that the formulated constraints ensure the realisation of a _UG_Solid by a GM_Solid. Nevertheless, the approach presented is incomplete. First, this must be said in terms of dimensions, since 0D–2D under-specified Geometries are not covered. Second, no relationship between the GM_Primitives has been modelled. This is surely a focus worthwhile for future research.

References

Ahnert F (1988) Modelling landform change. In: Anderson MG (ed) Modelling geomorphological systems. Wiley, Chichester, pp 375–400

Ahnert F (1996) Einführung in die Geomorphologie. Eugen Ulmer, Stuttgart

Brunsden D (1996) Geomorphological events and landform change. Z Geomorph NF 40:273–288 (Suppl.-Bd.)

Caine N (1974) The geomorphic processes of the alpine environment. In: Ives JD, Barry RG (eds) Arctic and alpine environments. Methuen, London, pp 721–748

Chorley RJ, Kennedy BA (1971) Physical geography: a system approach. Prentice-Hall, London
Chorley RJ, Schumm SA, Sudgen DE (1984) Geomorphology. Methuen, London
Church M, Slaymaker O (1989) Disequilibrium of Holocene sediment yield in glaciated British Columbia. Nature 337(2):452–454
Cooke RU, Doornkamp JC (1990) Geomorphology in environmental management. Oxford University Press, Oxford
Dalrymple JB, Blong RJ, Conacher AJ (1968) A hypothetical nine unit landsurface model. Z Geomorph NF 12:60–76
Dikau R (1989) The application of a digital relief model to landform analysis in geomorphology. In: Raper J (ed) Three dimensional applications in Geographic Information Systems. Taylor & Francis, London, pp 51–77
Dikau R (1996) Geomorphologische Reliefklassifikation und -analyse. Heidelberger Geographische Arbeiten 104:15–23
Egenhofer MJ, Herring JR (1990) A mathematical framework for the definition of topological relations. In: Proceedings of 4th international symposium on SDH, Zurich, Switzerland, pp 803–813
Evans IS (1972) General geomorphometry, derivates of altitude, and descriptive statistics. In: Chorley RJ (ed) Spatial analysis in geomorphology. Methuen, London, pp 17–90
Foley J, van Dam A, Feiner S, Hughes J (1995) Computer graphics: principles and practice, 2nd edn. Addison Wesley, Reading
Gröger G, Kolbe TH, Nagel C, Häfele K-H (2012) (eds) OGC city geography markup language (CityGML) encoding standard
Herring J (2001) The OpenGIS abstract specification, Topic 1: feature geometry (ISO 19107 Spatial Schema), Version 5. OGC Document 01–101
Hugget RJ (2003) Fundamentals of geomorphology. Routledge, London
Jordan P, Slaymaker O (1991) Holocene sediment production in Lillooet river basin British Columbia: a sediment budget approach. Géog Phys Quatern 45(1):45–57
Kada M, McKinley L (2009) 3D building reconstruction from Lidar based on a cell decomposition approach. Int Arch Photogrammetry, Remote Sens Spat Inf Sci XXXVIII((3/W4)):47–52
Kottman C, Reed C (2009) The OpenGIS® abstract specification topic 5: features, OGC document 08–126. (http://www.opengeospatial.org/standards/as)
Kugler H (1974) Das Georelief und seine kartographische Modellierung. Dissertation B, Martin-Luther-Universität Halle
Lake R, Burggraf DS, Trininic M, Rae L (2004) GML—geography mark-up language. Wiley, Chichester
Löwner M-O (2010) New GML-based application schema for landforms, processes and their interaction. In: Otto J-C, Dikau R (eds) Landform—structure, evolution, process control. Lecture notes in earth sciences, vol 115, pp 21–36
Löwner M-O, J-C Otto (2008) Towards an automatic identification of sediment cascades from geomorphological maps using graph theory. In: Proceedings of the international symposium on sediment dynamics in changing environments, Christchurch, Neuseeland, 1–5 Dezember 2008
Open Geospatial Consortium (2012) Abstract specifications (URL: http://www.opengeospatial.org/standards/as, last visited 23 Jan 2012)
Otto J-C, Dikau R (2004) Geomorphologic System Analysis of a high mountain valley in the Swiss Alps. Zeitschrift für Geomorphologie 48(3):323–341
Otto J-C, Schrott L, Jaboyedoff M, Dikau R (2009) Quantifying sediment storage in a high alpine valley (Turtmanntal, Switzerland). Earth Surf Proc Land 34:1726–1742
Rasemann S (2004) Geomorphometrische Struktur eines mesoskaligen alpinen Geosystems, Bonner Geographische Abhandlungen 111
Reid LM, Dunne T (1996) Rapid evaluation of sediment Budgets. Catena Verlag, Reiskirchen
Schrott L, Hoffmann T (2008) Seismic refraction. In: Hauck C, Kneisel C (eds) Applied geophysics in periglacial environments. Cambridge University Press, Cambridge, pp 57–80
Schrott L, Sass O (2008) Application of field geophysics in geomorphology: advances and limitations exemplified by case studies. Geomorphology 93:55–73

Schrott L, Hufschidt G, Hankammer M, Hoffmann T, Dikau R (2003) Spatial distribution of sediment storage types and quantification of valley fill in an alpine basin, Reintal, Bavarian Alps, Germany. Geomorphology 55:45–63
Summerfield MA (1997) Global Geomorphology: an Introduction to the study of landforms. Addison Wesley, Essex
Young A (1978) Slopes. Longman limited, London
Zheng W, Schenk T (2000) Building extraction and reconstruction from lidar data. Int Arch Photogrammetry Remote Sens XXX:958–964 (Amsterdam. Part B3)
Zlatanova S (2000) On 3D topological relationships. In: Proceedings of the 11th international workshop on database and expert system applications (DEXA 2000), Greenwich, London, pp 913–919, 6–8 Sept

Geometric-Semantical Consistency Validation of CityGML Models

Detlev Wagner, Mark Wewetzer, Jürgen Bogdahn, Nazmul Alam, Margitta Pries and Volker Coors

Abstract In many domains, data quality is recognized as a key factor for successful business and quality management is a mandatory process in the production chain. Automated domain-specific tools are widely used for validation of business-critical data. Although the workflow for 3D city models is well-established from data acquisition to processing, analysis and visualization, quality management is not yet a standard during this workflow. Erroneous results and application defects are among the consequences of processing data with unclear specification. We show that this problem persists even if data are standard compliant and develop systematic rules for the validation of geometric-semantical consistency. A test implementation of the rule set and validation results of real-world city models are presented to demonstrate the potential of the approach.

D. Wagner (✉) · J. Bogdahn · N. Alam · V. Coors
HFT Stuttgart—University of Applied Sciences, Faculty C, Schellingstraße 24, 70174, Stuttgart, Germany
e-mail: detlev.wagner@hft-stuttgart.de

J. Bogdahn
e-mail: juergen.bogdahn@hft-stuttgart.de

N. Alam
e-mail: nazmul.alam@hft-stuttgart.de

V. Coors
e-mail: volker.coors@hft-stuttgart.de

M. Wewetzer · M. Pries
Beuth Hochschule für Technik Berlin—University of Applied Sciences, Department II, Luxemburger Straße 10, 13353, Berlin, Germany
e-mail: mark.wewetzer@bht-berlin.de

M. Pries
e-mail: margitta.pries@bht-berlin.de

J. Pouliot et al. (eds), *Progress and New Trends in 3D Geoinformation Sciences*, Lecture Notes in Geoinformation and Cartography, DOI: 10.1007/978-3-642-29793-9_10,

1 Introduction

A steadily growing number of application fields for large 3D city models, such as navigation or solar potential analysis, have emerged in recent years. All of them are strongly relying on the quality of the underlying models. Besides semantic, topological or visualization aspects, geometric correctness is a major factor for the quality of virtual city models. So far, common standards defining correct geometric modeling are not precise enough to define a sound base for data validation. Depending on the application domain, different standards and guidelines allow various modeling alternatives. This fact causes difficulties for data suppliers as well as for customers due to the need of explicitly stating detailed requirements.

Furthermore, modeling and exchange formats like CityGML (Gröger et al. 2008) do not include formal constraints with respect to geometric-semantical consistency.

In this document we describe the validation of virtual urban models, mainly focused on CityGML format. We define basic checks to be performed on a given data set, where geometric requirements should be tested and validated to assure a user-defined level of quality. This approach focuses on common errors which are typically found in 3D city models. A prototype of these checks working with CityGML data sets is implemented in a system developed at HFT Stuttgart.

In the next step, validated geometry has to be checked for semantic consistency and plausibility. A concept built on a formal set of strict rules is presented.

2 Relevance of Quality Management Of Three-Dimensional Models

Data validation is the process of checking data against a predefined set of validation rules. The main goal is to ensure the correctness, integrity and consistency of the data set regarding certain requirements, which is *influenced by the field of application.* A second benefit is that validated data is standard compliant (if the standard is considered by the rule set) and has advantages with respect to interoperability on different systems. Data exchange should be based on validated data since both data supplier and customer have agreed on common underlying rules.

2.1 Quality Assurance of 3D models in Other Domains

It is acknowledged in the geoinformation domain that "the representation of 3D objects using CAD (Computer Aided Design/Drafting) is not new, and significant work has been done on ensuring that the computer-based model is valid" (Thompson and Oosterom 2011). It seems attractive to figure out if existing methods from the CAD, reverse engineering or computer visualization domain would be suitable for geodata as well, because quality issues have gained great importance since many years.

Some examples: Data transformation from one CAD-system to another could lead to "dirty geometry" caused by different tolerance values (Wöhler et al. 2009); if data for reverse engineering purposes is collected with a range scanner, it can be noisy or incomplete, because the scanner was not able to capture every wrinkle of the model correctly (Rocchini et al. 2004).

Quality issues of CAD-models have lead to the formulation of ISO standard ISO/PAS 26183 (International Organization for Standardization (ISO) 2006), also well known as SASIG PDQ[1] Guideline v2.1, defining and categorizing common errors, especially in the automotive industry. A lot of commercial software for checking and healing geometry has been developed, e.g. CADfix (International TechneGroup 2011), CADdoctor (Elysium 2008), VALIDAT (T-Systems International 2011) or Q-Checker (Transcat PLM 2011).

Quality related research of polygonal models is ongoing since more than 10 years for computational applications. Often these are restricted to surface models consisting of triangles, thus most of the healing algorithms are limited to triangles (Ju 2009). Because surface modeling in CAD applications is more error-prone than polygon based solid modeling, research was focused on healing free-form surfaces rather than polygons. Additionally, repairing complex models is not trivial and even today it is normally not possible to heal a corrupted model fully automatically (Wöhler et al. 2009; Butlin and Stops 1996).

Nonetheless the discussed problems are similar to those occurring in 3D city models: (Borodin et al. 2002) comment on the finding and filling of holes in surface models, or resolving self intersections. The latter topic is also discussed by Attene and Falcidieno (2006), Yamakawa and Shimada (2009) and Rocchini et al. (2004). For a broader overview we recommend the paper of (Ju 2009).

The efforts and achievements made so far are a good guidance for the development of similar documents and tools for 3D city model.

2.2 Validation of Geodata and 3D City Models

Although the advantages of assessing and managing spatial data quality are widely recognized, their application is not yet widespread (van Oort 2005). Today, the transition from traditional 2D mapping towards 3D modeling is in full progress for many real-world applications, and high data quality is becoming essential for applications such as cadastre or urban planning. Efforts to translate validation concepts from the 2D GIS world to 3D data are one of the approaches, eventually trying to link the two domains (Ghawana and Zlatanova 2010).

[1] Strategic Automotive Data Standards Industry Group—Product Data Quality.

ISO standards referring to spatial data quality exist (International Organization for Standardization (ISO) 2002; International Organization for Standardization (ISO) 2003a; International Organization for Standardization (ISO) 2003b), other standards for 3D data are well established, e.g. CityGML (Gröger et al. 2008). The OGC domain working group for data quality mentions a number of categories for quality measures, including accuracy, completeness, consistency and definition for semantic interoperability (Open Geospatial Consortium (OGC) 2011). However, *there is no generally acknowledged definition of quality so far.* As stated by (Bogdahn and Coors 2010), *quality aspects for 3D city models are highly dependent on user preferences and the field of application.* A comprehensive overview of geometry validation is described by Kazar et al. (2008) and implemented in an Oracle Spatial database system.

A series of vendor specific implementations in commercial software packages for the validation of 3D geometries came to the market in recent years. For example, Oracle integrated validation rules and algorithms for 3D geometries in their database system Oracle Spatial (Kazar et al. 2008), and ESRI's wide-spread shapefiles were extended to handle 3D geometries. As internal data models and processing rules are incompatible, interoperability becomes a challenge if not impossible. Thus, it became necessary to define the structure of the data in more detail, e.g. for geometry (Oosterom et al. 2005) or topological consistency and analysis (Ledoux 2011; Boguslawski et al. 2011). Specific rules and constraints based on user needs are developed to extend existing standards (Gröger and Coors 2010).

An overview of validation concepts and needs is presented by Karki et al. (2010). Their implementation is based on the decomposition of solids into tetrahedrons for validation of a 3D cadastre and describes entry-level validation rules in terms of continuity, reasonability and spatio-temporal aspects. All of the abovementioned approaches are dealing with geometry or topology, not considering semantic information.

Based on the semantic model of CityGML (Stadler and Kolbe 2007) distinguish six categories of model complexity and structure wrt. spatio-semantical coherence. Only models where semantic components correlate to geometric components on the same level of hierarchy can be fully coherent. Thus, "it becomes clear that besides the spatial and semantic complexity also the coherence of the two structures is an important quality aspect of 3D city models" (ibid.). CityGML support coherent representation of geometry and semantics. (Métral et al. 2009) underline the relevance of semantically enriched models in their ontology-based approach.

Currently, discussions in the German quality working group of SIG-3D[2] on correct usage of attributes and modeling alternatives are ongoing. The goal of the working group is to suggest "best-practice" rules for 3D city modeling. A handbook with modeling guidelines for typical elements is in preparation. Partially, these guidelines are directly derived from definitions in the standard. In

[2] SIG-3D—GDI-DE; Special Interest Group 3D of the Geo Data Infrastructure Germany.

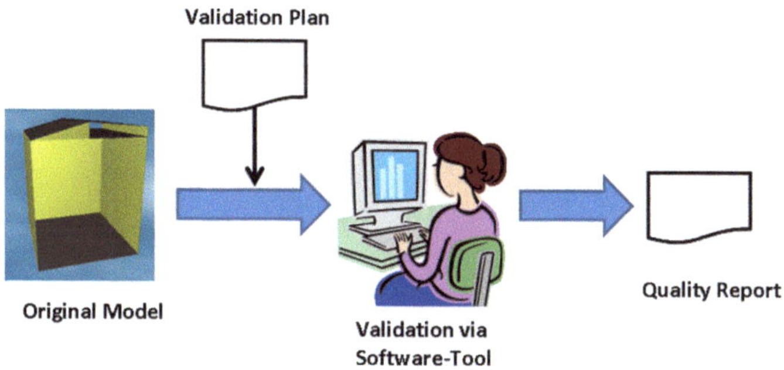

Fig. 1 Quality check process

many cases, however, the standard allows several alternatives. For example, the basic element `Building` can be modeled as `Solid` or as `MultiSurface` geometry. Some attributes might refer to different features without being clearly defined. A prominent example in this context is `MeasuredHeight` which reflects the building height which was measured by surveying methods. However, it is not clear if this value refers to the difference between ground surface and highest point of the structure, highest point of the roof, average roof height, etc.

Beyond the mere definition of these elements and attributes is their relation to the geometry. Intuitively, the geometric model height should be less or equal than the given attribute value of `MeasuredHeight`. Although attributes with geometric relevance encompass many obvious restrictions, geometric-semantical consistency is neglected so far. A reason might be the lack of recognized rules and software implementations.

In this paper, we present a concept to validate the consistency of geometry and semantics of CityGML models and define adequate rules and restrictions.

3 Concept of Quality Assurance for 3D City Models

A first concept of quality management for urban 3D models is presented by Coors and Krämer (2011). In this paper the implementation of a prototype for validation and healing of geometry errors is described (Fig. 1).

Gröger and Plümer (2009) developed a set of axioms to achieve geometric-topological consistency of 3D models. (Gröger and Coors 2010) defined valid geometry elements for CityGML in a more detailed approach. They add useful restrictions which are based on modeling guidelines discussed within the quality working group (SIG-3D Quality Working Group 2012). These extended rules are the base for the geometric model which we use for validation of CityGML data. The rules are applicable for data stored in XML documents as well as in data base solutions, which are customized adequately (Pelagatti et al. 2009).

Table 1 Modeling alternatives for the CityGML element `Building`

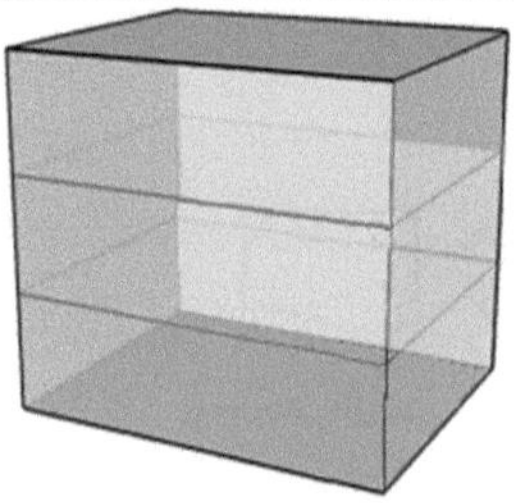	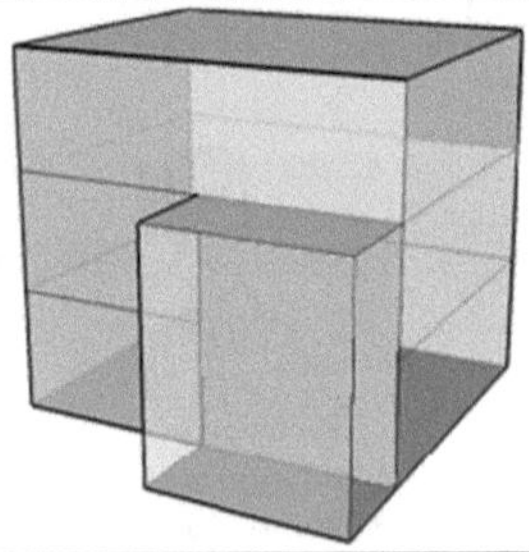	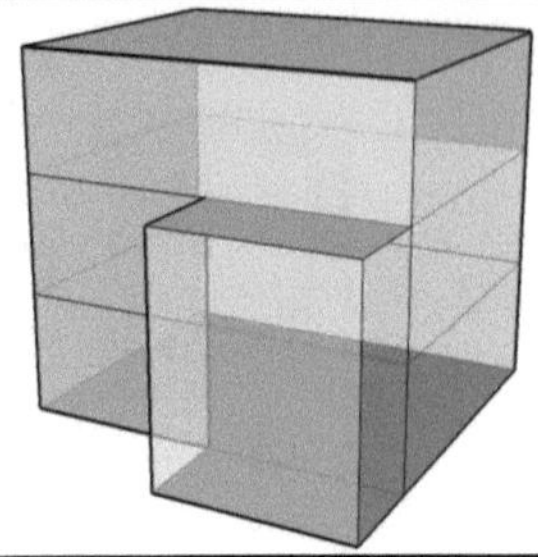

Alternative 1:
`1.Building (Solid)`
	Alternative 2:
	`1 Building (Solid)`
	`1 BuildingPart (Solid)`
	`combined geometry:`
	`CompositeSolid`
		Alternative 3:
		`1 Building`
		`(MultiSurface)`
		`1 BuildingPart`
		`(MultiSurface)`
		`combined geometry:`
		`Solid`

CityGML allows many different alternatives for modeling. This is an obstacle in the validation process, because it is not unambiguously defined what validity actually means without further specification. A verbal description of the modeling rules is a first step to formulate constraints for geometry, topology and semantics, but cannot be used for automated validation tools.

Simple examples are the CityGML elements `Building` and `BuildingPart`, which can be modeled in at least three different ways (Table 1):

- as single `Solid`
- as `CompositeSolid` (with XREF for shared faces)
- as `MultiSurface` geometry

All three alternatives are valid in CityGML. Validation of the models is only possible if the validation rules include information on the underlying modeling principles. Then `Buildings` which are modeled according to different rules can be detected as errors. It should be possible to allow several alternatives at the same time.

> To achieve geometric-semantical consistent models it is necessary to
>
> 1. specify allowed alternatives with formal rules, and
> 2. validate the geometry.

3.1 What is Valid Geometry?

Validation against the schema definition of CityGML is a basic requirement to ensure a standard compliant XML structure. However, no assertion on the geometric correctness of modeled objects is possible: A building can be described by a syntactically valid CityGML document but still have geometric errors like holes or non-planar faces, or contain inconsistent attribute values.

The following definitions are used in this paper. We stick quite close to the GML standard and limited the definitions to CityGML. Normally a polygon is bounded by a `gml:Ring`, but as there is only the linear ring as subtype of `gml:Ring` in CityGML we refer directly to `gml:LinearRing`.

3.1.1 Definitions of geometric elements

1. Point

   The element `gml:Point` is given as a triple of three real numbers:

$$p := (x, y, z), x, y, z \in \mathbb{R}$$

2. Edge

   A directed edge e is defined by an ordered pair of two points:

$$e_{k,l} := (p_k, p_l)$$

3. Linear Ring

   A sequence $R := (p_0, p_1 \ldots, p_n)$ of points p_i, with linear interpolation between the points (a sequence of connected edges) is called a `gml:LinearRing`, if $p_0 = p_n$ and $n \geq 3$. Note that the direction of an edge in a linear ring is determined by the order of the points.

4. Polygon

   By definition, a `gml:Polygon` P is a surface bounded by a `gml:LinearRing`. The boundary is coplanar and the polygon uses planar interpolation in its interior. Hence, a linear ring is planar if all points of the sequence are coplanar.

5. Solid

   A `gml:Solid` S is defined by a set $S := (P_1, \ldots, P_m)$ of connected polygons $P_i, i = 1, \ldots, m$. The set of polygons defines a 2-manifold surface.

3.1.2 Validity Axioms

It is quite clear, that the pure definitions of entities derived from the GML standard could lead to non-manifold, but still valid geometry. To avoid degenerated geometry we define additional validity axioms which the geometry has to satisfy to be valid in accordance with Gröger and Plümer (2009).

Linear Ring and Polygon

Let $\mathrm{R} = (\mathrm{p}_0, \mathrm{p}_1 \ldots, \mathrm{p}_\mathrm{n})$ be a linear ring and let P be the polygon defined by R. R and P are called valid, if

a. R consists of at least four ordered points (CP-NUMPOINTS);
b. The first and last point are identical: $p_0 = p_n$ (CP-CLOSE);
c. All points of the sequence besides the first and the last are different: $p_i \neq p_j,\ i,j = 0,\ldots n-1,\ i \neq j$ CP-DUPPOINT);
d. Two edges $e_{i,i+1}$ and $e_{j,j+1},\ i,j = 0,\ldots n-1,\ i \neq j$ of R do only intersect in one start-/endpoint. No other intersection is allowed (no self-intersection, CP-SELFINT);
e. According to definition (4), the boundary of P has to be planar (CP-PLANDIST, CP-PLANDISTALL, CP-PLANTRI).

Solid

Let $S = (P_1,\ldots,P_m)$ be a solid. S is called valid, iff

f. all *polygons* of S are valid (axioms a to e);
g. S *consists* of at least four polygons (CS-NUMFACES);
h. each edge e of a polygon of S is exactly referenced once by another polygon of S (CS-2POLYPEREDGE);
i. all *polygons* in S are oriented such that the normal vector of each polygon points to the outside of the solid. This means that an edge $e_{v,w}$ of a polygon P_k is directed in opposite direction in polygon P_l, which is sharing $e_{v,w}$: $e_{v,w} \in P_k \Rightarrow e_{w,v} \in P_l$ (CS-FACEORIENT, CS-FACEOUT);
j. the intersection of two polygonsP_k andP_l of S is either empty or contains only points p or edges e (points included), that are part of both polygons:

$$P_k \cap P_l = \begin{cases} \emptyset \\ \{p | p = (p_0,\ldots,p_c),\quad p_i \in P_k \wedge p_i \in P_l\} \\ \left\{ \begin{array}{c} e|e = \left(e_{v_0,w_0},\ldots,e_{v_d,w_d}\right), \\ \left(e_{v_i,w_i} \in P_k \wedge e_{w_i v_i} \in P_l\right) \wedge \\ (p_{v_i}, p_{w_i} \in P_k \wedge p_{v_i}, p_{w_i} \in P_l) \end{array} \right\} \end{cases}$$

(CS-SELFINT);

k. all polygons in S are connected (CS-CONCOMP);
l. each point is surrounded by exactly one cycle that is an alternating sequence of line segments and polygons, also called umbrella axiom (CS-UMBRELLA).

Based on these definitions geometry validation can be limited to the validity axioms of a solid for `Solid` geometries and the validity axioms of polygons for a `MultiSurface` geometry.

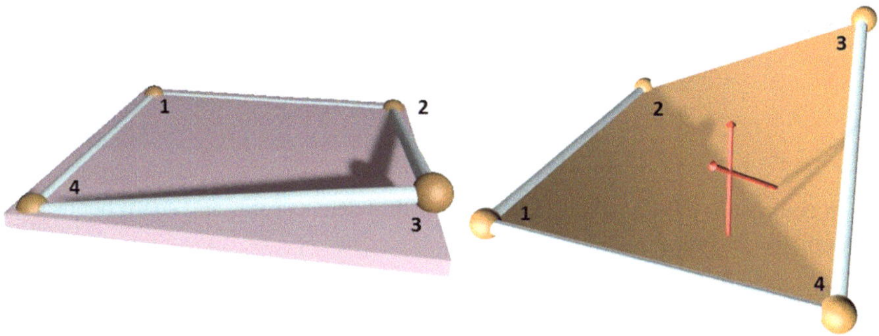

Fig. 2 Two examples of a non-planar `LinearRing`

3.2 Description of Check Routines

Five major check routines can be derived from the polygon definition above. The planarity checks are described in detail below. The other checks and the solid checks are defined in Coors and Krämer (2011). The relation of the axioms and the Check-IDs are given in Table 4.

Planarity can be defined in different ways as an internal discussion in the German Quality Working Group has shown and is referred to in Gröger and Coors (2010). Our implementation includes three different algorithms as a consequence of this discussion process, a fourth one is not yet implemented.

- Algorithm 1 (CP-PLANDIST)

Three non-linear points p_0, p_1, p_x of a linear ring R with $n + 1$ points describe a plane E with normal vector n. All other points of R must be situated in E (small deviations $\varepsilon \in \mathbb{R}$ are allowed):

$$|(p_0 - p_i) \cdot \boldsymbol{n}| < \varepsilon, i \in \{0, \ldots, n - 1\}$$

In `LinearRing` $R = \{4, 1, 2, 3, 4\}$ depicted in Fig. 2 (left), three points (4,1,2) would define a plane (visualized in gray), but Point 3 is not located in it. This polygon can still be considered planar if the distance d between Point 3 and the plane is less than a preset maximum deviation ε. Folds and sharp bends cannot be detected reliably, however.

- Algorithm 2 (CP-PLANDISTALL)

It tests if all points of a linear ring R with $n + 1$ points are situated in all possible planes E_l with normal vectors n_l, defined by three non-linear points p_i, p_j, p_k of R (small deviations $\varepsilon \in \mathbb{R}$ are allowed):

$$|(p_i - p_a) \cdot n_l| < \varepsilon, a \in \{1, \ldots, n - 1\}$$

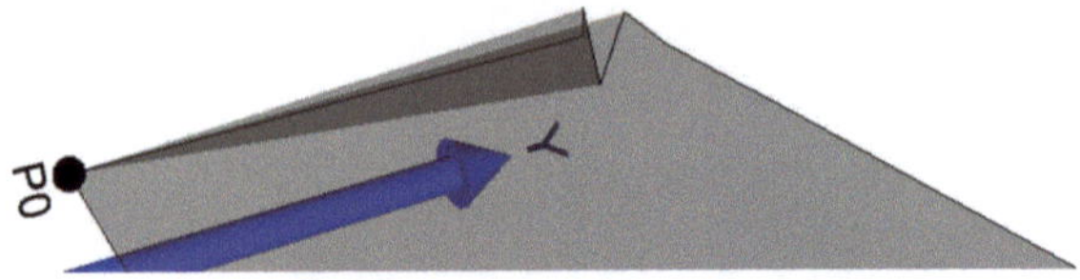

Fig. 3 `Linear Ring` with a fold

In contrast to CP-PLANDIST, this algorithm detects folds and sharp bends such as depicted in Fig. 3.

- Algorithm 3 (CP-PLANTRI)

Consider T as a triangulation of the polygon defined by R, containing m triangles. Additionally, let $n_l, l \in \{0, \ldots, m-1\}$ be the unit normal vector of the *l-th* triangle in T. R is planar, iff each scalar product of unit normal vectors n_a and n_b of two different triangles of T is less than a tolerance value ε:

$$\boldsymbol{n}_a \cdot \boldsymbol{n}_b < \varepsilon, a \neq b; \quad a, b \in \{1, \ldots, m-1\}$$

Folds and sharp bends are detected. Fewer problems with long polygons occur because the normal vector does not change its direction whereas point distance to a reference plane can increase compared to smaller polygons.

- Algorithm 4 (CP-PLANADJUST, not yet implemented in our validation tool, cf. Sect. 4)

An adjustment plane E with normal vector n for all points p_i of a linear ring is computed. All points p_i must be situated in E (small deviations $\varepsilon \in \mathbb{R}$ are allowed):

$$|(p_0 - p_i) \cdot n| < \varepsilon$$

However, as with CP-PLANDIST, folds and sharp bends cannot be detected reliably by CP_PLANADJUST.

Results of the error distribution of test models are given in Table 4. The highest number of non-planar polygons is detected by CP-PLANTRI, the lowest number is found by CP-PLANDIST. CP-PLANDISTALL is supposed to find all possible planarity errors as well; hence it is surprising that the number is different from the CP-PLANTRI result in some models. In both cases the tolerance value was set to 0.01. This reflects the maximum point distance for CP-PLANDISTALL and the maximum angle for CP-PLANTRI. Obviously, the thresholds are not comparable and CP-PLANTRI is more sensitive.

Another issue is that polygons with a long and narrow shape are expected to have a large deviation for certain point combinations in CP-PLANDISTALL. This leads to the question how to determine a useful value for the tolerance value ε. Our implementation is using a default value of 0.01, which can be changed by the user.

Table 2 Dependencies of geometry check routines

	CP-NUM-POINTS	CP-DUP-POINT	CP-PLAN[a]	CS-2POLY-PEREDGE	CS-FACE-ORIENT
CP-SELFINT	–	•	–	–	–
CP-PLAN[a]	•	•	–	–	–
CS-SELFINT	–	–	•	–	–
CS-FACEORIENT	–	–	–	•	–
CS-FACEOUT	–	–	–	–	•
CS-UMBRELLA	–	–	–	•	–

[a] CP-PLAN means either CP-PLANDIST, CP-PLANDISTALL or CP-PLANTRI

3.3 Dependencies in the Check Process

Some checks perform correctly only if the input has passed other checks before processing. This might be necessary to guarantee that certain assumptions are valid. For example, it would not make sense to test a polygon for planarity if there are less than 4 points, because planarity is not defined in that case.

In a strictly designed check process, all checks (except basic CP-NUMPOINTS) should depend on each other in a nearly linear way, according to the order as they are listed in Sect. 3.2. During testing we noticed that such strong restrictions prevent parts of the geometry to be checked by higher-order checks in case basic checks have not been passed. *The check process stops for the whole building geometry, even if there is only a CP-DUPPOINT error in one polygon. As a consequence, some checks at the end of the dependency chain are not executed for some buildings.*

To decrease the number of unchecked buildings we decoupled some checks from their former dependency. Table 2 gives a quick overview on all remaining dependencies which are underlying the current check process of our implementation. For example, CS-FACEOUT yields correct results only for solids which have passed CS-FACEORIENT successfully.

All other checks work correctly for geometric input extracted from CityGML files which can be validated against the current XML schema definition.

3.4 Concept of Semantic Validation

CityGML defines a semantic model that enables the user to add further information to spatial object. This goes far beyond the pure geometrical description of features. The idea is to enable a model to "know" what kind of features it contains and not only their location and shape. It can be enriched with domain-specific information which exceeds the default specification of CityGML. However, semantic information can be inconsistent with the geometry, e.g. if a building wall is defined as a `RoofSurface`.

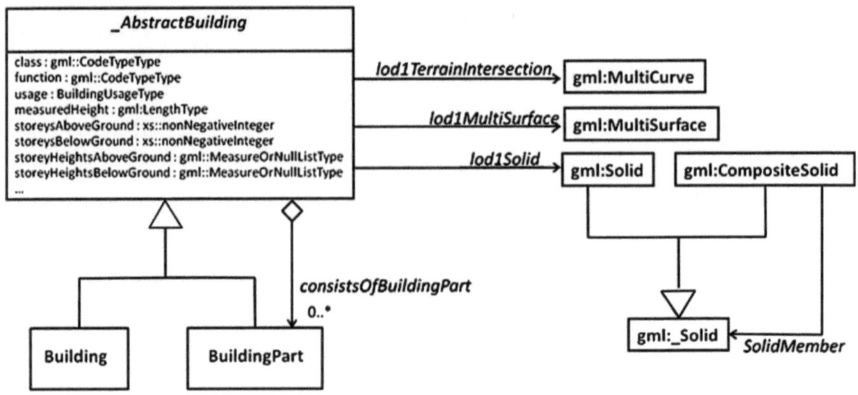

Fig. 4 UML diagram for CityGML element *_AbstractBuilding*

Based on validation of geometrical features as explained above, we introduce a concept for the validation of geometric-semantic consistency. Valid geometry is the base before other geometry-related information can be validated. A set of applicable and unambiguous rules is defined, mainly determined by the underlying data model (Stadler and Kolbe 2007).

For a start, the concept is focused on the element `Building` as specified in CityGML 1.1 standard (Gröger et al. 2008). A `Building` can have one geometry per level of detail (LoD) in CityGML. These geometries will be treated separately. In this paper, we limit the discussion to LoD 1, where the geometric-semantic themes volume part of the building shell (`gml:SolidType`), surface part of the building shell (`gml:MultiSurfaceType`), terrain intersection curve (`gml:MultiCurveType`) and building parts (`BuildingPartType`) are defined. The relevant part of the UML diagram is shown in Fig. 4.

3.5 Constraints for Building Models in CityGML

Geometry of LoD1 is either modeled as `MultiSurface` or as `Solid` and `CompositeSolid` respectively. A terrain intersection curve to define the intersection line of the building geometry with the terrain model may exist in addition.

A building can be modeled in several parts (`BuildingPart` elements), representing individual structural elements. The CityGML standard defines base requirements for conformance of `Building` elements:

"If a building only consists of one (homogeneous) part, it shall be represented by the element `Building`. However, if a building is composed of individual structural segments, it shall be modeled as a `Building` element having one or more additional `BuildingPart` elements. Only the geometry and non-spatial

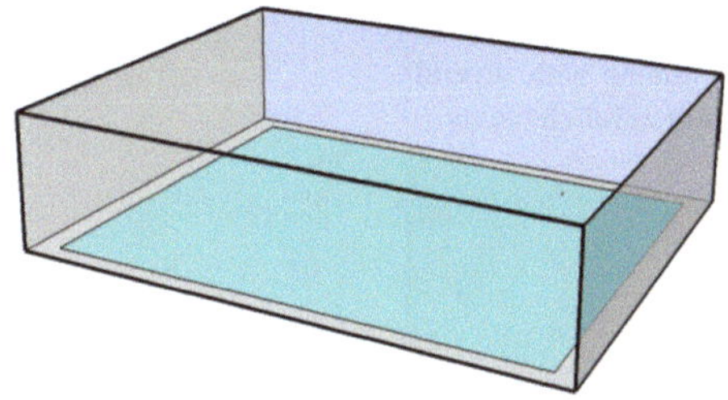

Fig. 5 Simple extrusion solid

properties of the main part of the building should be represented within the aggregating `Building` element" (Gröger et al. 2008).

Besides basic requirements no detailed rules are given in the standard how to model a `Building` correctly in a standardized form. Although there are comments or recommendations on the separation into several `BuildingPart` elements, different valid alternatives are possible according to the specification. However, a consistent modeling concept should be ensured throughout the actual city model, which is not included in a standard CityGML document. Hence, validation requires the formulation of rules in a standardized and machine-readable format.

3.6 Modeling Guidelines and Recommendations

In the simplest case a building geometry in LoD1 consists of an extrusion solid (Fig. 5). It should always be modeled as a `Solid`. The standard allows a `MultiSurface` in theory as well, which is widely used in practice. However, this way of modeling is not very helpful considering the consequences for potential applications of the model (e.g. watertightness).

3.7 Rules for Validation

In the following paragraphs a validation rule set is developed step by step, to ensure that a model is built of consistent geometric and semantic elements. The rules are given in colloquial language as well as in Object Constraint Language (OCL) as machine-readable format that can be processed automatically (Object Management Group 2011). The OCL statements given below are referring to the UML diagrams which define the CityGML elements in the standard. The relation of buildings and their parts is shown in Fig. 4, which includes the geometric elements for a valid model as well as the different alternatives for solids.

Additional rules are required to enable strict validation of models. We recommend a rule set on the basis of the modeling guidelines of SIG-3D (SIG-3D

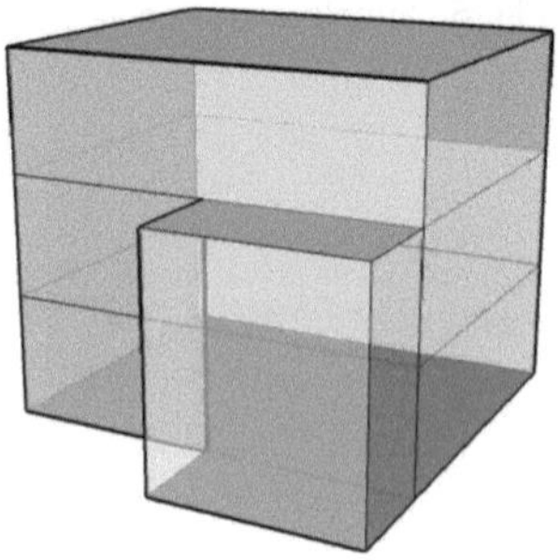
Fig. 6 Simple LOD 1 building with several structural elements

Quality Working Group 2012). Other alternatives could allow the users to deviate from these standards deliberately in case they require customized rules.

3.7.1 Simple Building Without Building Parts

Rule 1: If a building has no separable structural elements it is modeled as a simple `Solid` in LoD1:

```
context Building
inv: self.oclIsTypeOf(t:Building)
inv: self.lod1Solid->notEmpty() and
self.lod1MultiSurface->isEmpty()
inv: self.consistsOfBuildingPart->isEmpty()
```

3.7.2 Simple Building with Building Part

Different alternatives for modeling have to be considered in case a building is composed of several structural elements. This is illustrated in the following example building with different heights (Fig. 6).

This building can be modeled as a `Building` according to Rule 1, and no `BuildingPart` element is necessary in this case.

Just as well it is possible to model the building as `Building` with additional `BuildingPart`. The `BuildingPart` contains the lower structural segment at the front side of the building in Fig. 6. Different alternatives are considered subsequently:

Rule 2: If a building consists of several parts the main element is modeled as a `Solid`, as well as all other building parts. The combined geometry of all parts forms a `CompositeSolid`:

```
context Building
inv: self.oclIsTypeOf(t:Building)
inv: self.lod1Solid->notEmpty()and
self.lod1MultiSurface->isEmpty()
inv: self.consistsOfBuildingPart->notEmpty()->
```

```
forAll(b:BuildingPart | b.lod1Solid->notEmpty()
and b.lod1multisurface->isEmpty())
inv: self.lod1Solid->union(union
(self.consistsOfBuildingPart.lod1Solid))
= compositeSolid
```

The restriction to a CompositeSolid includes the requirement that all `Building` and `BuildingPart` elements are connected as stipulated by the standard.

The geometry according to rule 2 contains one or more common planes between adjacent solids (compare Table 1, alternative 2). This can be avoided by modeling the individual geometries as `MultiSurface` elements instead of `Solid` elements.

Rule 3: If a building consists of several parts the main segment is modeled as a `MultiSurface` element, as well as all other parts. The complete geometry of all building parts forms a `Solid`, i.e. the geometries are connected and the aggregated geometry could be defined by a single `Solid`:

```
context Building
inv: self.oclIsTypeOf(t:Building)
inv: self.lod1Solid- > isEmpty() and
self.lod1MultiSurface-> notEmpty()
inv: self.consistsOfBuildingPart- > notEmpty()->
forAll(b:BuildingPart | b.lod1Solid- > isEmpty()
and b.lod1multisurface- > notEmpty())
inv: self.lod1MultiSurface
- > union(union(self.consistsOfBuildingPart- >
any(exists(lod1MultiSurface))).isTypeOf(lod1Solid)
```

We recommend modeling according to rules 1 and 2. The rules can by combined: rule 2 OR rule 3 would allow using both alternatives in one model.

A `Building` can be modeled as a single `MultiSurface`, the remaining surfaces would be `BuildingPart` elements and the aggregated geometry would be a `Solid` (Fig. 7).

Rule 4: A Building or BuildingPart that is modeled as MultiSurface (according to rule 3) must have at least three faces.

```
context Building
inv: self.oclIsTypeOf(t:Building)
inv: self.lod1Solid->isEmpty() and
self.lod1MultiSurface->notEmpty()
inv: self.consistsOfBuildingPart->notEmpty()->
forAll(b:BuildingPart | b.lod1Solid->isEmpty()
and b.lod1Multisurface->notEmpty())
inv: self.lod1MultiSurface->size()≥3
and self.consistsOfBuildingPart.
lod1MultiSurface->size()≥3
```

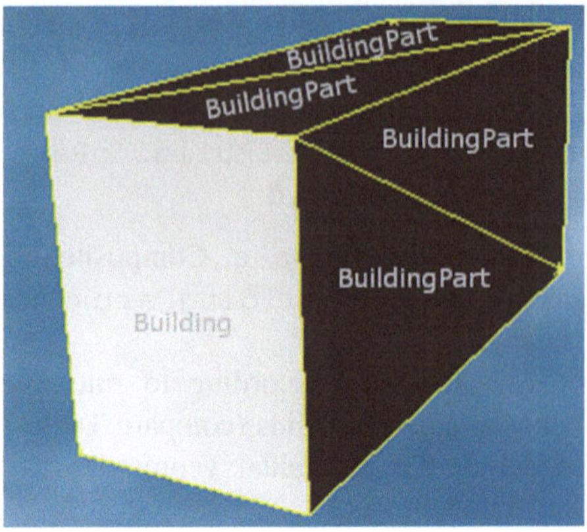

Fig. 7 `Building` and `Buildingparts` modeled as single surfaces

3.7.3 Attributes

Buildings as well as building parts can have further attributes. Some attributes describe geometry-related properties, e.g. `storeysAboveGround`, `storeysHeightsAboveGround` and `measuredHeight`. If the geometry includes underground structures, `storeysBelowGround` and `storeysHeightsBelowGround` can be present as well. In a correct model, their values should be consistent with the geometry. Since these attribute are not unambiguously defined their plausibility should be assessed.

Rule 5: If the number of storeys above ground is given, but not their heights, then a storey height between 2 and 3 meters is assumed. The height of the bounding box of the building geometry is expected to be within the calculated range.

```
context Building
inv: self.storeysHeightsAboveGround.oclIsUndefined()
inv: self.storeysAboveGround*2 <
self.lod1Solid.bboxHeight
< self.storeysAboveGround*3
def bboxHeight:Real
= self.lod1Solid.zCoordinate->maxValue()
- self.lod1Solid.zCoordinate->minValue()
```

The height of the bounding box is decreased by the extent of underground structures included in the model, or in case the `Solid` has been extended towards the ground in order to avoid gaps between the `GroundSurface` and the terrain model. In this case, the part below the lowest point of the `TerrainIntersectionCurve` is subtracted from the height of the bounding box.

```
context Building
inv: self.storeysHeightsAboveGround.oclIsUndefined()
and self.tic->notEmpty()
```

```
inv: self.storeysAboveGround*2 <
self.lod1Solid.zCoordinate- > maxValue()
- self.ticLowestPointHeight <
self.storeysAboveGround*3
def ticLowestPointHeight:Real
= self.tic.zCoordinate->minValue()
```

Rule 6: If `storeysHeightsAboveGround` is given, the height of the bounding box can be calculated more exactly by addition of all values of the list. The number of values should be equal to the number of `storeysAboveGround`.

```
context Building
inv: self.storeysHeightsAboveGround->notEmpty()
inv: self.storeysHeightsAboveGround->size()
= self.storeysAboveGround
inv: self.storeysHeightsAboveGround.sum() =
self.lod1Solid.bboxHeight ± tolerance
def bboxHeight:Real = self.lod1Solid.zCoordinate- >
maxValue() - self.lod1Solid.zCoordinate->minValue()
def: tolerance:Real = 0.5
```

4 Implementation of Test Tool and Validation Library

The presented tests and approaches to detect geometric errors in 3D city models are implemented by a library of check components. This library can be used in order to integrate the check functionality into different software tools and applications. It provides an interface to control which checks are actually performed and allows specifying a certain validation configuration.

The implementation is realized as a standalone Java tool for testing 3D city models for errors (Fig. 8). A Swing-based GUI is put on top of the functionality provided by the check library. Results are written to a log-file as well as displayed in the GUI (Fig. 9).

One example for the integration into an existing software tool is depicted in Fig. 10. Here the check library is integrated into FME (Safe Software 2011) as a custom transformer. The transformer takes *FMEFeatures* as input and checks the features using the check configuration specified by the user. Internally, *FMEFeatures* are converted into a data structure of the library and the checks are performed. As a result the transformer is providing two output channels: one for features without errors (original) and one for features not compliant with the validation rules (errors).

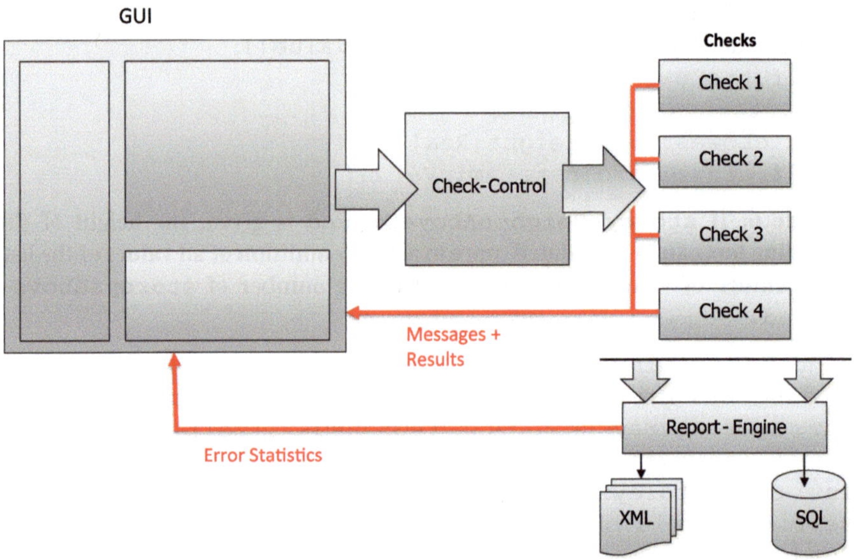

Fig. 8 Stand-alone implementation of the validation tool

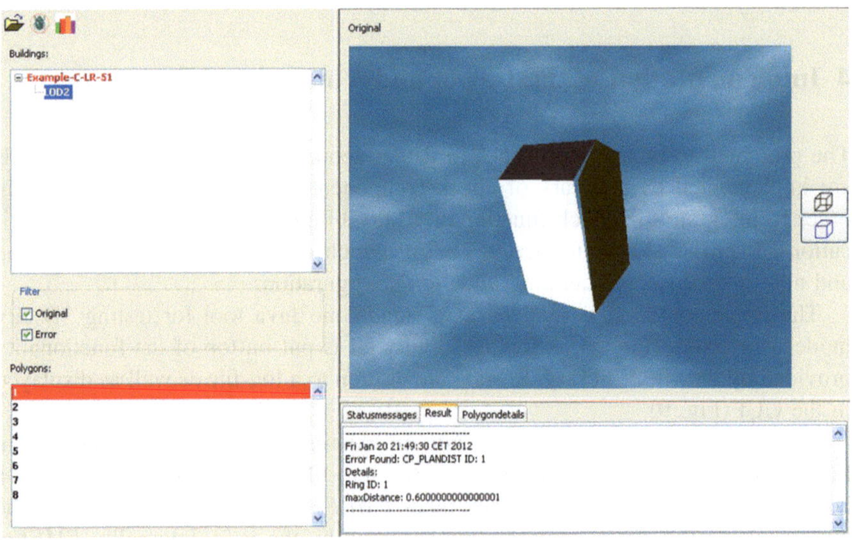

Fig. 9 Graphical user interface (GUI) of the stand-alone validation tool

Other plug-ins for test purposes have been developed for SupportGIS-3D of CPA Geo-Information (CPA-Systems GmbH 2011) and CityServer3D (Coors and Krämer 2011).

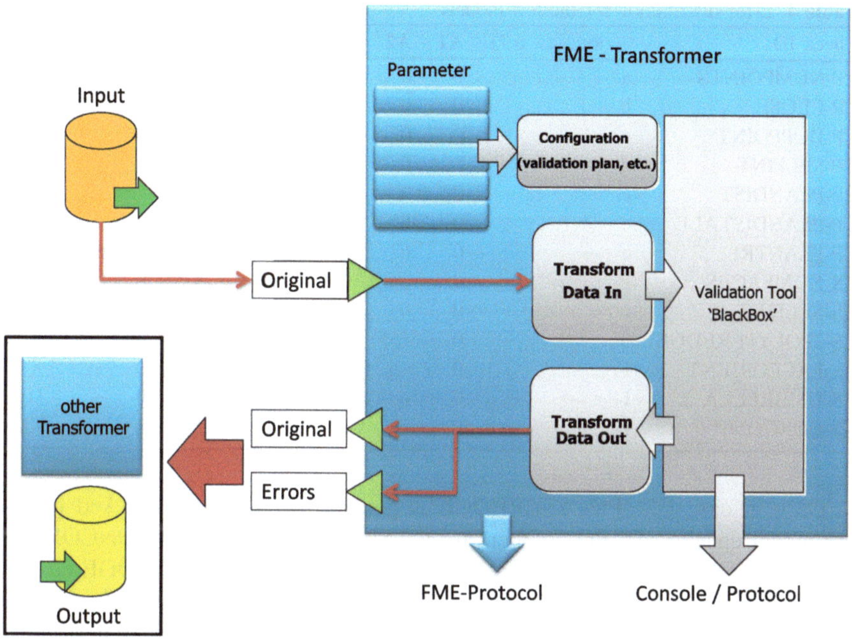

Fig. 10 Integration of the validation tool in FME

Table 3 Characteristics of test models

	A1	A2	H	J	K	L	Q	X
LOD	1	2	1	2	2	2	2	2
# buildings	1	1	61	61	551	4	7	1,922
# polygons	220	580	689	3,455	4251	69	403	32,546
# RoofSurface	–	273	–	1,483	772	10	73	4,957
# WallSurface	–	305	–	1,448	3122	54	25	24,067
# GroundSurface	–	2	–	524	360	5	305	3,522
# undefined	–	0	–	0	0	0	0	0
# edges	654	1402	1701	6,452		182	740	75,370
# vertices	436	779	1134	2,918		123	466	49,690
# holes	0	165	46	1,5504		0	664	0
# polygons per building	220	580	11.3	56.6	7.7	17.3	57.6	16.9

5 Test Results of Real-World Models

The outlined geometry checks of Sect. 3.3 are implemented with JAVA in a standalone application. It is tested against specially created models which contain certain types of errors as well as real-world models or extracts thereof. The models are characterized with certain key figures in Table 3; the validation results are given in Table 4.

Table 4 Error distribution for the test models

Check ID	Axiom (cf. p.7)	A1	A2	H	J	K	L	Q	X
CP-NUMPOINTS	a	0	0	0	0	0	0	0	0
CP-CLOSE	b	0	0	0	0	0	0	0	0
CP-DUPPOINT	c	0	0	0	0	46	0	4	0
CP-SELFINT	d	0	0	0	0	90	0	2	0
CP-PLANDIST	e	0	4	0	0	0	4	8	177
CP-PLANDISTALL	e	0	8	0	0	0	4	8	191
CP-PLANTRI	e	0	67	0	0	45	5	62	269
CS-NUMFACES	g	0	0	0	4,445	133	0	0	0
CS-SELFINT	j	0	0	22	–	647	2	–	4,575
CS-2POLYPEREDGE	h	0	155	46	15,484	6570	0	298	467
CS-FACEORIENT	i	0	–	–	–	–	0	–	–
CS-UMBRELLA	l	0	–	–	–	–	0	–	116
CS-CONCOMP	k	0	–	–	–	–	1	–	980

Only in case CS-2POLYPEREDGE is passed, a building is checked by the dependent checks, viz. the numbers for CS-UMBRELLA and CS-CONCOMP in models L and X refer to buildings which have no CS-2POLYPEREDGE errors.

Besides the different results for the planarity checks (cf. Sect. 3.3) the high numbers of CS-NUMFACES and CS-2POLYPEREDGE in model J are noticeable. A closer look at the details reveals that building installations are modeled as `MultiSurface` elements containing only one single polygon each. A `BuildingInstallation` element is currently treated as a `Solid` geometry which explains the high numbers of errors for CS-2POLYPEREDGE also.

6 Conclusion and Future Work

Rules for validation of geometry in 3D city models were defined and implemented in a JAVA library to enable automated processing of 3D city models. The functionality is available as standalone JAVA application and integrated as plug-in for three commercial systems. Validation tests with real-world models showed that nearly all models have geometric errors although they are visually in order. This fact confirms the importance of geometry validation for 3D city models to enable other applications to work correctly.

Error-free and standard-compliant geometry is a prerequisite for semantic validation. We presented a concept of validation rules for geometric-semantical consistency in LoD 1 of CityGML models. The rules are based on the standard and introduce additional useful restrictions which can help to increase the overall quality. They will be implemented as part of the validation tool in future. Furthermore, the rule set will be extended to LoD2 and above. Other CityGML elements with relation to geometry will be added.

CityGML can be extended by generic attributes (Application Domain Extensions—ADE). Some generic attributes have been defined by organizations such as the German AdV[3] and many municipalities. Initially, these attributes were intended to reflect additional feature properties which cannot be included in 2D data otherwise. For example, AdV has defined roof-type primitives like gable roof, hip roof, shed roof etc. In LoD 2 and higher, the geometry should reflect assigned roof type properties, thus a validation of certain generic attributes should be considered.

Development of consistency rules according to the concept above will be a major task to bring high-quality city models forward.

Acknowledgments The authors would like to thank the SIG-3D quality working group for fruitful discussion and the German Federal Ministry of Education and Research (BMBF) for funding of the project CityDoctor under provision number 17110B10.

References

Attene M, Falcidieno B (2006) Remesh: an nteractive environment to edit and repair triangle meshes. In: IEEE international conference on shape modeling and applications. p 41

Bogdahn J, Coors V (2010) Towards an automated healingof 3D urban models. In Kolbe TH, König G, Nagel C (eds) In: Proceedings of international conference on 3D geoinformation. International archives of photogrammetry, remote sensing and spatial information science. International conference on 3D geoinformation. Shaker Verlag, Aachen, Germany, pp 13–17

Boguslawski P, Gold C, Ledoux H (2011) Modelling and analysing 3D buildings with a primal/dual data structure. ISPRS J Photogram Rem Sens 66(2):188–197

Borodin P, Novotni M, Klein R (2002) Progressive gap closing for mesh repairing. In: Vince J, Earnshaw R (eds) Advances in modelling, animation and rendering, Springer pp 201–213

Butlin G, Stops C (1996) CAD data repair In: 5th international meshing roundtable. pp 7-12

Coors V, Krämer M (2011) Integrating quality management into a 3D geospatial server. In: UDMS 2011. Urban Data Management Society. Delft, p 8

CPA-Systems GmbH (2011) Available at: http://www.cpa-systems.de/

Elysium (2008) CADdoctor. Available at: http://www.elysiuminc.com/Products/caddoctor.asp. Accessed 20 Jan, 2012

Ghawana T, Zlatanova S (2010) Data consistency checks for building a 3D model: a case study of Technical University, Delft campus. Geospatial World, The Netherlands, p 4

Gröger G, Coors V(2010) Rules for validating GML geometries in CityGML. Available at: http://files.sig3d.de/file/20101215_Regeln_GML_final_DE.pdf. Accessed 18 Jan 2012

Gröger G, Plümer L (2009) How to achieve consistency for 3D city models. GeoInformatica 15(1):137–165

Gröger, G. et al (eds) (2008) OpenGIS city geography markup language (CityGML) encoding standard. Available at: http://portal.opengeospatial.org/files/?artifact_id=28802. Accessed 28 June 2011

International Organization for Standardization (ISO) (2002) ISO 19113 standard: geographic information – quality principles. Available at: http://www.iso.org/iso/iso_catalogue/catalogue_tc/catalogue_detail.htm?csnumber=26018

International Organization for Standardization (ISO) (2003a) ISO 19114 standard: geographic information—quality evaluation procedures. Available at: http://www.iso.org/iso/iso_catalogue/catalogue_tc/catalogue_detail.htm?csnumber=26019

[3] Working Group of the surveying agencies of the federal states.

International Organization for Standardization (ISO) (2003b) ISO/CD 19107, Geographic information—spatial schema

International Organization for Standardization (ISO) (2006) ISO/PAS 26183—SASIG product data quality guidelines for the global automotive industry

International TechneGroup (2011) CADfix. Available at: http://www.transcendata.com/products/cadfix/index.htm

Ju T (2009) Fixing geometric errors on polygonal models: a survey. J Comput Sci Technol 24(1):19–29

Karki, S., Thompson, R. & McDougall, K., 2010. Data validation in a 3D cadastre. In Neutens T, Maeyer P (eds) *Developments in 3D geo-Iinformation sciences*. Lecture notes in geoinformation and cartography, Springer Berlin Heidelberg, pp. 92-122

Kazar BM et al (2008) On valid and invalid three-dimensional geometries. In: Oosterom P et al (eds) Advances in 3D geoinformation systems. Springer, Berlin, pp 19–46 (Available at: http://www.springerlink.com/index/10.1007/978-3-540-72135-2_2. Accessed 22 June 2011)

Ledoux H (2011) Topologically consistent 3D city models obtained by extrusion. Int J Geogr Inf Sci 25(4):557–574

Métral C, Falquet G, Cutting-Decelle AF (2009) Towards semantically enriched 3D citymodels: an ontology-based approach. In: Academic track of geoweb 2009—cityscapes, international archives of photogrammetry, remote sensing and spatial information sciences (ISPRS). Vancouver, Canada

Object Management Group (2011) OMG object constraint language (OCL). Available at: http://www.omg.org/spec/OCL/2.3.1/. Accessed 20 Jan 2012

van Oort P (2005) Spatial data quality: from description to application. Optima Grafische Communicatie, The Netherlands

Oosterom P, Quak W, Tilssen T (2005) About invalid, valid and clean polygons. In Fisher PF (ed) Developments in spatial data handling. 11th international symposium on spatial data handling. Leicester, UK: Springer, Berlin, pp 1–16. Available at: http://dx.doi.org/10.1007/3-540-26772-7_1

Open Geospatial Consortium (OGC) (2011) Data quality DWG. Available at: http://www.opengeospatial.org/projects/groups/dqdwg. Accessed 19 Jan 2012

Pelagatti G et al. (2009) From the conceptual design of spatial constraints to their implementation in real systems. In Proceedings of the 17th ACM SIGSPATIAL international conference on advances in geographic information systems. Seattle, pp 448-451, Nov 4-6

Rocchini C et al (2004) The marching intersections algorithm for merging range images. Vis Comput 20:149–164

Safe Software (2011). FME desktop. Available at: http://www.safe.com/inc/vendors/elqNow/elqRedir.htm?ref=http://downloads.safe.com/fme/brochures/FME_Desktop.pdf

SIG-3D Quality Working Group (2012) Modellierungshandbuch Gebäude. Available at: http://www.sig3d.de/index.php?catid=2&themaid=8777960

Stadler A, Kolbe TH (2007) Spatio-semantic coherence in the integration of 3D city models. In: Proceedings of the 5th inter-national symposium on spatial data quality. Enschede. Available at: http://spirit.bv.tu-berlin.de/igg/htdocs/fileadmin/user_upload/ Stadler/SDQ2007_Stadler_Kolbe.pdf

Thompson R, Oosterom P (2011) Modelling and validation of 3D cadastral objects. In: Zlatanova S et al (eds) Urban and regional data management–UDMS annual 2011. CRC Press, Delft

Transcat PLM (2011) Q-Checker. Available at: http://www.transcat-plm.com/software/transcat-software/q-checker.html. (Accessed 24 Jan 2012)

T-Systems International (2011) VALIDAT. Available at: https://servicenet.t-systems.de/validat. Accessed 24 Jan 2012

Wöhler T, Pries M, Stark R (2009) Effiziente Verfahren zur Aufbereitung von Geometriemodellen für die virtuelle Absicherung. In 3. Symposium Geometrisches Modellieren, Visualisieren und Bildverarbeitung. Stuttgart

Yamakawa S, Shimada K (2009) Removing self intersections of a triangular mesh by edge swapping, edge hammering, and face lifting. In 18th international meshing roundtable

Advancing DB4GeO

M. Breunig, E. Butwilowski, D. Golovko, P. V. Kuper, M. Menninghaus and A. Thomsen

Abstract The analysis of complex 3D data is a central task for many problems in the geo- and engineering sciences. Examples are the analysis of natural events such as mass movements and volcano eruptions as well as 3D city planning and the computation of 3D models from point cloud data generated by terrestrial laser scanning for 3D data analysis in various domains. The volume of these data is growing from year to year. However, there is no geo-database management system on the market yet that efficiently supports complex 3D mass data, although prototypical 3D geo-database management systems are ready to support such challenging 3D applications. In this contribution we describe how we reply to these requirements advancing DB4GeO, our 3D/4D geo-database architecture. The system architecture and support for geometric, topological and temporal data are presented in detail. Besides the new spatio-temporal object model, we introduce new ideas and implementations of DB4GeO such as the support of GML data and the new WebGL 3D interface. The latter enables the direct visualization of 3D

M. Breunig (✉) · E. Butwilowski · D. Golovko · P. V. Kuper · M. Menninghaus
Geodetic Institute, Karlsruhe Institute of Technology, Karlsruhe, Germany
e-mail: martin.breunig@kit.edu

E. Butwilowski
e-mail: edgar.butwilowski@kit.edu

D. Golovko
e-mail: daria.golovko@kit.edu

P. V. Kuper
e-mail: kuper@kit.edu

M. Menninghaus
e-mail: mathias.menninghaus@kit.edu

A. Thomsen
Institute of Geosciences, Christian-Albrechts-Universität zu Kiel, Kiel, Germany
e-mail: athomsen@geophysik.uni-kiel.de

J. Pouliot et al. (eds), *Progress and New Trends in 3D Geoinformation Sciences*, Lecture Notes in Geoinformation and Cartography, DOI: 10.1007/978-3-642-29793-9_11,

database query results by a standard web browser without installing additional software. Examples for 3D database queries and their visualizations with the new WebGL interface are demonstrated. Finally, we give an outlook on our future work. Further extensions of DB4GeO and the support for the data management for collaborative subway track planning are discussed.

1 Introduction

The demand for modeling and handling large 3D and 4D data sets has been rapidly growing during the last decades (Breunig and Zlatanova 2011; Hashemi et al. 2009; Kolbe 2012; Kolbe et al. 2011; Mallet 2002; Raper 1989). Techniques and applications such as geodesy, terrestrial laser scanning (TLS), 3D city planning, geological modeling, geothermal reservoir modeling, and early warning of natural events strengthen this trend by generating large volumes of 3D data. However, the analysis of these data becomes a confusing task without the help of geo-databases, because the geo-expert has to access dozens or even hundreds of single data files. Furthermore, without documentation and long-time archiving of modeling and simulation (results) in a geo-database, many examinations become useless as soon as their authors are no longer available.

In this paper we argue that 3D geo-databases can be accessed by non-experts in a straightforward manner. The rest of this paper is organized as follows. In Sect. 2 the service-based system architecture of DB4GeO, our 3D geo-database kernel, is presented. Section 3 is dedicated to the geometric, topological, and temporal database support. Section 4 describes the support for GML data and implementation details of the REST communication interface. Section 5 presents the new WebGL interface of DB4GeO with various visualization examples. Finally, we give a conclusion and outlook on our future work concerning 3D and 4D geo-data support in DB4GeO, e.g. in the field of collaborative subway track planning.

2 System Architecture

DB4GeO (Bär 2007; Breunig et al. 2010; Thomsen et al. 2008b) has its roots in the development of GeoToolKit (Balovnev et al. 2004), an object-oriented library for 3D geometric data types for geo-databases. It has a service-based user interface and is exclusively implemented in the Java programming language. Hitherto REST (Fielding 2000) is used as communication platform, i.e., REST style web services are used to enable remote interaction of clients with the geo-database. The system architecture of DB4GeO is presented in Fig. 1. On the client side, GIS or mobile clients have access to 3D data managed by the DB4GeO server. On the server side, DB4GeO is accessed exclusively via its service infrastructure. The services are divided into simple and complex services (Breunig et al. 2010). The simple services

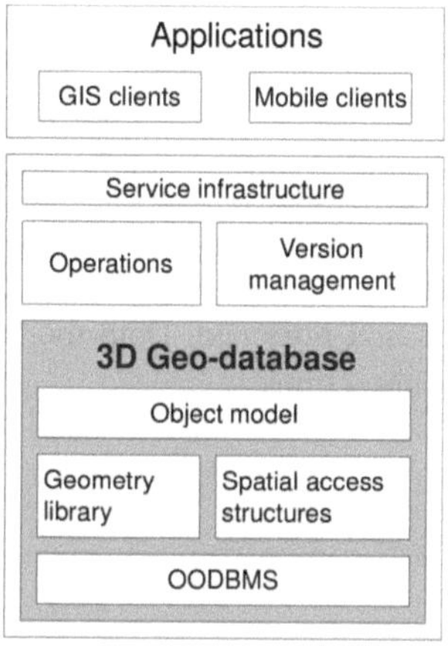

Fig. 1 DB4GeO system architecture

are equivalent to basic geometric and topological operations such as the distance between 3D objects or the determination if two 3D objects intersect etc. However, also computations such as the intersection between two surfaces belong to the simple services. A typical complex service is the so called "3D-to-2D service" that has been introduced in Breunig et al. (2010). It computes a vertical profile section of a geological subsoil model by intersecting all existing surface objects with a vertical plane. Finally, it projects all wells within a specified distance onto the vertical plane. The geological model has to be bounded by a 3D bounding box. Further examples of complex services are the "4D-to-3D service", which calculates the geometry of a spatio-temporal object at a given moment in time, and a triangulation service that creates a triangle net from a set of points.

The core of DB4GeO is its 3D geo-database which is based on a geometry library and the R-tree based spatial access structures. DB4GeO is implemented upon the open source object-oriented database management system db4o (Paterson et al. 2006; Versant Corp 2012).

Developed from a system designed primarily for geological applications, DB4GeO concentrates in the first place on the modeling of spatial and temporal characteristics of objects and their parts. Support of semantics has received less attention in our research so far but is expected to gain importance in our future work. Although DB4GeO has not had many users lately, we hope that extending the data model and the functionality of the geo-database will extend its user community. DB4GeO primarily aims at geoscientists who need to store and process 3D and 4D objects represented with simplicial complexes or who want to combine such data with non-simplicial geometries.

3 3D/4D Database Support

As a database management system, the main task of DB4GeO is to store and manage large sets of geo-data in an efficient way for a long period of time without loss and free of contradictions (i.e., consistent data storage). DB4GeO provides a tightly defined, extensive set of geometric, topological and spatio-temporal objects that can be stored, managed and retrieved. Data models of these objects will be presented in the following subsections.

3.1 Geometry

The object model of the geometric component in DB4GeO has been discussed in detail in Bär (2007). Roughly explained, the object model can be summarized as follows: at the root of the object model, a 3D object is located (`Object3D` class). Part of every 3D object is one three-dimensional spatial object (`Spatial3D`). The spatial objects can be of one of *four* abstract data types, namely either a *sample*, a *curve*, a *surface* or a *volume*. Any of the mentioned abstract data types has a concrete realization in the database. Currently, these are point nets (`PointNet3D`) as realization of the sample type, segment nets (`SegmentNet3D`) as realization of the type curve, triangle meshes (`TriangleNet3D`) as realization of the surface type and tetrahedral nets (`TetrahedronNet3D`) as realization of the volume type. All of these concrete classes for spatial data types are nets of the most simple geometric constructs (or geometric elements) of the respective dimension (so-called *simplices).*[1] A geometric net of any of these types in turn consists of an arbitrary number of disconnected net components. A net component is a contiguous geometric object, which consists entirely of geometric elements of one of the geometric types *point*, *segment*, *triangle* or *tetrahedron*.

3.2 Topology

The topology model of DB4GeO extends its geometry model and is closely linked to it. The topology model provides an additional construct which allows to go beyond the model of simplicial complexes and to manage more complex structures. Furthermore, it enables easy and efficient navigation in the meshes. The topology model of DB4GeO is based on the combinatorial concepts of *Generalized Maps* (abbreviated as G-Maps) and *cell-tuple structure* introduced by Lienhardt (1989)

[1] The entire geometric model of DB4GeO is based on the model of simplicial complexes introduced in the context of GIS by EGENHOFER and MOISE, cf. Breunig (2001).

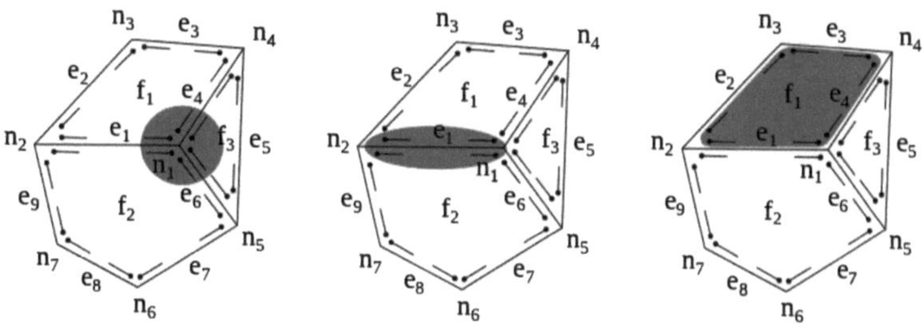

Fig. 2 Fragment of a face net with cell-tuples shown as darts and orbits of a node (*left*), an edge (*center*) and a face (*right*)

and Brisson (1989), respectively. Further research on the theory of G-Maps can be found in Fradin et al. (2005); Lévy and Mallet (1999), and Thomsen et al. (2008a).

The G-Map representation itself does not provide information about the spatial extent of modeled objects. The link between the topology model and the geometry model of DB4GeO is realized by constructing the topology representation on top of an already existing `TriangleNet3D` or `TetrahedronNet3D`. The created object is represented by the classes `FaceNet3d` or `SolidNet3d`, respectively. Similarly to the geometry model, it includes one or more components. Instead of points, segments, triangles and tetrahedrons of the geometry model, the topology model manages four types of cells: nodes, edges, faces and solids. The cells are not limited to simplices. A cell-tuple represents a unique combination of a node, an edge, a face and a solid. Cell-tuples that differ in only one cell are linked to each other by so-called *involutions*. Specific combinations of involutions form so-called *orbits* used to define cells and groups of cells. Cell-tuples and orbits of various dimensions are shown in Fig. 2.

Figure 3 demonstrates the principal classes of DB4GeO topology model used to manage face nets. Since the concept of G-Maps is not bound to a particular dimension, the handling of solid nets was designed similarly to the handling of face nets without great difficulties. Each component of a `FaceNet3d` or a `SolidNet3d` consists of a net level (C_{NL}, class `FaceNet3dCompNetLevel`) and an object level (C_{OL}, class `FaceNet3dCompLevel`). Both classes implement the interface `CellNet3dCompLevel`. The topology of C_{NL} exactly repeats the topology of the underlying triangle or tetrahedron net, i.e., every face or solid of C_{NL} is a simplex and points to the corresponding simplex of the triangle or tetrahedron net. Each node of C_{NL} is also linked to a particular point and can thus access its coordinates. C_{OL} describes the overall geo-object structure that is modeled by cells that are commonly composed of a large amount of simplices (so-called "big cells"). At the creation of the topological net, C_{OL} consists of exactly one cell whose boundary is identical to the boundary of the whole component. C_{OL} is linked to C_{NL} by the `higher` and `lower` attributes of their cell-tuples. The `higher` attribute of a cell-tuple of C_{OL} points to a corresponding cell-tuple of C_{NL}, and it is `null` for every cell-tuple of C_{NL}. The `lower` attribute of a cell-tuple of C_{NL} stores the link to its counterpart at C_{OL} if such a cell-tuple

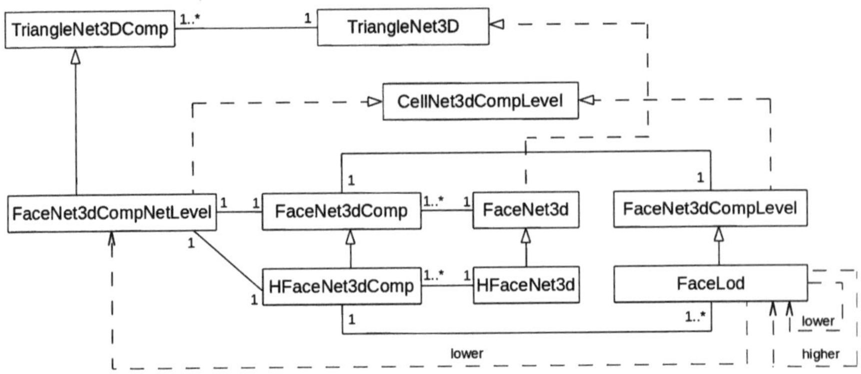

Fig. 3 Classes of DB4GeO topology model used to manage face nets

exists; otherwise the `lower` attribute is `null`. A special `OrbitIterator` allows retrieving cell-tuples of C_{NL} that are inside a particular cell of C_{OL}.

Such a structure permits separating geometry from topology, which is handy e.g. for the management of temporal changes that involve the object's coordinates but not its topology (cf. Sect. 3.3). Direct links to the geometry model enable access to the core functionality of DB4GeO at any time.

The topology module of DB4GeO offers the possibility to manage multiple levels of detail (LoDs) by using hierarchical G-Maps also referred to as HG-Maps (Lévy 2000; Fradin et al. 2005; Thomsen et al. 2008a). In DB4GeO, all LoDs have the same geometric extent and it is possible to relate particular locations of various LoDs to each other. This approach is different from e.g. the CityGML specification where various LoDs of the same object do not correspond to each other geometrically and can even be represented by different geometry types. For instance an object may be represented by a polygon at a more detailed LoD and by a line at a less detailed LoD. Furthermore, the geometry of objects in CityGML at each LoD is regarded as one entity and it is not always possible to determine which part of a more detailed LoD corresponds to a particular part of a less detailed LoD.

Similarly to cell-tuples, LoDs in DB4GeO also point to their `higher` and `lower` counterparts. The `higher` attribute of the most detailed LoD is C_{NL}. Note that, while a `FaceNet3dComp` has exactly one C_{OL}, an `HFaceNet3dComp` can have multiple LoDs (cf. Fig. 3). C_{OL} can be regarded as one of the LoDs.

Two ways of creating a new hierarchical face net (`HFaceNet3d`) are available to the user. First, it can be created from an already existing non-hierarchical face net (`FaceNet3d`). In this case, the underlying triangle net and C_{NL} of the original `FaceNet3d` are copied for the new `HFaceNet3d`, and C_{OL} of the original net is converted to be an LoD of the new hierarchical geo-object. This permits retaining the current subdivision of C_{OL}. Secondly, the user can create a new hierarchical face net from a set of triangles, similarly to the construction of a non-hierarchical net. In that case, the underlying triangle net is created first, then the net level, and finally the first LoD of the new hierarchical net with exactly one face. In order to ensure the geometric correspondence between LoDs, a new LoD can only be created as a copy of an already existing LoD. Besides, this allows the users to edit the hierarchical net more conveniently, because they can transfer already made changes to another LoD and continue editing.

The editing of LoDs in a `HFaceNet3d` and of C_{OL} in a `FaceNet3d` is possible via 3D Euler operations. DB4GeO offers methods for inserting and removing nodes and edges of a face net. Before the changes are applied on the net, a check of constraints takes place. For both types of nets, topological correctness within the edited C_{OL} or LoD has to be ensured. Hierarchical nets additionally require verifying that the hierarchy of LoDs is not violated, i.e., that every cell and every cell-tuple has its counterpart at the next more detailed LoD. If the constraints are fulfilled, the changes are carried out. Implementing 3D Euler operations and the management of hierarchies for solid nets are the objectives of future work.

3.3 Time

The modeling of time in geoscientific information systems (and in solid modelers in general) is a subject of widespread research. One of the research issues is the combination of continuous temporal geometry models with discrete temporal topology models. Polthier and Rumpf added some relevant work on this issue. In Polthier and Rumpf (1995) they propose the concept of *adaptive time-dependent discretization*. Thomsen and Rolfs designed a concept to handle and store the vertices of time-dependent simplex nets efficiently. This approach was termed *Point Tube model* (Rolfs 2005, p. 51). Another important approach in this context is the concept of *Delta-Storage* used by Strathoff during the development of GeoToolKit to reduce the storage volume of redundant data between two timesteps (Strathoff 1999).

In DB4GeO we implemented a combination of these three concepts to support spatio-temporal data. Therefore it is possible to create time series of 3D objects whose topology changes with time. Due to the concept of *Delta-Storage* and the *Point Tube model*, we were able to reduce the amount of required memory and increased the performance of operations such as the interpolation between time steps or the generation of snapshots at specified dates (Kuper 2010). Figure 4 shows a 4D object containing 5 timesteps with a change of the net topology at timestep 3.

However, a restriction of the implemented model is that the topology of the net configuration of the geo-object has to stay invariant throughout all time steps–only the geometry may change. Also Polthier and Rumpf make no statement on how to model the transition between two states of the object's topology (object level) of the same geo-object. In such a transition, the geometry generally stays unchanged, but the topology of the geo-object as a whole may change. The implementation of the space-time module of DB4GeO also excludes the issue of managing change in spatio-temporal topology. Raza and Kainz have done some relevant research in the

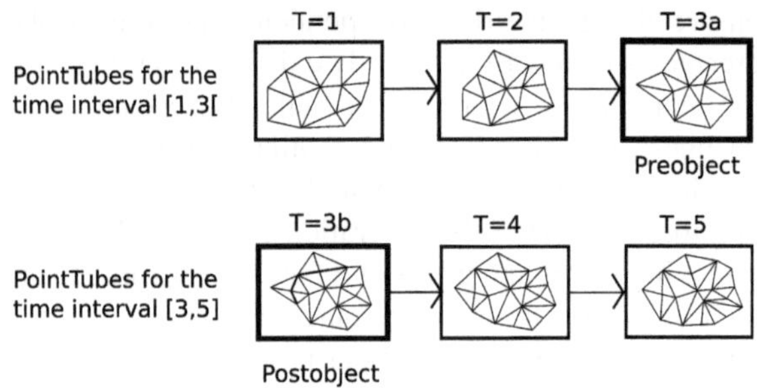

Fig. 4 Representation of a 4D object with the new 4D model of DB4GeO (Kuper 2010)

area of spatio-temporal topology. In Raza and Kainz (1999) they propose a temporal cell-tuple structure to manage *Spatio-Temporal-Attribute Objects* in generic temporal GIS (TGIS). The *temporal cell-tuple structure* used by Raza and Kainz is also based on the cell-tuple structure of Brisson and thus is comparable to the internal model of the topology module for DB4GeO. In contrast to the concept of Raza and Kainz, we use a graph representation and not a relational approach to manage and store the cell-tuple data. We also combine the concept of temporal cell-tuple structure with our previously discussed concepts of spatio-temporal geometry and hierarchical G-Maps to manage the temporal topology of "big cells" (cf. Sect. 3.2). Figure 5 shows a simple example application sketching the intended functionality of the proposed spatio-temporal topology module for DB4GeO. This module is currently in implementation process.

Figure 5 shows a similar case as demonstrated in Fig. 4. Both illustrations depict a geo-object that moves through time and changes its geometry (in time intervals) and the topology of its meshing (at time steps). The example of Fig. 5 consists of three time steps. The geo-object is created at time step $t = 1$. Between the time steps $t = 1$ and $t = 2$ the geometry of the spatial object changes (the object grows). At time step $t = 2$ the meshing of the object changes, while the geometry remains constant. The geometry then changes again in the period between time steps $t = 2$ and $t = 3$. In the spatio-temporal model of DB4GeO this temporal geo-object is internally managed as two sequential `PointTubes` (that describe the point geometry at each interval) and two sequential temporal triangle nets (`TriangleNet4D`) (that describe the meshing at each interval). This means that whenever a change in the meshing takes place (here at $t = 2$), there is a break in the continuity of the spatial object. At this point, all object identifications get lost. From this point on, the evolution of the spatial object cannot be clearly traced. An unambiguous assignment of all geometric elements between the pre- and post-object of $t = 2$ would not be possible due to the changed meshing.

However, even if an unambiguous assignment of all geometric elements is not possible, still some elements can be assigned (cf. coincidences of pre- and post-objects of $t = 2$ in Fig. 5). Such conditions can be used to trace the mentioned temporal big cells. In Fig. 5, the thick lines symbolize the boundaries of two face cells (big cells) that are part of a face net component at object level C_{OL}. While `TriangleNet4D` *a* ends at $t = 2$, the topology of the geo-object stays constant

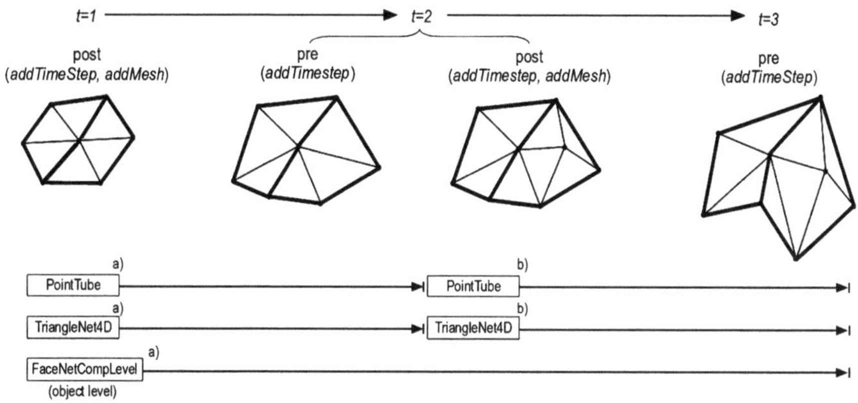

Fig. 5 Simple example of a temporal cellular network of net level (*thin lines*) and of object level (*thick lines*)

at the object level (cf. lifespan of `FaceNetCompLevel` in Fig. 5), thus the two faces can be traced throughout the whole lifespan of the geo-object.

4 Web-Based Geo-Data Access

One of the objectives of DB4GeO is to remotely provide geo-database services. Nowadays, web-based access to geo-data is closely associated with OGC standards. In the following, we discuss the handling of GML data in DB4GeO and afterwards present the implementation of the REST-based architecture of the geo-database (Fielding 2000).

4.1 Support for GML Data

Originally, DB4GeO was only capable of exchanging data in its own XML-format. However, as the OGC standards are developed and spread out among the providers and users of spatial information, the need for extending DB4GeO to offer data via OGC services is becoming more acute. To enable that, support for the Geography Markup Language (GML) is developed. This will make DB4GeO a handy tool that can at first import the results of 3D geomodeling software that does not offer OGC support, and then provide those results in a data format compatible with OGC standards. An example of such software is Gocad® (Gocad Research Group 2012), a widely used tool for 3D subsurface modeling. The geometry models of both DB4GeO and Gocad® are based on simplicial complexes making it easy to exchange data between the two. DB4GeO offers importers and exporters for various types of Gocad® geometrical objects (GObj).

A further reason to offer GML support in DB4GeO is the growing number of interdisciplinary projects with joint handling of data from different sources, e.g. of nature-formed and anthropogenic objects. While DB4GeO originated in the field of geosciences, particularly that of geology, integrating CityGML data will make the database interesting for instance for subsurface construction projects where building and infrastructure data are processed and visualized together with the information about geological structures (Breunig et al. 2011).

In GML, objects are modeled as *features* that have geometry as a property. The geometry model of GML is based on boundary representations, e.g. polygons are defined via their `exterior` and `interior` (the latter is used when polygons have holes). The `exterior` and `interior` are represented by ordered lists of point coordinates. In our work, GML 3.0 geometries only with linear interpolation between points have been considered. Those lists can include an arbitrary number of points. Planarity assumed by many GML geometries is well suited for modeling anthropogenic objects that are geometrically simple. This simplicity also allows to construct more complex hierarchies of geometries. However, nature-formed objects in most cases are more compex and far from planar, so they have often been modeled in geosciences using simplicial complexes native to DB4GeO (Breunig 2001; Balovnev et al. 2004). Because of the different assumptions of the two data models representing anthropogenic and nature-formed objects, data in most cases cannot be transferred between them without adjustments of geometry.

On the one hand, integration of non-simplex geometries of GML into DB4GeO requires a representation by simplices. A method that triangulates complex planar polygons has been implemented in DB4GeO for that purpose. On the other hand, importing GML data into DB4GeO just by triangulating non-simplicial geometric objects causes data loss and distortion. The initial structure of the data might be lost and thus the attributes related to it. For instance, if the original object with n polygons, each with the ?population density? attribute, is imported into DB4GeO, its polygons will be substituted with an even greater number of triangles, and the density attribute will lose its sense in the new geometric boundaries. Therefore, an additional construct is necessary to store the original geometric structure.

The topology model of DB4GeO provides such a construct (cf. Sect. 3.2). The G-Maps structure (Lienhardt 1989; Mallet 2002) enables managing non-simplex polygons (Thomsen et al. 2008a, 2008b). The geometric extent of the object is stored at the net level (C_{NL}) of the face net, enabling access to diverse geometric operations available in DB4GeO. The object level (C_{OL}) models how triangles of C_{NL} are aggregated to the original non-simplex geometries of GML.

Figures 6 and 7 provide examples of importing CityGML data into DB4GeO. After the data set of a part of Berlin's downtown (Fig. 6) obtained in CityGML format was imported into DB4GeO, it is possible to export it into one of the formats supported by the geo-database. For example, the data set can be exported into the

Fig. 6 GML data (© City of Berlin, obtained from citygml.org) converted to Gocad® format in DB4GeO, visualized with ParaViewGeo®

Gocad® format (.ts) and visualized in Gocad® or another 3D visualization tool, e.g. ParaViewGeo® (ParaViewGeo 2012), together with data illustrating the subsurface of the area. Figure 7 looks closer at how a building can be stored in DB4GeO. The building came as a part of a CityGML data set. Its surface is represented by the GML geometry type `CompositeSurface`, which is made up of 17 `Polygons`. In DB4GeO, the non-simplex polygons were triangulated, which resulted in 32 triangles for the whole building. Those triangles become faces of C_{NL}. The 17 original polygons are represented in DB4GeO by faces of C_{OL}. This enables, for instance, assigning a certain color or a texture to each wall of the building (cf. Fig. 7).

While additional classes for the modeling of topology are available in GML 3.0, GML offers an alternative which is simpler and almost as powerful. GML uses the XML concept of `Xlinks` that reference resources via their IDs avoiding redundant data storage. For instance, if two neighboring buildings share a wall, the wall can be stored just once and then be references via an `XLink` from all other objects that use it. Such references carry information about the topology of the objects: first, the neighbor relationship between the two buildings is stored; secondly, the relationship between the wall and each of the buildings is defined (Krimmelbein 2011, p. 14ff). We have chosen to use the G-Maps topology concept of DB4GeO to represent non-simplex geometry types of GML rather than the less widely used GML 3.0 topology classes. In the future, we also plan to use `XLinks` to avoid redundancy when providing data from DB4GeO in the GML format. The implementation of G-Maps in DB4GeO already takes care of storing each node, edge, face and solid in the net just once under a unique ID, even if the original data structure stored them redundantly. The advantage of topology management is that this approach is dimension-independent, i.e., it can be applied to 0D-, 1D-, 2D- and 3D-geometries. Furthermore, topology is modeled explicitly without the semantic information of the objects.

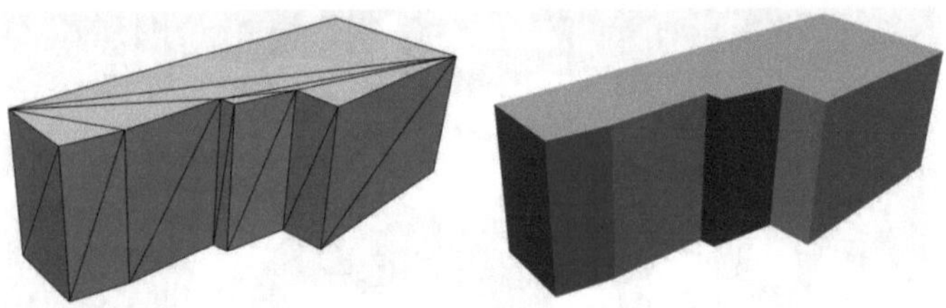

Fig. 7 Building (© Ordnance Survey Great Britain, obtained from citygml.org) represented by faces of the net level (*left*) and of the object level (*right*), visualized with ParaViewGeo®

4.2 REST-Based Access and Query Examples

DB4GeO is capable of providing its services via a RESTful web service architecture (Fielding and Taylor 2002). This architecture permits sending requests to the server via uniform resource identifiers (URIs). Each such request contains all the information necessary for the server to understand it and is independent of other requests that might have been sent to the server before. Such manner of communicating with the server is easier to interpret by humans without technical background, first, due to the familiarity of most people with URIs, and secondly, because of the isolation of requests. Furthermore, common browsers can serve as clients communicating with DB4GeO without installing additional software.

Manipulating resources using REST is possible via four commands: GET (retrieving a resource), POST (adding a new resource), PUT (updating an existing resource), and DELETE (removing a resource). DB4GeO uses two of them: GET and PUT. The latter can be used e.g. to add a new surface created by triangulation to the geo-database.

By default, when a request is sent to the geo-database, the response comes back in the DB4GeO-XML format and can be viewed in a browser. In DB4GeO, it is also possible to export the objects into other formats, e.g. the Gocad®, GML and VRML formats. In order to obtain an object in a certain format, the user should add the extension *.vrml*, *.gml*, *.ts*, *.vs*, *.so*, etc. at the end of the URI used to retrieve the object.

The REST-based services of DB4GeO are accessible from a web browser, via a Java application or an OpenJump plugin. In our future work, we plan to extend the number of operations available to the database users via the RESTful service and to create a user-friendly web interface to replace the XML-based representation in the browser.

Below are some examples of URIs used to query data from DB4GeO:

(1) http://server/projects/GeolProj
(2) http://server/projects/GeolProj/StructGeolSpace3D
(3) http://server/projects/GeolProj/StructGeolSpace3D/TestSurfaces
(4) http://server/projects/GeolProj/StructGeolSpace3D/TestSurfaces.ts

(5) http://server/projects/GeolProj/StructGeolSpace3D/TestSurfaces ?**intersects**(x1,y1,z1,x2,y2,z2)
(6) http://server/projects/GeolProj/StructGeolSpace3D/TestSurfaces ?**3dto2d**(CuttingPlane)
(7) http://server/projects/GeolProj/StructGeolSpace4D/TestSurfaces?**time**=0

Sets of objects in DB4GeO are managed by grouping them into spaces and projects. Spaces aggregate objects with the same dimension (e.g. 3D or 4D spaces), same coordinate reference, constraints, thematic information, etc. A project may contain multiple spaces. There is a defined way this hierarchical structure can be navigated via URIs. For example, the URI (1) returns information about the project *GeolProj*. That information includes the names of spaces that belong to the project. The user can easily add one of those names to the URI to access information about the corresponding space, like in the URI (2). In the same way, the user can go over to the object information in the XML format [(URL (3)]. If the object is to be retrieved in the Gocad® format for further processing in Gocad®, this is done by adding the extension *.ts* after the object name [(URI (4)].

Furthermore, it is possible to define operations to be carried out on objects via a URI. Operations include querying an object of a lower dimension, such as obtaining a 3D object for a certain point in time [(URI (7)] or the cross-section of a 3D object with a plane using the 3D-to-2D service of DB4GeO [(URI (6)]. Other operations include intersecting a given object with another object or a minimum bounding box [(URI (5)], projecting objects onto a plain and triangulating point sets.

5 Visualization with WebGL

Usually the visualization of 3D geodata takes place in various 3D modeling or visualization tools such as *Gocad®* or *ParaViewGeo®*. The user is responsible for the selection of one of these clients. For an alternative we developed a direct visualization of 3D database query results in a web browser based on WebGL (Khronos Group 2012).

WebGL (Web Graphics Library) is the mapping of OpenGL ES (Open Graphics Library for Embedded Systems) for web browsers and is now supported by almost all current browsers without installing an extra plugin. Since the graphical calculations runs directly on the hardware of the client, a high performance compared to traditional solutions such as VRML (Virtual Reality Modeling Language) and X3D (Extensible 3D) is provided. Thereby the user does not need to install any additional software apart from a modern web browser.

Thus with the help of our WebGL viewer it is possible to visualize geo-objects of the geo-database directly in a browser in 3D including lighting, colors, and controls.

For the visualization in WebGL we are using the free library *Three.js* (Three.js 2012). This JavaScript based library provides various types of cameras, lights and various shading concepts (e.g. Flat, Gouraud, Phong). We developed an exporter

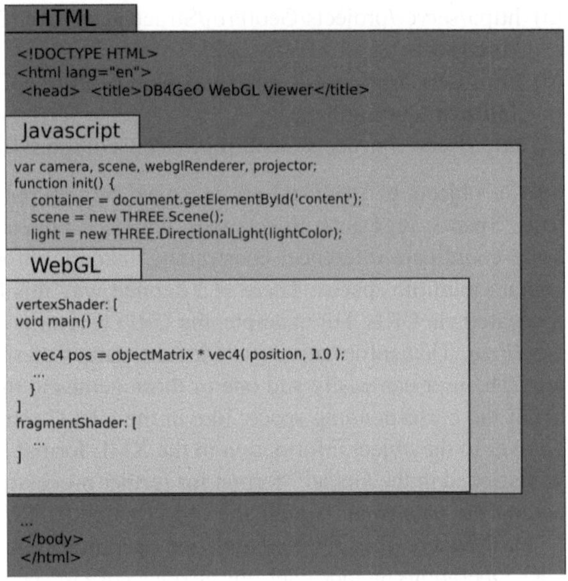

Fig. 8 Structure of the HTML file generated by DB4GeO calling WebGL

which creates a 3D scene by using the geometry of the DB4GeO database. The following criteria were relevant for this intent:

- A suitable lighting.
- An intuitively controllable camera to view around the 3D object.
- Suitable shaders, colors, etc.

We decided to use `THREE.DirectionalLight` lights and a `THREE.PerspectiveCamera` camera. The DB4GeO WebGL Exporter creates an HTML file that executes JavaScript code with the use of *Three.js*. Within the JavaScript code different *Three.js* objects are created, adapted and transferred from the geometries of the spatial database into a format suitable for *Three.js*. For the DB4GeO objects `Point3D` and `Triangle3D` we use `THREE.Vertex` and `THREE.Face3`, respectively.

An overview of such an HTML file is shown in Fig. 8 and the result of an export in the Chrome® browser is shown in Fig. 9.[2] The viewer supports zooming, panning, and the possibility to rotate the object. The implemented WebGL viewer also shows the numbers of triangles for the visualized object. For instance the object demonstrated in Fig. 9 has 98,740 triangles.

The whole 3D model is transmitted from the database at the start of the viewer. Even larger amounts of data (test data sets with up to 2.5 million points) are represented efficiently due to the use of WebGL. In order to avoid further loading while viewing the model, caching is dispensed. To reduce the amount of data to be transferred, we plan to optionally provide a reduced 3D model.

---

[2] 3D-Model of "The Thinker" by Simon Schuffert (KIT), all rights reserved.

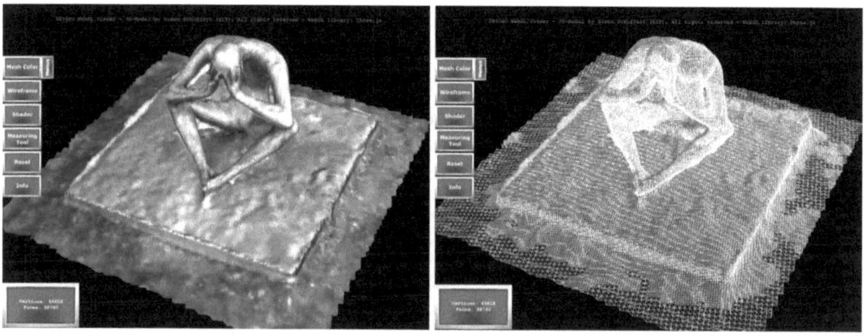

Fig. 9 WebGL viewer of DB4GeO with colored mesh (*left*) and wireframe representation (*right*)

As one of the additional spatial operations the distance function has been implemented. In the future, various operations and queries should be directed back to the geo-database. After their executions, the results should again be displayed visually in the browser.

6 Conclusion and Outlook

In this contribution we have presented our last steps advancing DB4GeO, our service-based geo-database architecture. Besides the new object model for spatio-temporal data, new features are the support for GML data (implementation started) and the WebGL-interface (implementation completed) enabling direct geo-database access and visualizing 3D objects via a standard web browser without installing any additional software.

In our future work we intend to query geo-database operations directly from the WebGL viewer. Our focus here is on a BBox query, the query of meta-data, and the comparison of different time steps of a 4D object. Therefore we need to develop some additional spatio-temporal operations in DB4GeO and extend our WebGL viewer. Additionally we intend to examine how DB4GeO can be accessed via extended OGC services.

Finally, we intend to develop a new branch of DB4GeO supporting spatio-temporal data used for multi-scale subway track planning.

In the research group "Computer-Aided Collaborative Subway Track Planning in Multi-Scale 3D City and Building Models" (Breunig et al. 2011) a spatio-temporal database will be used to store several states and versions of building plans. As DB4GeO by now has been used to store geological data in 3D and time, it should be enhanced to support a data model that is commonly used to describe building models (Eastman 1999). Such parametric models stand in contrast to the simplicial complexes which are a core concept of DB4GeO. Re-using the simplicial model would lead to a loss of data when converting the parametric data into the discrete triangle nets which can be stored by DB4GeO. Converting and storing the data in higher resolution would only increase the memory consumption, but not decrease the data loss in an equal manner. We will test two solutions. First, we intend to implement a hybrid model, which means that the database will convert the parametric geometries into discrete triangle nets handled by DB4GeO and we store the original file-based data linked to these converted geometries. By doing this we will be able to use the high performance queries of DB4GeO and its topology module. This does not cause any data losses, because both the parametric data is stored in the database and changes are simultaneously made on the converted geometries. Secondly, we intend to implement a complete new data-structure keeping the core concepts in mind. Therefore, we do not need to build the database from scratch and can use an already stable and proven system.

Another issue in advancing DB4GeO within the research group is the usage of spatio-temporal data within the database. In the geosciences, spatio-temporal data often consists of moving objects. Construction plans, however, may be modeled in two different ways. First, the temporal axis can show the construction progress of a building, i.e., it can be modeled which part of a building will appear at first and which one will appear later on. Therefore, the geometry of an object will not change in time, but only appear at a certain time step and may disappear later. Secondly, if several people develop a plan, there will be different versions of that plan, but it is unlikely to have one definite plan all the time. These different versions should be stored in the database in order to compare them with each other and re-use them later on, if needed. By using two qualities of time, *valid-time* and *transaction time*, we need to extend DB4GeO with a Bi-Temporal model (Worboys 1994). Experiences gained in handling the two qualities of time might be useful to integrate further time dimensions in DB4GeO later in the future. In order to perform efficient spatio-temporal queries, we also need to implement an efficient indexing technique. At first, we will extend the R*-Tree (Beckmann et al. 1990), which also is used in DB4GeO, to a Spatio-Temporal R-Tree (Saltenis and Jensen 1999). After that, we will concentrate on suitable and efficient query techniques (Snodgrass 1995).

Acknowledgments We thank Thomas Kolbe from the Technical University of Berlin, the City of Berlin and Ordnance Survey of Great Britain for the CityGML data sets. Furthermore, we are grateful to Simon Schuffert from Karlsruhe Institute of Technology for providing the 3D model of the figure "The Thinker". This research has been funded by the German Research Foundation (DFG), grant no. BR2128/12-1 and BR2128/14-1.

References

Balovnev O, Bode T, Breunig M, Cremers AB, Möller W, Pogodaev G, Shumilov S, Siebeck J, Siehl A, Thomsen A (2004) The story of the GeoToolKit–an object-oriented geodatabase kernel system. GeoInformatica 8(1):5–47 (Kluwer Academic Publishers, Hingham)

Bär W (2007) Management of geoscientific 3D data in mobile database management systems. In German. PhD thesis, University of Osnabrück, Germany

Beckmann N, Kriegel HP, Schneider R, Seeger B (1990) The r*-tree: an efficient and robust access method for points an rectangles. In: Proceedings of the ACM SIGMOD international conference on management of data (SIGMOD'90), pp 322–331

Breunig M (2001) On the way to component-based 3D/4D geoinformation systems. Springer, New York

Breunig M, Zlatanova S (2011) 3D geo-database research: retrospective and future directions. Comput Geosci 37(7):791–803. doi:10.1016/j.cageo.2010.04.016

Breunig M, Schilberg B, Thomsen A, Kuper PV, Jahn M, Butwilowski E (2010) DB4GeO, a 3D/4D geodatabase and its application for the analysis of landslides. Lecture notes in geoinformation and cartography for risk and crisis management, pp 83–102

Breunig M, Rank E, Schilcher M, Borrmann A, Hinz S, Mundani RP, Ji Y, Menninghaus M, Donaubauer A, Steuer H, Vögtle T (2011) Towards computer-aided collaborative subway track planning in multi-scale 3D city and building models. In: Proceedings of the 6th 3D geoinfo conference, p 17.

Brisson E (1989) Representing geometric structures in d dimensions: topology and order. In: Proceedings of the 5th ACM symposium on computational geometry, ACM Press, Washington, pp 218–227

Eastman CM (1999) Building product models, 1st edn. CRCPress, Taylor& Francis group, Boca Raton, Florida

Fielding RT (2000) Architectural styles and the design of network-based software architectures. PhD thesis, University of California, Irvine

Fielding RT, Taylor RN (2002) Principled design of the modern web architecture. ACM Trans Int Technol 2(2):115–150

Fradin D, Meneveaux D, Lienhardt P (2005) Hierarchy of generalized maps for modeling and rendering complex indoor scenes. Signal Image Communication laboratory, CNRS, University of Poitiers (Tech. rep.)

Gocad Research Group (2012) http://www.gocad.org. Accessed 29 Feb 2012

Hashemi L, Mostafavi MA, Pouliot J, Therrien R (2009) Developing an adaptive topological tessellation for 3D modeling in geosciences. J Can Inst Geom Geomat (GEOIDE students special issue) 63(4):419–431

Khronos Group (2012) http://www.khronos.org/webgl/. Accessed 17 Jan 2012

Kolbe TH (2012) CityGML. http://www.citygml.org. Accessed 12 Jan 2012

Kolbe TH, Konig G, König G, Nagel C (2011) Advances in 3D geo-information sciences. In: Lecture notes in geoinformation and cartography, Springer

Krimmelbein A (2011) Topologie in CityGML (Topology in CityGML). Diploma thesis, Karlsruhe Institute of Technology, Germany

Kuper PV (2010) Development of 4D object management for the geo-database DB4GeO. Diploma thesis, University of Osnabrück, Germany (in German)

Lévy B (2000) Computational topology: combinatorics and embedding. PhD thesis, National Polytechnic Institute of Lorraine (in French)

Lévy B, Mallet JL (1999) Cellular modeling in arbitrary dimension using generalized maps. http://alice.loria.fr/publications/papers/1999/gmaps/gmaps.pdf. Accessed 14 Dec 2011

Lienhardt P (1989) Subdivisions of n-dimensional spaces and n-dimensional generalized maps. In: Proceedings of the fifth annual symposium on computational geometry, ACM Press, Washington, pp 228–236

Mallet J (2002) Geomodeling. Oxford Press, New York

ParaViewGeo (2012) http://sites.google.com/a/objectivity.ca/paraviewgeo/. Accessed 29 Feb 2012

Paterson J, Edlich S, Hörning H, Hörning R (2006) The definitive guide to db4o. Apress Series, APress

Polthier K, Rumpf M (1995) A concept for time-dependent processes. Visualization in Scientific, Computing, pp 137–153

Raper J (1989) Three dimensional applications in geographical information system. Taylor& Francis, London

Raza A, Kainz W (1999) Cell tuple based spatio-temporal data model: an object oriented approach. ACM-GIS, pp 20–25

Rolfs C (2005) Design and implementation of a data model for the management of 3D models in geoscientific applications. Diploma thesis, University of Osnabrück, Germany (In German)

Saltenis S, Jensen CS (1999) R-tree based indexing of general spatio-temporal data. Tech. rep, TimeCenter

Snodgrass RT (1995) The TSQL2 temporal query language, 1st edn. Kluwer Academic Publishers, Dordrecht

Strathoff F (1999) Memory-efficient management of time-dependent geometries for GeoToolKit. Diploma thesis, University of Bonn, Germany (In German)

Thomsen A, Breunig M, Butwilowski E (2008a) Towards a G-Map based tool for the modeling and management of topology in multiple representation databases. Photogrammetrie, Fernerkundung. Geoinformation (J Photogram Rem Sens Geoinf Process) 3:175–186

Thomsen A, Breunig M, Butwilowski E, Broscheit B (2008b) Modelling and managing topology in 3D geoinformation systems. In: Advances in 3D Geoinformation Systems. Springer, Heidelberg, pp 229–246

Three.js (2012) http://github.com/mrdoob/three.js/. Accessed 17 Jan 2012

Versant Corp (2012) Db4o. http://www.db4o.com. Accessed 12 Jan 2012

Worboys MF (1994) A unified model for spatial and temporal information. Comput J 37(1):26–34

Glob3 Mobile: An Open Source Framework for Designing Virtual Globes on iOS and Android Mobile Devices

Agustín Trujillo, Jose Pablo Suárez, Manuel de la Calle, Diego Gómez, Alfonso Pedriza and José Miguel Santana

Abstract The widely development of mobile devices is contributing to a high demand in 3D graphics, as they have also become a very important requirement of modern applications. Virtual Globes integrating environmental data at any time or place, remain a challenge within the technical constraints imposed by mobile devices. We present Glob3 Mobile, an open source framework for the development of virtual globes on familiar iOS and Android mobile devices. The paper discusses the design and development choices for each platform. The aim of this work is twofold. First, to provide an efficient Virtual Globe application, testable and freely accessible from the web and providing a truly 3D navigation experience with smooth flying. Second, to provide the main software components to easily design and implement 3D Virtual Globes based applications, on both iOS and Android platforms.

Keywords Virtual globe · Smartphone application · iOS · Android

A. Trujillo · J. M. Santana
Imaging Technology Center (CTIM), University of Las Palmas de Gran Canaria, Canary Islands, Spain
e-mail: atrujillo@dis.ulpgc.es

J. P. Suárez (✉)
Division of Mathematics, Graphics and Computation (MAGiC), IUMA, Information and Communication Systems. University of Las Palmas de Gran Canaria, Canary Islands, Spain
e-mail: jsuarez@dcegi.ulpgc.es

M. d. la Calle · D. Gómez
Departamento de I+D, IGO SOFTWARE, Santiago Caldera 4, 10004 Cáceres, Spain
e-mail: mdelacalle@igosoftware.es

A. Pedriza
COTESA, Área de Sistemas de Información, C/Luis Proust, 17. Boecillo, 47151 Valladolid, Spain
e-mail: alfonsopedriza@grupotecopy.es

J. Pouliot et al. (eds), *Progress and New Trends in 3D Geoinformation Sciences*,
Lecture Notes in Geoinformation and Cartography, DOI: 10.1007/978-3-642-29793-9_12,

1 Introduction

Today, the popularity of smart devices, such as phones and tablets, and mobile networks has considerably changed the way people access computers and remote services.

A Virtual Globe is a geographic information system that provides graphical access to huge amount of imagery and elevation models of the Earth. Virtual Globes on smart devices come to make it free and easy 3D Earth exploration and mobility. They enhance human capacity by providing geo-location facilities, collaboration and ultimately decision support.

Over last years, smart devices have made important advances. Increasing processor speed, emerging platforms with dual processors, improved memory resources, power graphics and more than acceptable programming capabilities have opened these platforms to the development of general purpose applications. Of increasing interest on mobile application are Virtual Globes, as they play a central role in GIS applications, map browsers etc. The development of Virtual Globes began in 2001 with companies that pursued to effectively communicate their research and results to a broad audience worldwide. It was Google Earth project that brought Virtual Globes in 2004 to world-wide attention. At the same time, NASA World Wind, Bell et al. (2007) also emerged with a power open source implementation and versatile possibilities. It is written in Java, so it can be run as an applet, or embedded in a web page. World Wind still does not work on mobile devices, and moreover, in Apple devices it is not expected, since there is no Java interpreter on these devices. World Wind can also be expanded to include additional imagery and data whereas Google Earth is very limited here, as it uses commercial satellite imagery. If we look for truly open source options to develop Virtual Globes, Google Earth is not surely the best option. World Wind, however is open source and this has led to a proliferation of add-ons and plugins which are enriching and increasing the power of Virtual Globes.

At recent times, many other appreciated Virtual Globes engines can be found on desktop computers. WebGLEarth is also a valuable option, a free software project that focus on web browsers, using JavaScript and WebGL technology. It shows neither dependency on native languages nor computer platforms. It run on any device with a standard HTML5 compliant browser. A major drawback is that it still does not work properly on mobile devices. And when it does, the obtained performance is still very poor, maybe caused because it is a very recent project. A different alternative deserving attention is osgEarth, a C++ terrain rendering toolkit that uses OpenSceneGraph. Although osgEarth does not consider mobile devices yet, OpenSceneGraph has been released in 2011 to be used with handheld devices, supporting OpenGL-ES enabling both iOS and Android platforms. An increasing interest has also arisen to the research community. A recent work of terrain navigation in mobile devices, although in prototype stage, has been presented in Noguera et al. (2011), which handles efficiently the computational resources and network bandwidth. Several World Wind based frameworks for desktop computers with powerful capabilities have emerged in the last years, for example, iGlobe, Chandola et al. (2011), and Glob3.[1] Glob3 is an open source 3D GIS multiplatform framework and it is the precursor of the Glob3 Mobile presented in this work.

1.1 iOS and Android Platforms

Android is an open source software project and operating system developed by Google for mobile devices.[2] It is build on the open Linux Kernel and allows developers to write programming code in Java language. However, it does not support the standard Java JME libraries and then the specific Google-developed Java libraries must be used instead. JME was designed some years ago for small handsets, whereas the Google Java libraries have been developed with the aim of running in more powerful modern smartphones. In terms of graphics capabilities, Android currently supports OpenGL-ES 2.0 specification. Lastly, Android has been extended to other devices, not only smarphones, as for example to notebooks and ebooks, providing so to be a reliable and efficient open source operative system.

On the other hand, iPhone is a smartphone designed by Apple Inc. It is powered by the iPhone Operating System (iPhone OS) which is based on a variant of the same basic kernel that is found in Mac OS X.[3] Objective-C together with Cocoa API, Davidson (2002) is the development framework provided by Apple on iPhone OS. This language is a superset of C, which includes both syntactic and semantic features to support object-oriented programming. iPhone OS also provides the OpenGL-ES framework which conforms to the OpenGL-ES v2.0 specification.

Contributing with a third party application onto the iOS is only possible after paying a membership fee, whereas in Android is free. Today, a surprisingly amount of applications are being developed for mobile devices, free or paid they are served to the user in an easy and fast way.

1.2 3D Graphics Overview for Mobile Devices

Computer graphics are rapidly advancing in mobile devices as a consequence of improvement of mobile operating systems, i.e. iOS and Android. However, there are still some technical limitations in mobile devices that make it difficult to translate algorithms and programming strategies from desktop computers as they are, Xiao et al. (2010). In general handheld devices lack of the CPU power and memory capabilities to manage large data as in terrain representation. Fortunately,

[1] Glob3: An open source 3D GIS multiplatform framework (2011) http://glob3.sourceforge.net

[2] Android Mobile Operating System (2012) http://www.android.com

[3] iOS Mobile Operating System, (2012) http://www.apple.com/iphone/ios

programming capabilities for mobile devices are improving considerably. Classical programming paradigms as Java and C can be used, together with specific libraries as for 3D Graphics as OpenGL-ES, Khronos Group (2004) and M3G, Pulli et al. (2007). OpenGL-ES (Embedded Systems) is a well-defined subset of the familiar OpenGL 3D graphics API. It is specially designed for embedded devices including mobile phones, video game consoles and other handheld devices. OpenGL-ES provides a flexible and powerful low-level interface that facilitates the software development through graphics acceleration. For a review of OpenGL-ES and other graphics library options for mobile devices see Pulli et al. (2005).

A salient advance in the hardware of handheld devices is the inclusion of specific 3D graphics hardware. This feature is being a differentiating factor for the fabricants of mobile devices. This permits GPU-based solutions with a increasing expected performance of applications.

With the lack of one unique proposal for implementing Virtual Globes on the most familiar mobile devices iOS and Android, in this paper we present Glob3 Mobile, an open source solution for Virtual Globes, enabling a true 3D navigation experience and scalable possibly with enriched extensions as plugins. Moreover, user may benefit of a viable methodology enabling to reproduce own Virtual Globe enjoying with user specific features over it. Instantly access to Glob3 Mobile can be done at http://ami.dis.ulpgc.es/glob3m. Figure 1 shows Glob3 Mobile running in Apple iPhone and Samsung Tab Android.

The paper is organized as follows. Section 2 introduces the software aspects of the Virtual Globe, focusing on the architecture, object classes and details of the implementation. In Sect. 3 we define the policy to represent the Earth globe in handheld devices, and outline de LOD strategy. Section 4 is devoted to 3D view interaction with multitouch devices, and gives details of the implemented gestures to interact with the globe. Accessing and handling of images are described in Sect. 5, where we focus on accessing to imagery data from public and WMS related servers, and explaining texture handling in mobile GPU. In Sect. 6 we give final conclusions and outline future work.

2 Globe3 Mobile Framework Architecture

Nowadays, with the fragmented and rapidly emerging mobile platforms, the development of applications easily portable to different platforms is a must. With this aim, we have developed a multilayer object oriented architecture.

Figure 2 shows the software layers model. Lowest two layers are platform dependent, whereas higher layers are designed to be decoupled from specific platforms. Glob3 Mobile framework, at the user view level, is located at the uppermost layer. With the aim to provide extra capabilities to the globe, a new layer is identified where developers may find a way to program user-specific tasks as Plugin Extensions.

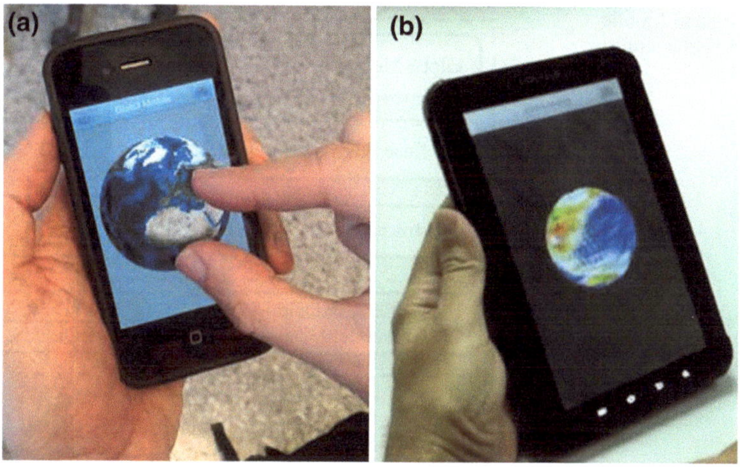

Fig. 1 Glob3 mobile running in **a** Apple iPhone and **b** Samsung Tab Android

The main decision regarding the software architecture is the separation of cross-platform independent programming from the issues close related to iOS and Android. A base engine is constructed that permits abstracting the virtual classes for file management, network petitions, image processing and event handling. Then such virtual classes are implemented on each platform using its own native language.

The Android dependent classes have been written in Android Java, which is slightly different to standard Java. The same classes have been implemented in Objective-C and linked to the kernel to generate the Apple version of the globe. Besides, the kernel makes use of a virtual graphic class, which is implemented using OpenGL-ES technology, following two different versions: 1.1 for older devices, and 2.0 for newer ones.

The methodology followed to develop the framework can be organized in three steps:

1. Develop a full engine in C++ that serves as base of the API architecture and supports the core functionality for the 3D graphic visualization. By using virtual classes, iOS and Android implementations are facilitated afterwards.
2. An iOS version of the Virtual Globe is built, using the base engine and specific iOS implementation using Objective-C.
3. An Android version is finally built, based upon the C++ to Java straightforward conversion of the base engine together with the new classes developed in Java.

The C++ kernel used for iOS is converted automatically with a Java converter to generate a Java kernel with exactly the same functionality. It uses the same OpenGL classes than in the iOS version, but using Java instead of C++.

Figure 3 shows the main components of the engine, organized in four different blocks. *Layers* block deals with WMS protocol to connect to servers and ask for

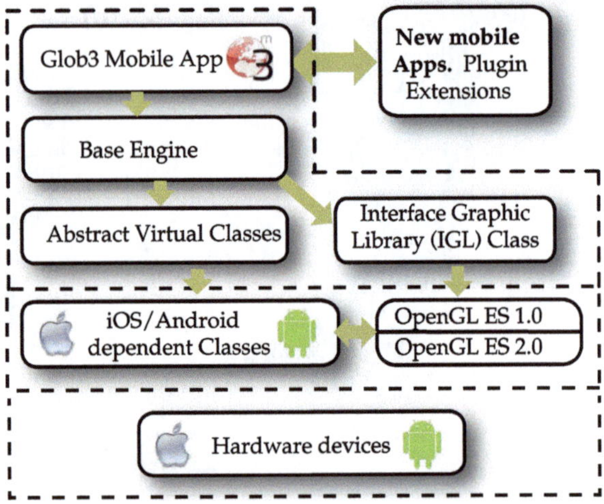

Fig. 2 Glob3 Mobile software architecture

textures. *LUA* block (probably will be substituted in the future versions by JavaScript) will enable users to write plugins applications by means of LUA[4] script language. By means of plugins, developers could add features to Glob3 Mobile without changing the program's source code, so expanding the possibilities of our Virtual Globe. Plugins are small programs written in a scripting language which could be loaded and compiled by Glob3 Mobile at startup. LUA is chosen as it is an embeddable, fast and simple scripting language. Separately, *TinyXML*[5] block, a third party library, is devoted to read and write XML files needed when connecting to WMS servers.

Engine is the bigger block, and it is organized in two separate modules, *Core* and *Renderers*. *Core* controls the whole scene currently displayed by handling user interaction with screen. Virtual classes for file handling, network petitions and events are defined within Interfaces. Precise camera movement and scene visualization are elaborated in the View submodule. Globe information as triangle meshes, vertices, ellipsoidal data and LOD, together with geometric implementations of the navigation are also grouped in submodule Planet. The rendering process is one of the main task of the Virtual Globe and so programming objects are encapsulated separately in several modules named *Renderers*. The main components to be rendered in the globe are the tiles that compose the earth surface (*TileRenderer* object). However, enriched scenes of the Virtual Globe may include the rendering of atmosphere, stars, 3D objects on the terrain, etc. and so these objets are programmed in different *Renderer* submodules. *RendererList* contains all the objects that must be rendered in the scene. Finally, the *GLU* module

[4] www.lua.org

[5] www.grinninglizard.com/tinyxml/

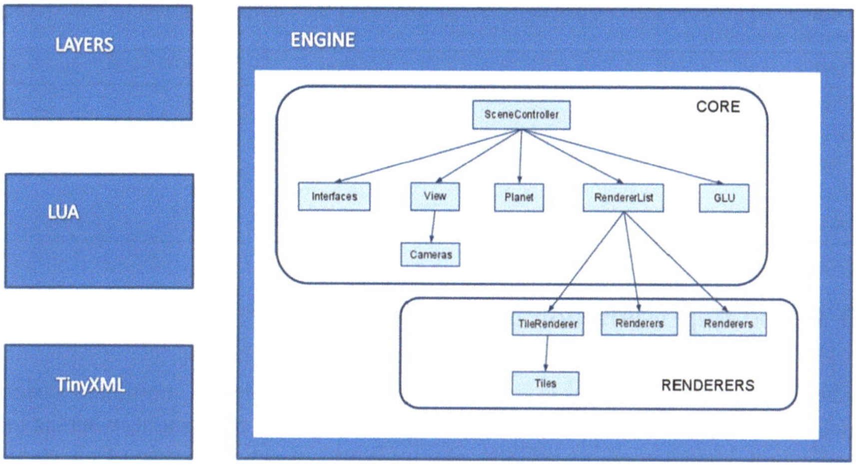

Fig. 3 Framework architecture for the virtual globe

includes some geometric and mathematical utilities such as projection and unprojection computation, matrix and vector operations, etc.

2.1 *Adding Functionality by a Plugin Mechanism*

An user API or plugin mechanism is under development on Glob3 Mobile. As an open source project, users can access to provided engine functions which permit adding WMS layers, moving the camera, and also drawing geometry on the globe. This can be viewed as a scripting capability to personalize the appearance of the virtual globe. For example, users may perform a camera movement by entering following high level command:

```
globe.camera.position=Position.create(lat,lon,height);
```

Ideally, this API permits writing code once and then run such code in all smartphones platforms. A preliminary version of the plugin mechanism can be tested in the Glob3 Mobile page project.[6]

[6] http://ami.dis.ulpgc.es/glob3m/

Table 1 iOS and Android mobile devices resolution and values adopted for number n of vertices ($n \times n$) per mesh

Device	Resolution	n
iPhone 3	480 × 320	8
Galaxy S-SII	800 × 400	10
iPhone 4	960 × 640	10
iPad-iPad2	1024 × 768	12
Galaxy Tab 7″–10.1″	1280 × 800	12

3 Globe Representation and LOD Strategy

The Earth model widely used is the ellipsoid. The geographic coordinate system defines each position on the globe by longitude, latitude and height, whereas the projection system used is WGS84. As in World Wind application Bell et al. (2007), due to restrictions of NASA World Wind server (explained more in detail in Sect. 5), we initially construct the ellipsoid with a tessellation of 10×5 patches (tiles), each one obtained each $36 \times 36°$. It should be noted that many terrain visualization tools, specially those dealing with very large elevation data uses adaptive triangle meshes to model the surface, for example Pajarola (1998). As in Losasso et al. (2004) we adopt a regular mesh for representing the earth surface. We generate a regular triangle mesh per patch of $n \times n$ vertices, where n depends upon the screen resolution of the device. See Table 1 for a list of some familiar mobiles devices, screen resolution and our choice for n values.

Using a LOD strategy is crucial to avoid bottle-necks in terrain navigation. In the last decade, several strategies have appeared to cope with this problem, see a review in Cozzi and Ring (2011). For example, Geometry Clipmapping, Losasso et al. (2004) renders terrain data as a mipmapped height map. We use a Chunk LOD system, Cline et al. (2001) that incrementally renders the surface of the globe. We break the terrain into a quadtree of tiles, named *chunks*. The root of that quadtree is a low-detail representation of the globe and the successive child chunks are new divisions of the globe into four-equal-sized areas that provide higher-detail of the terrain. A geometric error dictates which portion in the quadtree is displayed at the screen scene.

When the observer is located close to ground, the number of visible tiles could be very high (LOD value is close to 18). We must use a LOD strategy that keeps the maximum number of visible tiles under a limited value.

Many of the LOD methods use a threshold that depends on the projected area of each tile on the screen. This works properly when camera view direction is normal to the globe surface, but when this is not the case, all the tiles located behind the camera, have a very low LOD level. In a handheld device user can quickly rotate the camera with the fingers to see what is behind. This produces several frames with a very low detail of that part of the scene, until all the correct LOD levels are uploaded in GPU. Other methods use the distance from the tile to the camera, but again, when view direction is not normal to the surface, we have a lot of tiles located just below the camera with a maximum level of detail when it is not necessary.

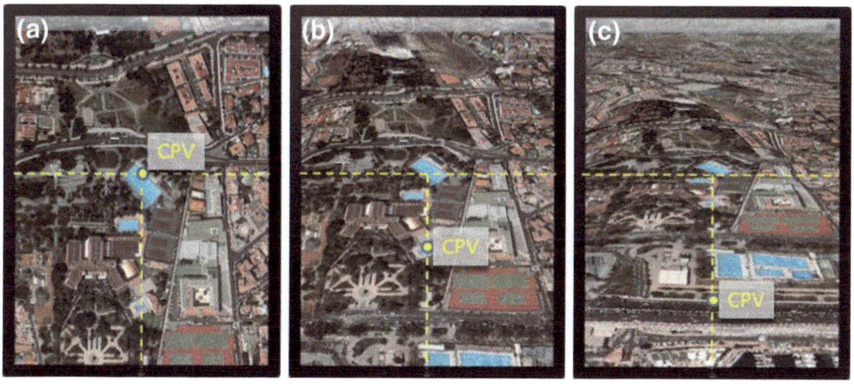

Fig. 4 The location of CPV depends on the camera tilt. **a** View direction normal to the surface. **b** 30°. **c** 60°

To verify if the current LOD level of the tile is correct or not we propose a different test that is computed in every frame for each visible tile. The test dictates if current LOD should be changed to a more detailed level, subdividing it into four children, or to a less detailed level grouping it with its three brothers.

The first step of the test is computing, at each frame, the estimated central point of user view, CPV. Usually, when the camera view direction is normal to the surface, CPV is located on the center of the screen, but when this is not the case, this point is moved towards the bottom of the screen, as seen in Fig. 4. A ray is casted from the eye to CPV point, and the intersection with the globe is obtained. This surface point should be the one with more detail on the screen.

After obtaining the searched point, we compute for every tile T the geodesic distance d_T from this point to the center of the tile. The criteria for deciding whether a tile T should be divided or not is that the relationship between this geodesic distance and the current width of the tile, w_T be lower than a threshold ε. This value, that is not constant, is given by:

$$\varepsilon(T) = \varepsilon_{\max} + (\varepsilon_{\min} - \varepsilon_{\max}) \frac{w_T - D/4}{\pi R - D/4} \tag{1}$$

where w_T is the current width of the tile, D is the distance from the eye to CPV, and R is the radius of the globe. After several experimental tests, chosen values for $\varepsilon_{\max}$ and $\varepsilon_{\min}$ have been delimited to 1.1 and 0.9 respectively.

Then, a tile is subdivided if the following two conditions are true:

$$\begin{gathered} \frac{d_T}{w_T} < \varepsilon \\ w_T > D/4 \end{gathered} \tag{2}$$

where Eq. 2 is forcing the highest LOD in the scene (for the tile including CPV). Two different situations of sample tiles are seen in Fig. 5.

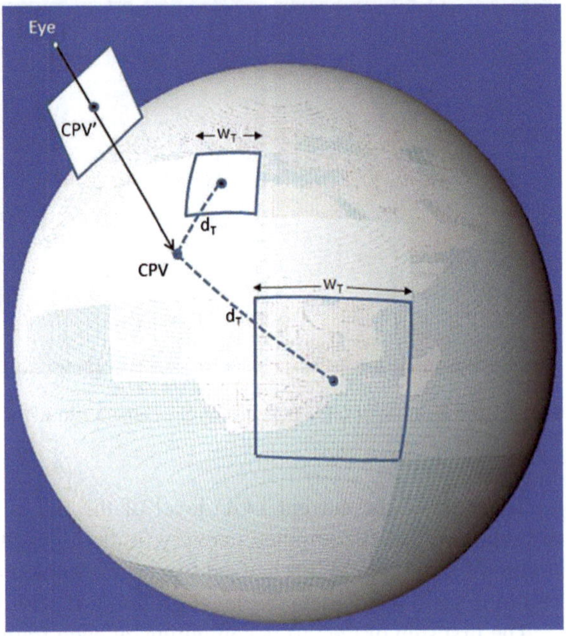

Fig. 5 The CPV point on the globe indicates the terrain point with highest detail. The criteria to decide if a visible tile T must be subdivided or not depends on its width an the geodesic distance from its center to CPV

By zooming in toward the globe, the distance from the eye to the globe is decreased, and new child partitions come into the new scene with higher detailed resolution. In a reverse manner, when zooming out, the engine will capture previous lower-detail portions of the globe. A common problem of this technique is to preserve coherence between adjacent tiles, as the two chunks may not have the same number of supported terrain vertices. We adopt here a simple but efficient solution, Ulrich (2002) that fills gaps between chunks using *skirts*, a triangle strip covering the four perimeters of the chunk. This triangle strip is then mapped together with the wrapping texture of the chunk, as showed in Fig. 6.

This skirt solution is a very fast technique, because it only depends on each tile, and not on its neighbors. On the other hand, the mesh is not readapted which may induce a very high consuming time.

Such skirts cover the holes formed by these gaps avoiding undesirable visual artifacts, as illustrated in Fig. 7.

4 3D View Interaction in the Multitouch Mobile System

Realistic navigation and smooth flying over the Virtual Globe is high demanded. For this reason, a careful programming of the user interaction with 3D ellipsoid is required, implying a very precise control of the geometry involved in globe movements.

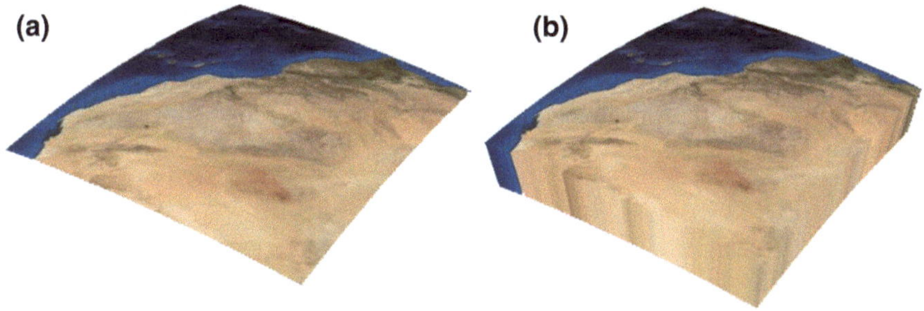

Fig. 6 A terrain chunk **a** without skirts and **b** with skirts

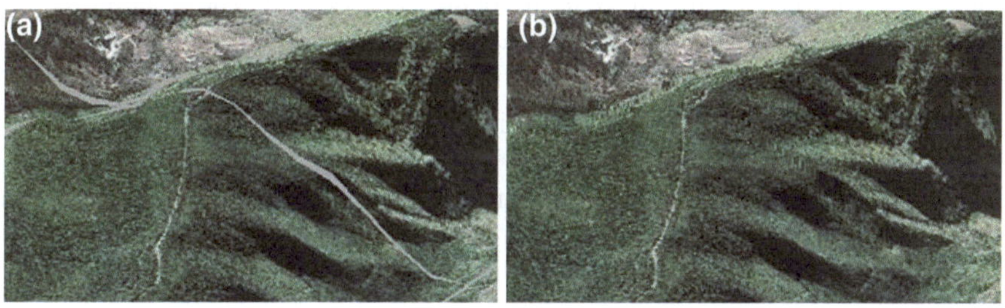

Fig. 7 Quality of earth visualization. **a** Without using skirts. **b** Using skirts

It should be noted that the devices we are dealing with use finger gestures for the interaction with the touchscreen. A prominent task in the programming of our globe is to detect finger gestures that users make on the screen and then accordingly modify the globe. A reference guide to touchscreen gestures can be found at Villamor et al. (2010).

To help applications detect gestures, iOS introduces gesture recognizers. However, Android systems lack of gesture event handling similar to iOS that facilitate the gesture detection. Then, to normalize the programming on both platforms we developed our own control of gesture events, using only two low level event gestures: *fingerdown* and *fingermove*. We describe next the types of gestures introduced in Glob3 Mobile. Readers may test the performance and quality of such gestures by downloading the free application at the page project online.

4.1 Tapping

To clearly describe the geometric meaning of the gestures, we show in Fig. 8 detailed explanations of each gesture type. The common notation in this figure is: c is the point that represents the observer eye. When user makes a quick up-and-down touch on the screen with one finger, point P'_0 is defined. If the ray starting from c and passing through P'_0 is projected back directly to the earth globe, we have point P_0.

Normally, tapping is the first event before other gestures, like dragging with one or more fingers on the screen. The first step is always finding the unprojected point on the globe surface. This option is also useful for future functionalities, as getting information about that location (height, geographic coordinates), or picking up some object on the terrain.

4.2 Panning

The user moves fingertip over the screen without losing contact. Globe is continued rotated whereas the finger is in contact, see Fig. 8b. The rotation axis is given by the cross product of vectors $\overrightarrow{OP_k}$ and $\overrightarrow{OP_0}$ whereas the rotation angle is obtained from the dot product.

If the finger is released while dragging on the screen, a small animation is executed during the following seconds, continuing the rotation with decreasing angle until the globe is stopped, or until the screen is touched again. This result in a pleasant and attractive visual effect.

4.3 Pinching

In this event, user touches surface with two fingers and brings them closer together or moves them apart. A zooming effect is attained by moving the camera along the view direction. The goal is displacing a magnitude making that those points resulted when the user fingers touched down on the screen, keep in contact approximately with the fingers. The idea is obtaining the similar effect that when stretching a 2D image. The distance to displace the camera is given by the length $\overline{O_0O_k}$, see Fig. 8c. This magnitude is calculated from the two rays given by c and passing through the points given respectively by fingers P'_0 and Q'_0. These rays are projected back onto the globe giving points P_0 and Q_0. Moreover, two other points P'_k and Q'_k are known while user keeps dragging the fingers. Then, the new camera position is estimated along its view direction, searching for the position that verifies that the projection of original points P_0 and Q_0 after the displacement are close to P'_k and Q'_k.

4.4 Double Tapping

If the user quickly touches the screen twice with one finger, similar to the mouse double clicking, the resulting action will be a combination of panning and pinching. The visual effect is a smooth animation of two seconds, where the terrain

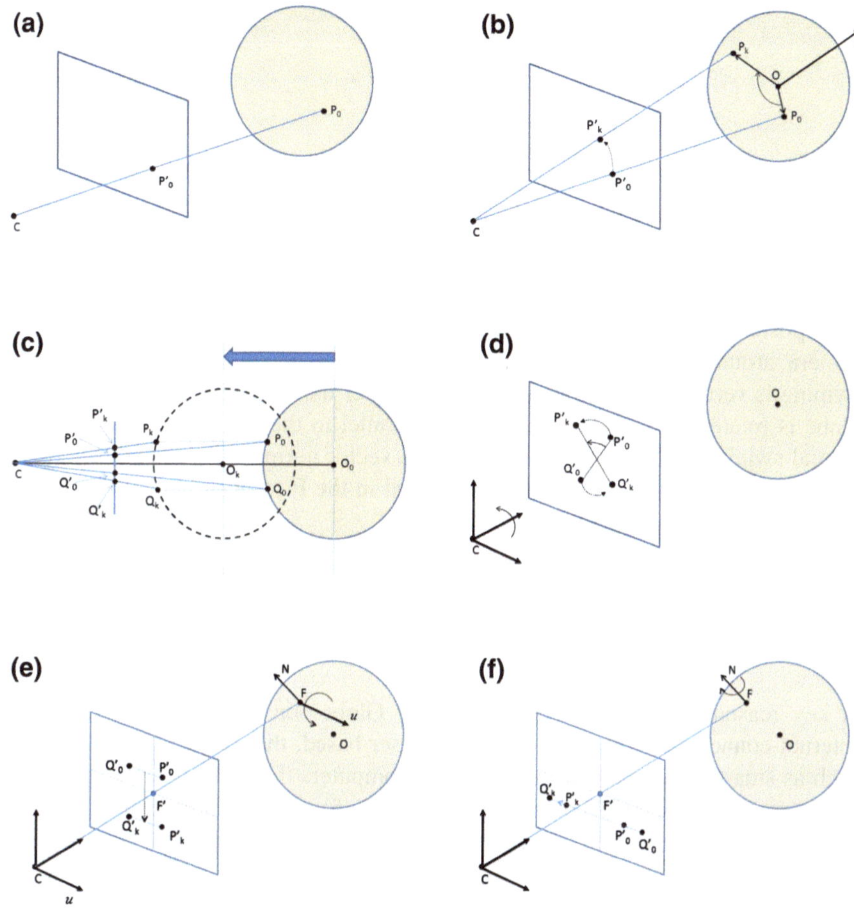

Fig. 8 Multitouch gestures implemented in Glob3 Mobile. **a** Tapping. **b** Panning. **c** Pinching. **d** Rotating. **e** Vertical swiping. **f** Horizontal swiping

point touched by the user moves toward the center of the screen, and simultaneously, the camera approaches a little bit.

4.5 Rotating

To rotate the globe, user touches the screen at two points, and then rotate them around themselves on the screen, as seen in Fig. 8d.

Starting and finishing points of the rotation define two rays and then it is straightforward to obtain the rotation angle by the formula:

$$\frac{|\overrightarrow{P'_0Q'_0} \cdot \overrightarrow{P'_kQ'_k}|}{|\overrightarrow{P'_0Q'_0}| \cdot |\overrightarrow{P'_kQ'_k}|}$$

4.6 Swiping

This gesture is reproduced when user touches the screen with two fingers, giving points P'_0, Q'_0 and drag them parallel to each other without losing the contact to the final position P'_k, Q'_k. The desired globe movement is obtained by rotating the camera around the globe with respect to an axis which depends weather the swiping is vertical or horizontal. When the finger movement is on the vertical the globe is rotated through the horizon axis parallel to the screen, whereas the horizontal swiping implies rotation through the vector normal N to the surface earth, see Fig. 8f. Both rotation axes are positioned in the F point on the terrain, that is the unprojected point of the screen center.

5 Image Access and Handling

A key reason for the advances of Virtual Globes on handheld devices is the Internet-connected applications, often browser-based, that run on mobile devices such as smartphones, tablets or embedded computers that have wireless access to the Internet. Through GPS or other positioning information gathered from internet servers, mobile devices can report their position to applications that deliver location services. Next sections deal with remote access to textures and height maps in the Virtual Globe.

5.1 Access to Web Map Services

The Open Geospatial Consortium (OGC), an international voluntary consensus standards organization, works to enable geographic information more usable and useful in GIS related applications. For example users can access the enormous amount of available internet data layers by means of the Web Map Service (WMS), as it provides map images from almost everywhere on Earth.

Glob3 Mobile supports access to remote data repositories from any public WMS server. The interface available in Glob3 Mobile enables a list of proved WMS servers and also the user may provide the URL of the web service to use, see Fig. 9. By accessing to this menu, we can select a WMS server, or just add a new one.

Working simultaneously with the LOD algorithm, every time a tile must be subdivided in four new children tiles, four new petitions are submitted to the WMS

Fig. 9 User dialog to choose the WMS texture servers on Glob3 Mobile

server. To detect which server to use for each petition, the list is visited sequentially, until a server is reached that includes completely the bounding box of the tile. By using the WMS selection dialog, users may insert at first server positions, such URLs of web services corresponding to imagery data more close to the ground surface, while other global terrain imagery is recommended to be added at last positions. The last server in the list is always the NASA World Wind server, because that server covers the whole world with a good resolution.

The NASA World Wind server actually consist of two different servers. The first one is the Blue Marble WMS Global Mosaic, that includes the first levels of detail for images of the whole globe (from LOD 0 to LOD 3). The second one is Virtual Earth Tile Server, that is used for higher detailed images. This server is the backend for Microsoft Map Web services. It is not a standard WMS server, but works with fixed location images of predefined size. The first level of this server if composed of a mosaic of 160×80 images covering areas of $2.25 \times 2.25°$. For this reason, our globe starts with an initial mosaic of 10×5 tiles in the LOD 0, in order to match the resolution of the Virtual Earth tiles.

To avoid that the application stop while a network petition is not completely downloaded, all the petitions are made in an asynchronous way using multi-threading. In this way, several network petitions could be sent at the same time, while the user is navigating through the globe. Obviously, each tile won't be subdivided into its four children until all of their textures have arrived. Experiments have validated this technique for several bandwidth connections, leading so a good navigation experience.

Some WMS servers do not allow to send many petitions simultaneously. For this reason, network petitions are not made in every frame, and in every case, only tiles with highest priority, depending on the distance to the observer, are sent to the server. Moreover, no more than 50 tiles are allowed to be sent simultaneously.

All the images are saved to a disk cache after downloading, using a very simple caching strategy. Thus if a tile is removed from the quadtree, because the observer has navigated to a different place, when coming back to same place, the texture is first searched in the cache before sending the request to the WMS server. So simply, this caching mechanism has proven to be efficient, allowing a faster visualization for those areas previously visited.

5.2 Dealing with Textures in Mobile GPU

Some limitations of graphic hardware in mobile devices avoid using usual compressed image formats as PNG or JPG, requiring instead own compressed format that is hardware dependent. For example, iPhone only supports a specific type of compression called PVRTC, supported by PowerVR chip, the iPhone's graphic processor, see Rideout (2010). The problem is that a desktop computer is needed to compress a JPG image to this format, because there is no available source code to make the compression task within the iPhone hardware. The iPhone SDK only comes with a command-line program that must be used to generate PVRTC data. In this way, image compression can be done firstly, and then attached to the iPhone project.

Obviously, the solution of such command-line compression is not efficient at all. The WMS server only responds with well-known formats, as JPG and PNG, and we would be forced to compress images on the fly. For this reason, we read the pixel data with its original uncompressed 24 bit color values, and convert them to a 16 bit representation. This simple image compression has proven to be successful in the of tested devices.

We are using 256×256 textures for each tile, with 16 bits/pixels, that must be stored in GPU memory (130 Kb approximately). After several tests in different smartphones, a number of 300 textures in GPU memory is proven to be acceptable for most of them. We have included this parameter in the LOD strategy, in order to keep the maximum number of visible tiles close to this value.

Transparency or alpha channels are used frequently and then are supported in our framework. Some WMS textures include this channel to indicate that some areas on the maps must not be displayed, see Fig. 10. In these cases, two textures are required for each tile. In a desktop computer, this could be done using multitexture in the GPU, but the problem is that we need double memory requirements

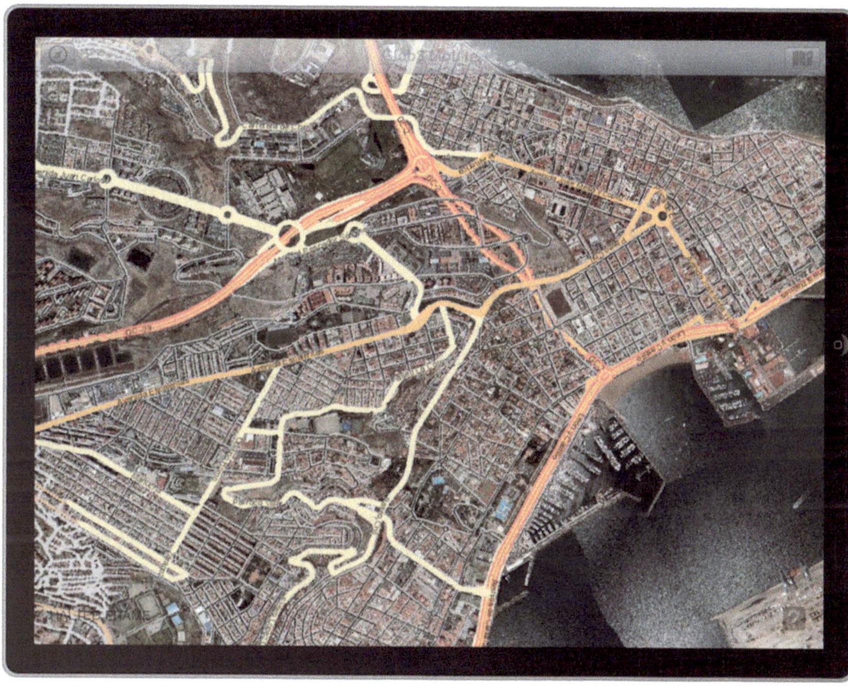

Fig. 10 Displaying a transparent texture (street map of a city) merged with an aerial photo

in the GPU. For this reason, the merging of textures is done at CPU, to maintain the maximum value of 300 textures in GPU memory.

5.3 Including Elevation Data

Elevation data is obtained using another WMS server from NASA World Wind. This server uses a specific format (BIL) to return elevation data. Every pixel in the resulting image (named height map) is a 16 bit signed value, that indicates the height in meters of the corresponding point. Then, each tile is composed by a triangle mesh of $n \times n$ vertices, and a BIL image of $n \times n$ pixels. With this image, each vertex in the mesh is translated in the normal direction to the globe surface as many meters as indicated in the height map.

After several experimental results, we have seen that the precision of height maps is not very accurate. To obtain more precision in the elevation values close to the tile borders, it is better to send petitions that include a wider area. For this reason, if the tile has a mesh resolution of $n \times n$, we ask for a height map with a resolution of $3n \times 3n$, that includes an area three times larger (in both dimensions). In this manner, the obtained accuracy in the $n \times n$ inner area is more precise.

6 Discussion, Future Work and Conclusions

In this work we contribute with the development of an open source Virtual Globe that can be freely testable from the Glob3 Mobile page project.[7] We describe here the main software components to easily design and implement a Virtual Globe and also user-specific application over it, on both iOS and Android platforms.

Our server-client approach provides a highly scalable architecture, capable of dynamically balancing the workload of the textures and height fields. Careful attention has been paid to the 3D user interaction in multitouch iOS and Android platforms. We have tested the application in a dozen of different mobile devices, leading to analogous (real-time) performance for an average of internet bandwidth wifi connection between 10 and 100 kb/s. Another interesting feature of our Virtual Globe, is that it could work without internet access, if the geographical data (textures and elevation data) of those areas to be visited have been stored previously in the cache.

A major challenge in this project is extending the globe to the emerging technology WebGL, executing the embedded 3D application in an HTML page. Some nearest extensions to the globe which are coming up very soon are:

- Display buildings and other 3d models on the globe.
- Display photos, panoramic pictures and videos on the terrain.
- Include gateways to open Internet API's such as Climate, Geonames, Twitter, Facebook.

Some other long-term promising implementation goals include:

- Display large clouds of points or other geometry (vectors, polylines) using streaming.
- Including Streaming video.
- Access to a scene server.
- Display different graphics geographic formats (KML, WFS, GML).
- Work with a distributed model view controller client/server architecture.
- Integration with Euclid 3D geometry library, developed within the project.

Acknowledgments This work has been supported in part by Spanish firms IGO software and COTESA, by CYCIT Project MTM2008-05866-C03-02/MTM from Ministerio de Educación y Ciencia of Spain and by AECID Project A/030194/10 of Ministerio de Asuntos Exteriores y de cooperación of Spain.

[7] http://ami.dis.ulpgc.es/glob3m/

References

Bell DG, Kuehnel F, Maxwell C, Kim R, Kasraie K, Gaskins T, Hogan P, Coughlan J (2007) NASA world wind: opensource GIS for mission operations. In: Proceedings of the 2007 IEEE aerospace conference, vol 3, issue 10, pp 1–9

Cline D, Parris KE (2001) Terrain decimation through quadtree morphing. IEEE Trans Visual Comput Graph 7(1):62–69

Cozzi P, Ring K (2011) 3D engine design for virtual globes. CRC Press

Chandola V, Vatsavai RR, Bhaduri BL, (2011) iGlobe: an interactive visualization and analysis framework for geospatial data. In: Proceedings of the 2nd international conference and exhibition on computing for geospatial research& application, COM.Geo, Washington, DC. ACM International Conference Proceeding Series. doi:10.1145/1999320.1999341

Davidson AJ (2002) Learning cocoa with objective C. O'Reilly& Associates

Khronos Group (2004) OpenGL ES: the standard for embedded accelerated 3D graphics. http://www.khronos.org/

Losasso F, Hoppe H (2004) Geometry clipmaps: terrain rendering using nested regular grids. ACM Trans Graph (SIGGRAPH) 23(3)

Noguera JM, Segura RJ, Ogáyar Joan-Arinyo R (2011) Navigating large terrains using commodity mobile devices. Comput Geosci 37(9):1218–1233

Pajarola R (1998) Large scale terrain visualization using the restricted quadtree triangulation. In: Proceedings of the conference on visualization '98. IEEE Computer Society Press, Los Alamitos, CA, pp 19–26

Pulli K, Aarnio T, Miettinen V, Roimela K, Vaarala J (2007) Mobile 3D graphics with OpenGL ES and M3G. Morgan Kaufmann

Pulli K, Aarnio T, Roimela K, Vaarala J (2005) Designing graphics programming interfaces for mobile devices. IEEE Comput Graph Appl 25(6):66–75

Rideout P (2010) iPhone 3D programming: developing graphical applications with OpenGL ES. O'Reilly Media, Inc., USA

Ulrich T (2002) Rendering massive terrains using chunked level of detail control. In: Proceedings of SIGGRAPH2002. ACM Press

Villamor C, Willis D, Wroblewski L (2010) Touch gesture reference guide. http://www.lukew.com/ff/entry.asp?1071

Xiao J, Zhu M, Wang X, Wan W (2010) Analysis of mobile graphics pipeline with real-time performance. In: Proceedings of the international conference on audio language and image processing (ICALIP), pp 489–493

(α, δ)-Sleeves for Reconstruction of Rectilinear Building Facets

Marc van Kreveld, Thijs van Lankveld and Maarten de Rie

Abstract We introduce the concept of (α, δ)-sleeves as a variation on the well-known α-shapes. The concept is used to develop a simple algorithm for constructing a rectilinear polygon inside a plane; such an algorithm can be used to delineate a building facet inside a single plane in 3D from a set of points obtained from LiDAR scanning. We explain the algorithm, analyse different parameter settings on artificial data, and show some results on LiDAR data.

1 Introduction

In recent years, public interest in the use of virtual cityscapes has drastically increased. Applications in a variety of fields like navigation, urban planning, and serious games, increasingly use building models for visualization and simulation purposes. Simultaneously, the quality, complexity, and availability of urban datasets are increasing. Smart-phones are becoming ubiquitous, making photo and video images very easy to obtain, while modern LiDAR devices can capture hundreds of data points per square meter (John Chance Land Surveys and Fugro 2009).

This wide interest requires efficient automation of urban scene reconstruction to process the vast datasets. The general goal of urban reconstruction is recreating the geometry and visual likeness of the buildings in the scene. Whether applying structure from motion and dense stereo reconstruction to image data (Furukawa

M. van Kreveld · T. van Lankveld (✉) · M. de Rie
Department of Information and Computing Sciences, Utrecht University, Utrecht, The Netherlands
e-mail: T.vanLankveld@uu.nl

J. Pouliot et al. (eds), *Progress and New Trends in 3D Geoinformation Sciences*, Lecture Notes in Geoinformation and Cartography, DOI: 10.1007/978-3-642-29793-9_13,

et al. 2009; Seitz et al. 2006), or directly using LiDAR data, urban geometry reconstruction is usually aimed at identifying the shapes of buildings from a point cloud.

Earlier methods in photogrammetry would use a predefined collection of parametric models of complete buildings and either try to determine the model that best fits the data (Brenner 2005; Schwalbe et al. 2005), or only model roofs supported by vertical walls (Rottensteiner 2003; You et al. 2003; Zhou and Neumann 2008). While these methods are able to construct scenes from very sparse data sets, they are inherently limited by the versatility of the building models in their collection.

Other methods reconstruct free-form triangular meshes that interpolate the data points (Carlberg et al. 2009; Marton et al. 2009; Tseng et al. 2007). While these methods can reconstruct buildings of any shape, most have difficulty dealing with the artifacts inherent in the point data like measurement error and outliers. Additionally, most mesh-based reconstruction methods reconstruct smooth surfaces, removing the sharp edges and simple shapes widely present in urban scenes.

We present a method that is partially parametric and partially free-form to deal specifically with the shapes of urban scenes. While most parametric methods reconstruct the scene per building, we assume that these buildings are composed of planar surfaces and reconstruct these individual surfaces.

Most point clouds measured in an urban scene have inliers and outliers, points measured from a planar surface and the remaining points respectively. The data also contains noise, a small displacement in the point locations. We use Efficient RANSAC (Schnabel et al. 2007) to cluster the points per individual surface, although methods based on region growing (Tseng et al. 2007) could also be used.

Both dense stereo and LiDAR produce point sets densely covering the viewed surfaces. Our method estimates the shape that the points were measured from per individual surface. A surface can be reconstructed by computing a polygon that contains all its points while not containing large empty regions. A popular method for computing such a polygon is the α-shape (Edelsbrunner et al. 1983).

Another prevalent feature of the surfaces in urban data sets is rectilinearity, caused by the predominant use of right angles. In recent work (van Lankveld et al. 2011), we showed that roughly a third of the surfaces in many city scenes are rectangles and we presented a method for reconstructing these surfaces. Here, we broaden this to general rectilinear shapes that tightly bound a point set. To determine these shapes, we present a simple variation of the α-shape, called the (α, δ)-sleeve. This structure creates a buffer around the shape and we search for a rectilinear shape within this buffer. Figure 1 shows an overview of our method.

Finding a rectilinear shape that is close to a given shape is a problem that has been studied in different contexts. For example, restricted—orientation line simplification has been studied for the purpose of schematized map computation (Buchin et al. 2011; Swan et al. 2007; Wolff 2007). Another example is squarifying, an operation that occurs in ground plan generalization (Mayer 2005; Regnauld et al. 1999; Ruas 1999). In these cases, the starting point is a polygonal line or planar subdivision, whereas (initially) we start with a set of points. Furthermore,

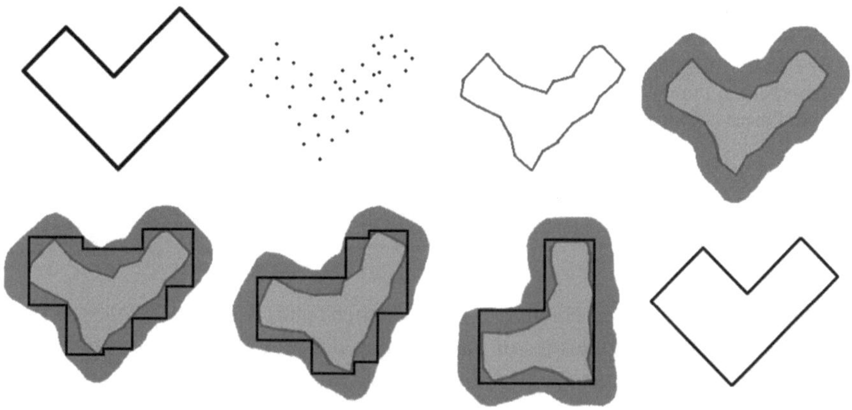

Fig. 1 An overview of our method. From *top left* to *bottom right*: the surface, a point sample of the surface, the α-shape, the (α, δ)-sleeve, *three* rectilinear minimum-link paths within the (α, δ)-sleeve for different rotations, and the path with the fewest links over all rotations

since sampled points are assumed to be inside the rectilinear polygon to be found, we wish to find an outer approximation of the boundary of the point set. Hence, an area-preserving method like in (Buchin et al. 2011) does not appear suitable. Our method guarantees that the rectilinear polygon is within a specified distance of the α-shape but still outside it; other methods do not have this feature.

2 (α, δ)-Sleeves and Minimum-Link Paths

In this section we describe the approach for computing a rectilinear polygon that corresponds to the shape of a set of points well. We first define the α-shape (Edelsbrunner et al. 1983), then we introduce a new structure called the (α, δ)-sleeve. We show properties of this new structure and give an efficient algorithm for its construction. Finally, we show how we can use the (α, δ)-sleeve to determine a suitable rectilinear polygon, and give an algorithm to compute it.

Our objective is to bound the point set by a rectilinear shape with few edges. This shape must have all points to the inside or on it, but we must allow the shape to cover some area outside of the α-shape to accomodate a rectilinear shape with few edges.

Like was done for finding rectangles to fit a set S of points in (van Lankveld et al. 2011), we will use the δ-coverage concept. There we defined the δ-coverage region to be the union of the radius-δ disks centered on the points of S. Any point in the plane not in the δ-coverage region is at least at distance δ from all sample points. We required the approximating rectangle to contain all points of the sample, but not be outside the δ-coverage region. The value of δ should be chosen small enough so that the rectangle cannot be too far away from the sampled points

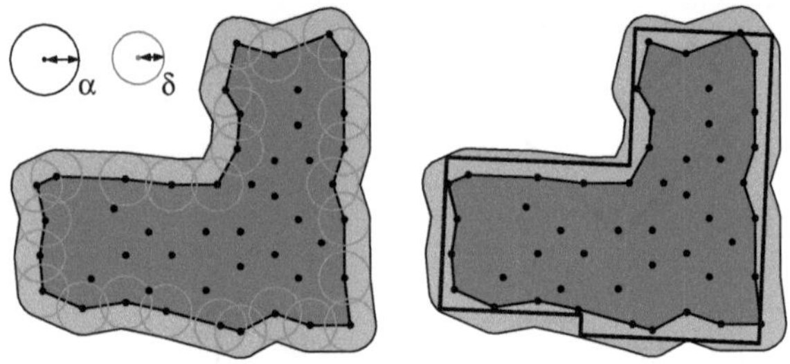

Fig. 2 *Left*, values of α and δ shown by disks, and the (α, δ)-sleeve of the points shown. *Right*, a minimum-link rectilinear path in the sleeve

and therefore from the likely shape. On the other hand, δ should be chosen large enough so that the irregularities in the sampling do not exclude the existence of a rectangle in the δ-coverage region that encloses the points as well.

2.1 Definition and Properties of (α, δ)-Sleeves

We adopt these ideas from rectangles to rectilinear shapes. Let S be a set of sampled points in a plane, and let P denote a desired rectilinear polygon. Let $\alpha > 0$ be a real parameter related to the sampling density, typically between 20 and 50 cm. Let $\delta > 0$ be another such parameter, also related to the sampling density.

Definition 1 (Edelsbrunner et al. 1983) Given a point set S, a point $p \in S$ is α-extreme if there exists an empty open disk (i.e., not containing any point from S) of radius α with p on its boundary. Two points $p, q \in S$ are called α-neighbors if they share such an empty disk. The α-shape of S is the straight-line graph whose vertices are the α-extreme points and whose edges connect the respective α-neighbors.

We will compute the α-shape A of S and require that P contains A completely. If P was a rectangle, there is no difference between requiring P inside or A inside, but for other rectilinear shapes it can make a difference. Our main reason for using the α-shape is the guarantee that P is not self-intersecting if the α-shape consists of one connected component. Under regular sampling conditions of an individual surface, we can make sure that the α-shape has only one component by carefully choosing α based on the sampling density.

From the α-shape A of S we will compute the (α, δ)-sleeve, defined as follows.

Definition 2 For set S of points in the plane, the (α, δ)-sleeve is the Minkowski sum of the α-shape of S with a disk of radius δ, where only the part outside the α-shape is taken.

The Minkowski sum with a disk creates a buffer region around the shape. The (α, δ)-sleeve is an outer proximity region of the α-shape.

Not all values of α and δ, or combinations of values, give nice properties to the (α, δ)-sleeve. Let us assume that α is such that the α-shape is in principle a good

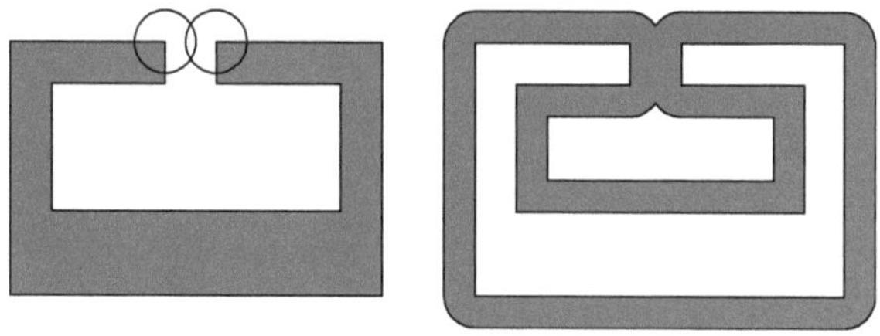

Fig. 3 If the α-shape is the *grey* interior shown *left*, and δ corresponds to the radius of the disks shown *left*, then the (α, δ)-sleeve (shown in *grey* on the *right*) has more than one inner boundary

approximation of the underlying shape. In particular, let us assume that it is connected and has no holes.

This implies that the inner boundary of the (α, δ)-sleeve is the boundary of a simple polygon. If δ is sufficiently small, the outer boundary of the (α, δ)-sleeve will be the boundary of a simple polygon as well, but with circular arcs. For some shapes and larger values of δ, the (α, δ)-sleeve can have several inner boundaries: not just those created by subtracting the interior of the α-shape, but also ones where opposite sides on the outside of the α-shape are close. Figure 3 shows an example.

It turns out that if we set $\delta < \frac{4}{5}\alpha$, then the (α, δ)-sleeve will have the desired topology with one outer and one inner boundary. We prove this after proving some more properties of the α-shape.

Lemma 1 *For a given $\alpha > 0$, let A be an α-shape of a set S of points in the plane. Any two points $p, q \in S$ that share the boundary of an empty open disk d of radius $\leq \alpha$ are connected by a line segment inside A.*

Proof According to Edelsbrunner et al. (1983), the interior of the α-shape is the union of all Delaunay triangles with a circumcircle of radius $\leq \alpha$.

The Delaunay triangulation on S must contain the edge p, q as witnessed by d. Now consider the Delaunay triangle t incident to p, q on the side of the center of d; we distinguish two cases. Either (a) this triangle has a circumcircle of radius $\leq \alpha$ and is part of the interior of A, or (b) there is an empty radius-α disk with p, q on its boundary and p, q are α-neighbors. In both cases, p, q must be either in the interior or on the boundary of A. □

Lemma 2 *For a given $\alpha > 0$, let A be an α-shape of a set S of points in the plane. Any two points $p, q \in S$ at distance at most 2α are connected by a path that lies both inside A and inside the disk with p, q as its diameter*

Proof In this proof, we use a property of α-shapes that is easily inferred from Lemma 1: for two points $p, q \in S$ at distance at most 2α, either (1) they are α-neighbors, or (2) the connecting line segment is interior to the α-shape, or (3) the disk with p, q as its diameter contains another point of S.

In cases (1) and (2), it is clear there is a connecting path within A and within the disk with p, q as its parameter. In case (3), we consider the collection of empty disks with their center on p, q and touching two points of S, as shown in Fig. 4. These disks impose an ordering $(p, s_1, s_2 q)$ on a subset of the points inside the disk. For any two

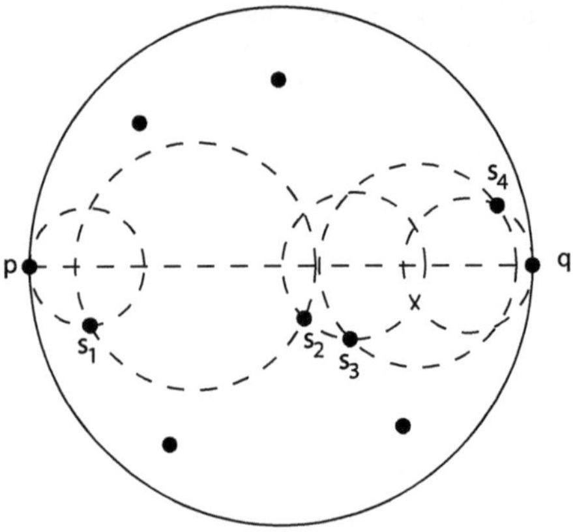

Fig. 4 Illustration of the proof of Lemma 2

subsequent points x, y in this ordering, the line segment connecting x, y is contained in A, according to Lemma 1. □

Lemma 3 *For a given $\alpha > 0$, let A be an α-shape of a set S of points in the plane, and assume that A is the boundary of a simple polygon. Then for $0 < \delta \leq \frac{4}{5}\alpha$, the (α, δ)-sleeve has the topology of an annulus.*

Proof Since for very small δ, the topology of the (α, δ)-sleeve is an annulus, we can imagine growing δ to the first (smallest) value δ' where the topology is no longer an annulus. This happens only when there are two points p, q, not necessarily in S, on the α-shape at distance $2\delta'$. Let m be the midpoint of p, q. Then m is at distance δ' from p and q, and no point of the α-shape is closer to m. Hence, the disk Δ centered at m with radius δ' does not intersect the α-shape in points other than p and q.

Assume first that the two points are two vertices $p, q \in S$. According to Lemma 2, Δ contains an edge inside the α-shape. This contradicts the assumption that no point of the α-shape is closer to m than p or q.

Assume next that $p, q \notin S$. Then they lie in the interior of two different α-shape edges. If they are not parallel, it is impossible that the first topological change occurs at m. If they are parallel, then the topological change will occur simultaneously along a stretch of the two edges, and we can choose p and q such that at least one of them is an endpoint and therefore in S. So this case is treated together with the final case.

Assume finally that only one point is in S, say, $p \in S$ and $q \notin S$. Let q_1 and q_2 be the endpoints of the α-shape edge that q lies on, so $q_1, q_2 \in S$. The diametral disk Δ of p and q is tangent to the edge q_1, q_2 at q. Assume without loss of generality that $\overline{q_1 q_2}$ is vertical, that q_1 is the lower endpoint of $\overline{q_1 q_2}$, that q_1 is closer to p than q_2, and that p is to the right of $\overline{q_1 q_2}$, see Fig. 5.

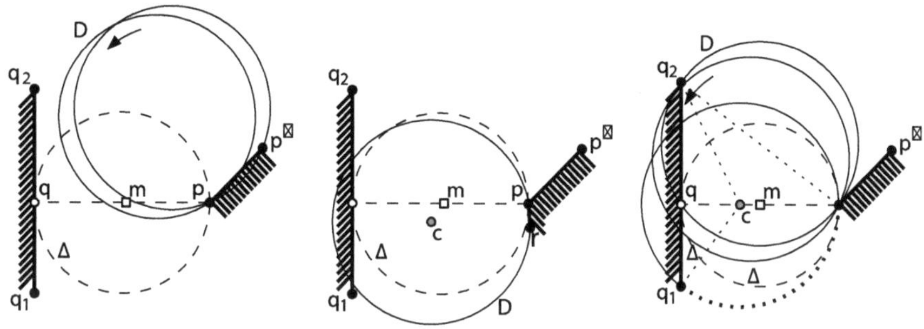

Fig. 5 Illustration of the proof of Lemma 3

Since the topology of the (α, δ)-sleeve changes for the first time (smallest δ') due to the contact at m, the α-neighbors of p cannot lie to the left of the vertical line through p. Let p' be the α-neighbor of p that makes the smallest angle with the vertical upward direction. Since p and p' are α-neighbors, an empty α-disk D exists that has p and p' on its boundary and its center to the left of the directed edge from p to p'.

We now rotate D in contact with p in counterclockwise direction, see Fig. 5. Initially D does not contain any point of S inside. Let r be the first point of S that is reached by the boundary of D, such that r would be inside if we were to rotate D further. We distinguish several cases.

If $r = q_2$, then $\overline{pq_2}$ are α-neighbors (as witnessed by the emptiness and current position of D). The implied α-shape edge will intersect Δ because q_2 lies left of p, a contradiction. The same contradiction is obtained when $r = q_1$, or when r is any point that lies left of the vertical line through p. Hence, r lies to the right of this vertical line or on it. This implies that the disk D has rotated beyond the situation where it has a vertical tangent. This is equivalent to stating that the center c of D lies below the horizontal line through p. Also, this center must lie right of the line through q_1 and q_2, because q_1, q_2 is an α-shape edge with the interior to its left.

Because q_1, q_2 are α-neighbors, $\|q_1q_2\| \leq 2\alpha$. We argue that $\|pq_2\| > 2\alpha$. According to Lemma 2, if $\|pq_2\| \leq 2\alpha$, there is a path connecting p, q_2 inside their diametral disk. Because $\|pq_1\| \leq \|pq_2\|$, the same holds for p, q_1. This means that if $\|pq_2\| \leq 2\alpha$, then either there is a point in A closer to m than p, or A is not a simple polygon; both cases contradict an assumption.

We are interested in the threshold case, where $\|pq\| = 2\delta'$ is as small as possible, and the (α, δ)-sleeve is an annulus for $\delta \leq \delta'$, but not if δ is infinitesimally larger than δ'. Because $\|q_1q_2\| \leq 2\alpha$, $\|pq_2\| > 2\alpha$, and $\|pq_1\| \leq \|pq_2\|$ and $\angle pqq_2 = \frac{\pi}{2}$, the smallest $\|pq\|$ occurs when the angle $\angle q_1q_2p$ is as small as possible. At the same time, the radius-α disk D touching p, r cannot contain q_1 and cannot have its center above $\overline{pq}$. Finally, because r was the first point encounterd by D during its rotation, either $\|pq_1\| \geq 2\alpha$ or c is above $\overline{pq_1}$. If $\|pq_1\| \geq 2\alpha$, then $\|pq\| \geq \sqrt{3}\alpha$ so $\delta' \geq \frac{1}{2}\sqrt{3}\alpha > \frac{4}{5}\alpha$. We continue to show that the other case can give a smaller lower bound for δ. Here $\|pq_1\| < 2\alpha$ and c is above $\overline{pq_1}$. By Lemma 2, A has a sequence of edges connecting p with q_1 via r inside the disk that has $\overline{pq_1}$ as its diameter.

The threshold case is shown in Fig. 5(right). If D stays empty while rotating beyond the point where c lies on p, q, it is possible for the boundary of A to connect q_1 to r and p through a series of edges strictly outside Δ. This would result in a hole in the (α, δ)-sleeve just below m, meaning that the topology of the (α, δ)-sleeve is not an annulus.

In the threshold case, c lies on p, q, and then Δq_2cq_1 and Δq_2cp are mirrored triangles, so the angles $\angle q_2q_1c = \angle q_2pc$, and Δq_2qp and Δcqq_1 are similar triangles. This implies that $\frac{2\alpha}{\alpha} = \frac{\|q_2q\|}{\|cq\|}$ or $\|q_2q\| = 2\|cq\|$. If we combine this with the Pythagorean theorem on Δq_2qp we can derive that $2\delta' = \|pq\| = \frac{8}{5}\alpha$:

$$\begin{aligned}
(2\alpha)^2 &= \|q_2q\|^2 + (\|cq\| + \alpha)^2 \\
4\alpha^2 &= 4\|cq\|^2 + \|cq\|^2 + 2\|cq\|\alpha + \alpha^2 \\
3\alpha^2 - 2\|cq\|\alpha - 5\|cq\|^2 &= 0 \\
(\alpha + \|cq\|)(3\alpha - 5\|cq\|) &= 0 \\
\|cq\| = -\alpha \text{ or} \|cq\| &= \frac{3}{5}\alpha
\end{aligned}$$

The first option leads to a degenerate triangle Δq_1q_2p. The second option leads to $\|pq\| = \|cq\| + \alpha = \frac{8}{5}\alpha$.

The threshold case presented results in an (α, δ')-sleeve with the topology of an annulus: the value of δ' does not allow D to rotate beyond a vertical tangency at p when it reaches r but before it reaches q_1. Hence, r is not right of p. If r is vertically below p (a degenerate situation), then we can repeat the whole construction with r instead of p, which means we can ignore this case. Hence, in order to get a different topology, r must be strictly right of p, and c must be strictly below p, q. Because D cannot contain q_1, $\|q_1q_2\| \leq 2\alpha$, and $\|pq_2\| > 2\alpha$, this implies that $\|pq\| = 2\delta' > \frac{8}{5}\alpha$. □

We can use known algorithms to compute the (α, δ)-sleeve for given values of α and δ. For a set S of n points in the plane, the α-shape can be computed directly from the Delaunay triangulation of S (Edelsbrunner et al. 1983). Then we use a buffer computation algorithm on the α-shape. Such an algorithm can be based on computing the Voronoi diagram of the line segments of the α-shape first (de Berg et al. 2008; Yap 1987). Then the buffer boundary can be found in each Voronoi cell, and these can be merged into the boundaries of the δ-buffer of the α-shape. Converting this to the (α, δ)-sleeve is then straightforward. This procedure takes $O(n \log n)$ time in total.

Fig. 6 The bottom part of an (α, δ)-sleeve, the lowest vertex v of the α-shape (*left*), and the conversion of the sleeve to a simple polygon (*right*)

2.2 Minimum-Link Paths in (α, δ)-Sleeves

The (α, δ)-sleeve gives a region in which we want to determine a rectilinear shape. The rectilinear shape should separate the inner boundary from the outer boundary. We will show how to use a minimum-link path algorithm to find the shape. We will assume that the (α, δ)-sleeve has the topology of an annulus.

For any simple polygon and start- and endpoints s and t inside, we can find a minimum-link path that uses only horizontal and vertical edges. This problem has been well-studied in computational geometry, and a linear-time algorithm exists that finds such a path (Hershberger and Snoeyink 1994). Our problem is different in three aspects: (a) We do not have a simple polygon but a shape with the topology of an annulus. (b) We do not have a start- and endpoint but we want a rectilinear cycle. (c) We do not know the orientation of the edges beforehand, we only know that the angles on the path are 90°.

Lemma 4 *In an (α, δ)-sleeve, there exists a minimum-link axis-parallel cycle that passes through the lowest point of the inner boundary of the sleeve (the α-shape).*

Proof Consider any minimum-link cycle in the sleeve, and let e be its lowest horizontal edge. If e contains a vertex of the α-shape the lemma is true, otherwise we can move e upwards while shortening the two adjacent edges of the cycle. During this move two things can happen: (a) An adjacent edge reduces to length 0, but then a cycle with two fewer links is found. (b) Edge e is stopped by a vertex of the α-shape, which must be its lowest vertex because all vertices of the α-shape have a y-coordinate at least as high as the lowest edge of the cycle. This proves the lemma. □

Another property of the minimum-link rectilinear path is that it is non-selfintersecting. Otherwise, we could remove the extra loop and obtain a cycle with fewer links.

Lemma 5 *Any minimum-link rectilinear path in an (α, δ)-sleeve is non-selfintersecting.*

Suppose that we know the orientation of the minimum-link path. Then we can rotate the (α, δ)-sleeve so that this orientation becomes the axis-parallel orientation. By the lemma above, we now know a point that we can assume to lie on the minimum-link cycle. To convert the (α, δ)-sleeve to a polygonal region we do the following. We find the lowest vertex v of the α-shape and insert a new edge vertically down from v, until it reaches the outer boundary, see Fig. 6. This edge will split the annulus into a shape with the same topology as a simple polygon. We duplicate the new edge including its endpoints, splitting v into a left copy v_l and a right copy v_r. Now our shape does not have doubly used edges, and it can be treated as a normal simple polygon. We will find a minimum-link axis-parallel path from v_r to v_l using the algorithm in (Hershberger and Snoeyink 1994). The output path can easily be converted back into a minimum-link cycle.

3 Experiments

In the previous section we presented a method to construct a rectilinear minimum-link path that approximates the α-shape. Here, we evaluate this method on synthetic and real data with the main goal of checking whether the method is suitable for piecewise planar urban scene reconstruction. In particular, we wish to discover whether for a given point density, values of α and δ exist that lead to a rectilinear minimum-link path that is close to the true building facet shape.

This section describes both the setup of these experiments and the results. The first two subsections describe our experiments on synthetic data to determine a value for δ to get the correct number of edges or the best overlap with the ground truth, respectively. The last subsection describes an experiment on urban LiDAR data.

3.1 Universal δ

For the evaluation on synthetic data, we have constructed fifteen test cases of varying shape and complexity. Each test case comprises a ground truth polygon T that should represent a realistic urban surface shape with an area of between 30 and 60 m^2, as shown in Fig. 7. To counter bias towards a certain initial orientation, T is rotated by a random real-valued angle for each test. Each rotated T then yields a point set of predetermined density by uniform sampling from its interior.

For each case, we compute the (α, δ)-sleeve using a fixed α based on the sampling density ρ and a varying δ. We have chosen a fixed α such that the α-shape is connected and without holes irrespective of test case. We vary δ to determine whether there is a single value of δ at each ρ such that our method produces polygons similar to the ground truth in all test cases. To measure the impact of ρ on the results of our method, we have repeated the experiments for three different densities. The chosen α for each density ρ is shown in Table 1. We vary δ using increments of 1 cm.

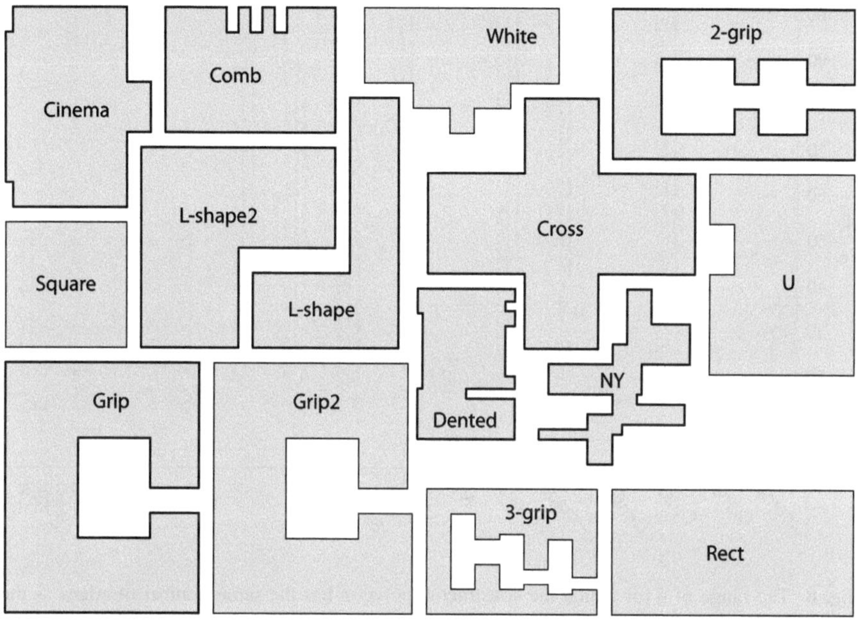

Fig. 7 The different test cases used for the experiments on synthetic data

Table 1 The different densities used for the point sampling and the associated values of α

ρ (points/m^2)	α (cm)
50	60
100	20
200	17

A rectilinear polygon $\hat{P}$ is constructed inside the (α, δ)-sleeve, according to the algorithm given in Sect. 2, such that $\hat{P}$ has the minimum number of edges over all rotation angles. We use an angular step size of 1 degree, meaning that we run the minimum-link algorithm 90 times for each (α, δ)-sleeve. To get a canonical result for each angle, we 'shrink' $\hat{P}$ to a smaller polygon P by moving each edge inward until it touches the α-shape. We move edges in descending order of edge length. If there are multiple polygons with the minimum number of edges, created for different rotation angles, P is chosen as the one with the smallest area.

In urban reconstruction, the shape of a surface is not known a priori. For this reason, we have analyzed whether our method reconstructs the correct shapes. We measure the correctness of a shape by its number of edges and its angle of rotation, by comparing them to the ground truth. In the optimal case, there is a value for δ at which our method always produces the correct shape. However, it is very likely that the optimal value of δ depends on ρ. Because we want our method to be

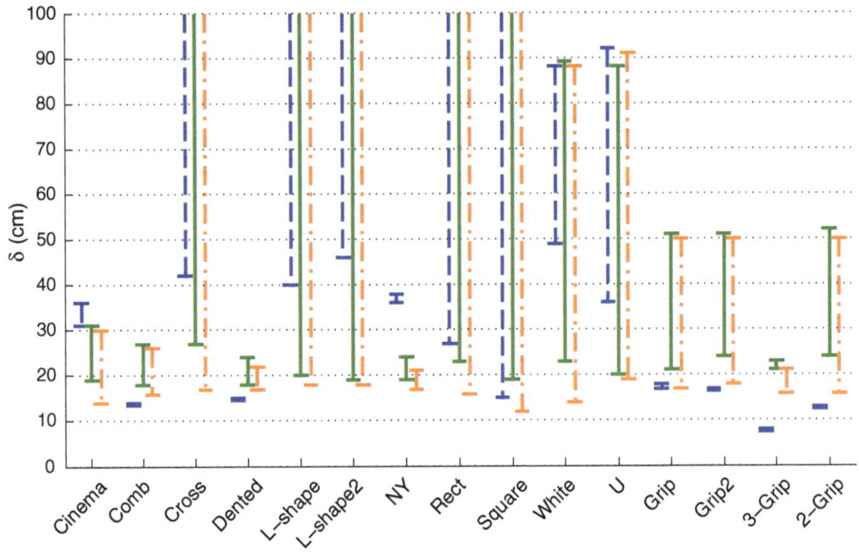

Fig. 8 The range of δ for which the constructed polygon has the same number of edges as the ground truth at point densities 50 (*dashed*), 100 (*solid*), and 200 (*dash-dotted*) points/m^2

shape-invariant, we are looking for a δ for which our method performs well, irrespective of ground truth case.

Our experiments showed that the chosen polygons P always have a rotation within 1 degree of the rotation of the ground truth. Figure 8 shows the ranges of δ values for which the constructed polygon has the same number of edges as the ground truth. At the higher densities (100 and 200 points/m^2), we can choose the value of δ at 24 cm and 20 cm respectively to construct a shape with the same number of edges for most cases.

At the lowest density (50 points/m^2), there is no δ value that consistently results in a correct polygon. Additionally, Fig. 8 shows that in half of the cases the range of δ resulting in the correct number of edges is very small. These cases all have some small features that add edges to the ground truth while being difficult to make out in the rough point samples. Visual inspection showed that the openings of the "grip" cases were closed off in their α-shape, explaining the problems with constructing the correct shape. For the cases without a small feature, a value of δ between 50 and 85 results in the correct polygons.

Considering the values of α and δ together, we observe that we should choose δ slightly larger than α. However, by Lemma 3, we are not guaranteed to get an (α, δ)-sleeve with the topology of an annulus in this case. Especially in "grip" cases, finding ways to correctly deal with (α, δ)-sleeves that do not have an annulus topology may yield better results.

3.2 Data-Dependent δ

The shapes encountered in urban scenes vary greatly. Small features of the shape combined with insufficient sampling density may make it very difficult to correctly estimate the number of edges of the shape, even for a human modeler. Additionally, the sampling density may vary greatly between surfaces. For this reason, it may be interesting to estimate the best value for δ from the point data itself instead of choosing some fixed value.

One way to determine which δ is best is to analyze how P changes as δ increases. At the smallest δ, P will approximate the α-shape and from some large δ onwards, P is a rectangle. Recall that P is the shrunken version of $\hat{P}$, so we can expect the largest changes in P when its number of edges change. Additionally, given a fixed number of edges, it is likely that the polygon that approximates T best is found immediately after a jump to that number of edges. While growing δ, we use the polygon just after each jump as representative for the polygons with the same number of edges. This leads to a succession of representative polygons $\{R_0 R_k\}$, where each consecutive polygon has fewer edges, culminating in four edges at R_k.

The best R should strike a balance between complexity and approximation of the shape of T. The complexity can again be measured in the number of edges, with a preference for low complexity. How well R approximates T could be measured from their area of symmetric difference. Unfortunately, for real-world data the ground truth is not known, so the symmetric difference cannot be used to determine the best δ. However, because the area of T is fixed, a simple approximation for the symmetric difference is to use the area of R.

Because R always covers the α-shape, we can assume that the reason for large jumps in the area of R is the removal of a large feature of T from R. The goal then becomes to determine when the process of increasing δ stops reducing the complexity of the shape and starts removing large features. If we denote by χ_i the change in area between consecutive representative polygons R_i and R_{i+1}, we search for a threshold value $\bar{\chi}$ for χ_x. The idea is that if the area of R makes a jump of $\chi_i > \bar{\chi}$ due to a decrease in the number of edges, then the polygon loses a key feature of T so R_i is the simplest polygon that contains all important features.

We determine $\bar{\chi}$ by selecting which reference polygon R_i best approximates T by visual inspection and computing χ_i. Choosing $\bar{\chi}$ such that it is smaller than χ_i but larger than χ_j for all $j < i$ would result in terminating the automatic search for δ at the preferred polygon. Because we want the method to be shape-invariant, we are looking for a value of $\bar{\chi}$ for which our method performs well, irrespective of ground truth case.

The ranges of $\bar{\chi}$ resulting in the polygon selected by visual inspection are shown in Fig. 9. At the higher densities (100 and 200 points/m^2), there are values for $\bar{\chi}$ that result in the correct polygon in all the test cases. At a density of 50 points/m^2, there is no $\bar{\chi}$ that produces the correct polygon in all cases, although 0.8 m^2 is a good value for most cases.

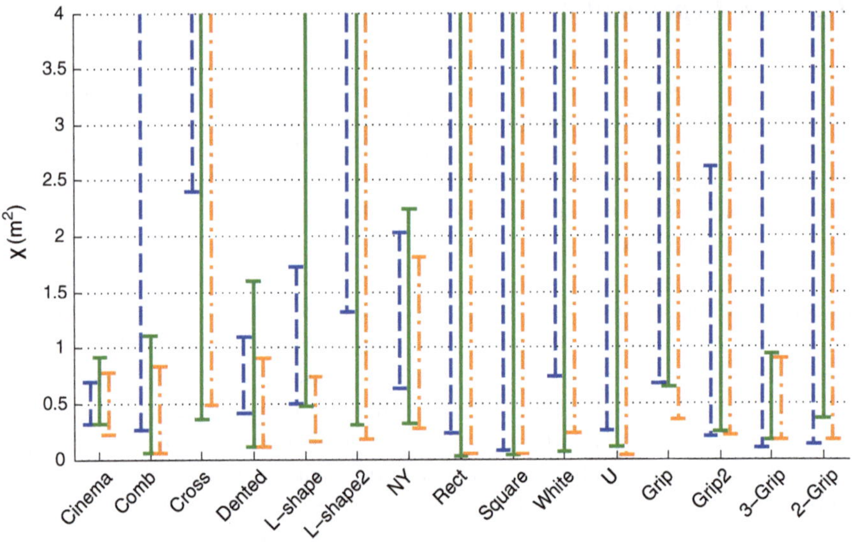

Fig. 9 The range of $\bar{\chi}$ for which the polygon selected by visual inspection is constructed at point densities 50 (*dashed*), 100 (*solid*), and 200 (*dash-dotted*) points/m^2

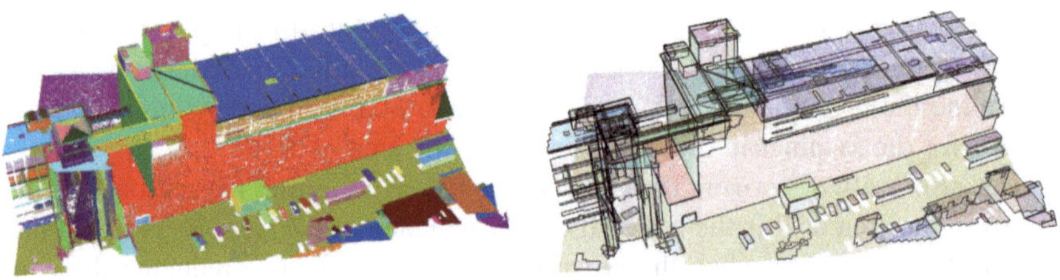

Fig. 10 A LiDAR data set with points colored per surface and the rectilinear boundaries of those surfaces

3.3 LiDAR Data

Apart from synthetic data, we have also applied our method to an airborne LiDAR data set of a building, as shown in Fig. 10. The point density varies greatly between surfaces, because of the scanning method. However, based on our earlier work on the same data (van Lankveld et al. 2011), we set α to 125 cm for surfaces close to vertical and 60 cm for the other surfaces.

The building contains many rectilinear surfaces, which we have reconstructed using $\delta = 60$ cm. The surfaces near the edge of the scene were given jagged edges because the buildings are not aligned with the scene. However, our results do favour long straight edges for the remainder of the shape.

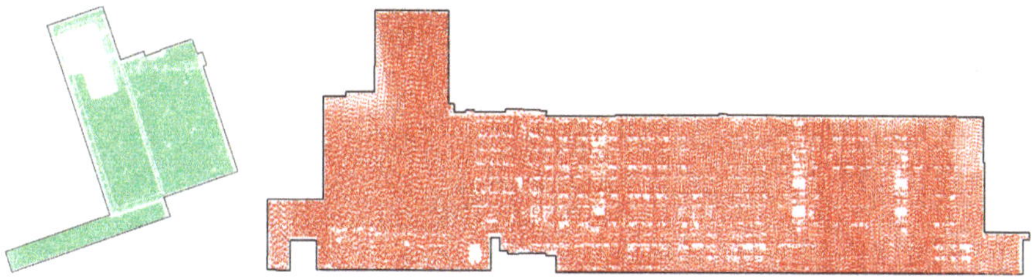

Fig. 11 Two interesting rectilinear surfaces

Figure 11 shows two interesting surfaces and their rectilinear boundaries. Note how the rotation of the reconstructed polygons matches the neighboring surfaces. A human modeler may construct a similar shape with fewer edges for these surfaces. It seems that these faults in our results are caused by missing data in the input sets.

4 Conclusions

We have presented a novel concept, the (α, δ)-sleeve, which contains a proximity zone around a point set. When combined with a minimum-link path algorithm, this structure can be used to reconstruct simple shapes that contain the point set. We have presented a method aimed at reconstructing surfaces in urban scenes from a point set by combining the (α, δ)-sleeve with a rectilinear minimum-link path. Our experiments showed that when the surfaces are sampled sufficiently dense, there are parameter settings for α and δ that lead to correct reconstructions for artificial data. Finally, we have shown on an urban LiDAR data set that our method produces plausible results.

A number of interesting possibilities for extensions and improvements remain. In most urban scenes there are some surfaces that are not rectilinear. By changing the rectilinear minimum-link path algorithm to allow a few edges that do not follow one of the principal directions, the (α, δ)-sleeve can be used to reconstruct such surfaces as well. Alternatively, we could use a post-processing method that replaces long stair-like parts in the rectilinear surface boundary by one line.

Another recurring problem we encountered is the rounding of concave corners. As Fig. 12 shows, the α-shape can "round off" a concave corner, often going outside the original polygon. In such cases, constructing a polygon within the (α, δ)-sleeve requires either more edges or a larger δ. This problem may be overcome by using pieces of the α-hull (Edelsbrunner et al. 1983) as inner boundary of the (α, δ)-sleeve, instead of the α-shape. The α-hull uses concave circular arcs between its boundary vertices, allowing the polygon to go deeper into concave corners. Unfortunately, using α-hull arcs everywhere may cause the inner boundary to self-intersect, which in turn can result in a self-intersecting rectilinear

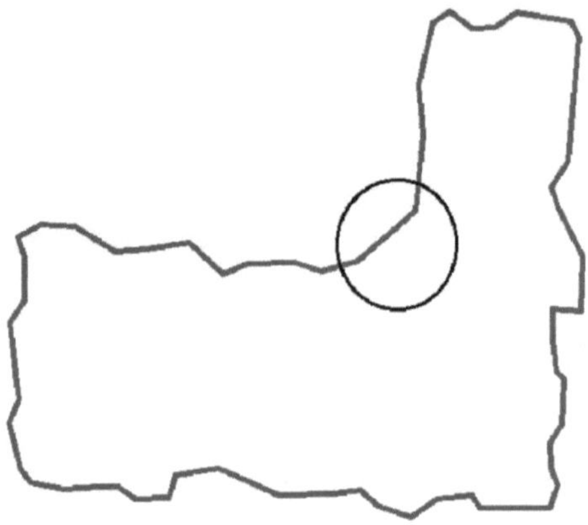

Fig. 12 An example α-shape where a concave corner is rounded off

path. Hence, this solution would require additional steps to ensure that the constructed polygon does not have self-intersections.

Finally, for this work we have ignored holes in the α-shape. Reconstructing those holes can be done in a fashion similar to the method described in this paper. The (α, δ)-sleeve would consist of several components, each one with the topology of an annulus. We can run minimum-link path algorithms in each annulus with the same orientations in order to find an orientation that is best overall. This way we can easily ensure that the rectilinear directions of the outside boundary and of the hole boundaries are the same.

Acknowledgments This research has been supported by the GATE project, funded by the Netherlands Organization for Scientific Research (NWO) and the Netherlands ICT Research and Innovation Authority (ICT Regie).

References

Brenner C (2005) Building reconstruction from images and laser scanning. Int J Appl Earth Obs 6(3–4):187–198

Buchin K, Meulemans W, Speckmann B (2011) A new method for subdivision simplification with applications to urban-area generalization. In: Proceedings SIGSPATIAL, pp 261–270

Carlberg M, Andrews J, Gao P, Zakhor A (2009) Fast surface reconstruction and segmentation with ground-based and airborne lidar range data. Technical Report ADA538860, University of California at Berkeley

de Berg M, Cheong O, van Kreveld M, Overmars M (2008) Computational geometry-algorithms and aplications. Springer, Berlin

Edelsbrunner H, Kirkpatrick DG, Seidel R (1983) On the shape of a set of points in the plane. IEEE Trans on Inf Theory 29(4):551–559

Furukawa Y, Curless B, Seitz SM, Szeliski R (2009) Manhattan-world stereo. In: IEEE Computer Society, pp 1422–1429

Hershberger J, Snoeyink J (1994) Computing minimum length paths of a given homotopy class. Comp Geom 4:63–97

John Chance Land Surveys, Fugro (2009) Fli-map specifications. http://www.flimap.com/site47.php
Marton ZC, Rusu RB, Beetz M (2009) On fast surface reconstruction methods for large and noisy point clouds. In: Proceedings ICRA, pp 2829–2834
Mayer H (2005) Scale-spaces for generalization of 3D buildings. Int J Geogr Inf Sci 19:975–997
Regnauld N, Edwardes A, Barrault M (1999) Strategies in building generalisation: modelling the sequence, constraining the choice. In: Proceedings ICA
Rottensteiner F (2003) Automatic generation of high-quality building models from lidar data. IEEE Comput Graphics Appl 23(6):42–50
Ruas A (1999) Modèle de généralisation de données géographiques à base de contraintes et d'autonomie. PhD thesis, Université de Marne la Vallée
Schnabel R, Wahl R, Klein R (2007) Efficient RANSAC for point-cloud shape detection. Comput Graphics Forum 26(2):214–226
Schwalbe E, Maas HG, Seidel F (2005) 3D building model generation from airborne laser scanner data using 2D GIS data and orthogonal point cloud projections. In: Proceedings ISPRS, pp 12–14
Seitz S, Curless B, Diebel J, Scharstein D, Szeliski R (2006) A comparison and evaluation of multi-view stereo reconstruction algorithms. In: Proceedings CVPR, vol 1, pp 519–528
Swan J, Anand S, Ware M, Jackson M (2007) Automated schematization for web service applications. In: Web and Wireless GISystems. LNCS 4857:216–226
Tseng YH, Tang KP, Chou FC (2007) Surface reconstruction from LiDAR data with extended snake theory. In: Proceedings EMMCVPR, pp 479–492
van Lankveld T, van Kreveld M, Veltkamp RC (2011) Identifying rectangles in laser range data for urban scene reconstruction. Comput Graph 35(3):719–725
Wolff A (2007) Drawing subway maps: a survey. Inform Forsch Entwickl 22(1):23–44
Yap CK (1987) An $O(n \log n)$ algorithm for the Voronoi diagram of a set of simple curve segments. Discrete Comput Geom 2:365–393
You S, Hu J, Neumann U, Fox P (2003) Urban site modeling from LiDAR. In: Proceedings ICCSA. LNCS, vol 2669, pp 579–588
Zhou QY, Neumann U (2008) Fast and extensible building modeling from airborne LiDAR data. In: Proceedings SIGSPATIAL, pp 1–8

A 3D-GIS Implementation for Realizing 3D Network Analysis and Routing Simulation for Evacuation Purpose

Umit Atila, Ismail Rakip Karas and Alias Abdul Rahman

Abstract The need for 3D visualization and navigation within 3D-GIS environment is increasingly growing and spreading to various fields. When we consider current navigation systems, most of them are still in 2D environment that is insufficient to realize 3D objects and obtain satisfactory solutions for 3D environment. One of the most important research areas is evacuating the buildings with safety as more complex building infrastructures are increasing in today's world. The end user side of such evacuation system needs to run in mobile environment with an accurate indoor positioning while the system assist people to the destination with support of visual landscapes and voice commands. For realizing such navigation system we need to solve complex 3D network analysis. The objective of this paper is to investigate and implement 3D visualization and navigation techniques and solutions for indoor spaces within 3D-GIS. As an initial step and as for implementation a GUI provides 3D visualization of Corporation Complex in Putrajaya based on CityGML data, stores spatial data in a Geo-Database and then performs complex network analysis under some different kind of constraints. The GUI also provides a routing simulation on a calculated shortest path with voice commands and visualized instructions which are intended to be the infrastructure of a voice enabled mobile navigation system in our future work.

U. Atila
Directorate of Computer Center, Gazi University, Ankara, Turkey
e-mail: umitatila@gazi.edu.tr

I. R. Karas (✉)
Department of Computer Engineering, Karabuk University, Karabuk, Turkey
e-mail: ismail.karas@karabuk.edu.tr

A. A. Rahman
Department of Geoinformatics, Universiti Teknologi Malaysia, Johor, Malaysia
e-mail: alias.fksg@gmail.com

J. Pouliot et al. (eds), *Progress and New Trends in 3D Geoinformation Sciences*, Lecture Notes in Geoinformation and Cartography, DOI: 10.1007/978-3-642-29793-9_14,

Keywords 3D-GIS · Network analysis · Navigation · Evacuation

1 Introduction

The need for 3D visualization and navigation within 3D-GIS environment is increasingly growing and spreading to various fields. Most of the navigation systems use 2D or 2.5D data (e.g. road layer) to find and simulate the shortest path route which is lacking in building environment (Musliman and Rahman 2008). When we consider current navigation systems, most of them are still in 2D environment and there is a need for different approaches based on 3D aspect which realize the 3D objects and eliminate the network analysis limitations on multilevel structures (Cutter et al. 2003; Pu and Zlatanova 2005; Kwam and Lee 2005; Zlatanova et al. 2004).

Passing from two-dimensional (2D) GIS toward 3D-GIS, a great amount of 3D data sets (e.g. city models) have become necessary to be produced and satisfied widely. This situation requires a number of specific issues to be researched, e.g. 3D routing accuracy, appropriate means to visualize 3D spatial analysis, tools to effortlessly explore and navigate through large models in real time, with the correct texture and geometry (Musliman et al. 2006).

One of the most important research area is evacuating the buildings through the shortest path with safety in a case of extraordinary circumstances (i.e. disastrous accidents, massive terrorist attacks) happening in complex and tall buildings of today's world which is the subject of 3D network analysis applications for indoor.

In research environments, two main approaches to indoor evacuation systems are currently accepted. One is 3D modeling environment that this study follows and the other is fire simulation models. Originating from 3D modelling environment, evacuation and routing is based on graph networks (Karas et al. 2006; Jun et al. 2009), while 3D visualization problems achieved by CityGML (Kolbe 2008). Researchers following the network approach generally modify the existing 2D routing algorithms to the 3D aspect (Musliman and Rahman 2008). Initial requirements of 3D-GIS and navigation (Musliman et al. 2006), concepts, frameworks and its application from a bigger scope of view were widely represented (Pu and Zlatanova 2005), but there is currently still a lack of implementation of 3D network analysis and navigation for evacuation purpose.

Compared to outdoor space there is more tendency to be lost in building environment because of a more complicated recognition of landmarks in the indoor environment (Gartner et al. 2009). An ideal evacuation system should have an engine for route calculation and an engine for adapting the routing presentation. The most advanced possibility is to use mobile devices like cell phone, PDA, etc. for presenting the calculated route to the user. Mobile devices can supply people with voice commands and graphical evacuation instructions (Pu and Zlatanova 2005). Therefore, providing such evacuation systems needs to solve some complex 3D topology, 3D modeling, 3D network analysis and so on.

This paper investigates 3D network analysis within an evacuation system and presents how to manage 3D network analysis using Oracle Spatial within a Java based 3D-GIS implementation. All experiments highlighted in this paper are on a 3D model of Corporation Complex in Putrajaya, Malaysia. Section 2 gives some examples on visualization of 3D building and network model from CityGML format. Section 3 gives some information on storing spatial data in a Geo-DBMS and explains how to create Network Model in Oracle Spatial. Section 4 gives visualized results of some 3D network analysis performed by 3D-GIS implementation. Section 5 elaborates the routing engine integrated in simulation module of 3D-GIS implementation and gives some visualization samples. Finally, we conclude the work with some future tasks that needs to be addressed.

2 Visualization of 3D Building and Network Model

Visualization of 3D building model is performed by Java based 3D-GIS implementation. The implementation reads data from CityGML format which is a common semantic information model for the representation of 3D urban objects that can be shared over different applications. CityGML is designed as an open data model and XML-based format for the storage and exchange of virtual 3D city models (Gröger et al. 2008).

For visualization of 3D spatial objects, OpenGL graphic library is used. The implementation supports 4 different types of view mode for spatial objects. These modes are Wireframe (Fig. 1a), HiddenLine (Fig. 1b), Shaded (Fig. 1c) and Shaded with Texture (Fig. 1d).

CityGML supports different Levels of Detail (LOD). LODs are required to reflect independent data collection processes with differing application requirements(Gröger et al. 2008). The prepared 3D-GIS implementation uses *citygml*4*j* Java class library and API for facilitating work with the CityGML and JOGL Java bindings for OPENGL to carry out visualization. CityGML datasets from LOD0 to LOD2 are supported. Building model is represented in LOD2 described by polygons (Fig. 2) and network model is represented as a linear network in LOD0 using Transportation Module of CityGML (Fig. 3).

3 Managing 3D Network in Geo-DBMS

Using Geo-DBMS in 3D modeling and spatial analysis has a lot of advantages. Beside the standart advantages of DBMS with respect to centralized control, data independence, data redundancy, data consistency, sharing data, data integrity and improved security, geo-DBMS brings efficient management of large spatial data

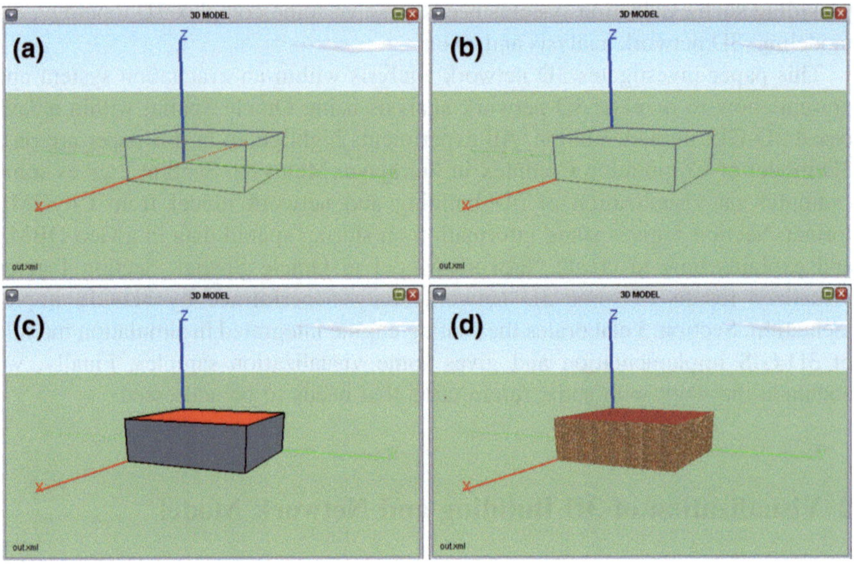

Fig. 1 Viewing modes of 3D spatial objects

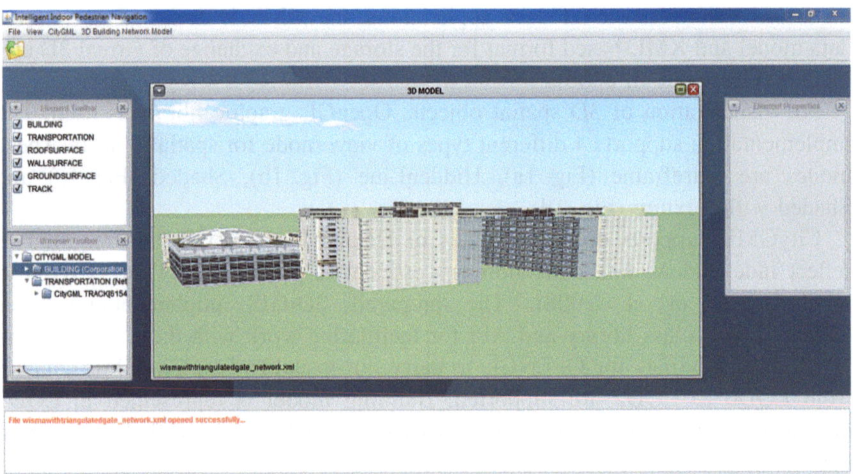

Fig. 2 Building model

sets. The management of a 3D network requires usage of graph model in DBMS. While CityGML is used to store and visualize 3D spatial objects, the graph model is used to perform network analysis.

A network is a type of mathematical graph that captures relationships between objects using connectivity. A network consists of nodes and links. Oracle Spatial maintains a combination of geometry model and graph model within Network

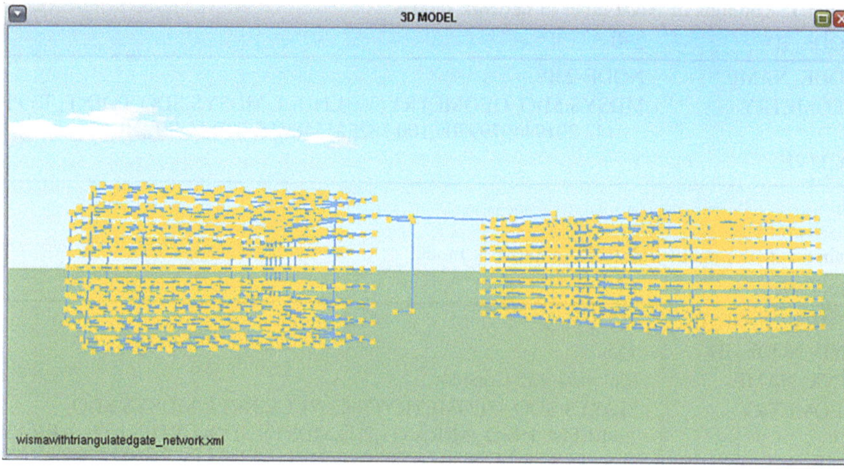

Fig. 3 Network model

Data Model. Network elements (links and nodes) may have geometric information associated with them. A logical network contains connectivity information but no geometric information. A spatial network contains both connectivity information and geometric information. In a spatial network, the nodes and links are SDO_GEOMETRY objects representing points and lines, respectively. A spatial network can also use other kinds of geometry representations. One variant lets you use linear referenced geometries. Another lets you use topology objects.

To define a network in Oracle Spatial, at least two tables should be created: A *node* and a *link* table. These tables should be provided with the proper structure and content to model the network. A network can also have a *path table* and a *path link table*. These tables are optional and are filled with the results of analyzes, such as the shortest path between two nodes.

Node table (see Table 1) describes all nodes in the network. Each node has a unique numeric identifier (the NODE_ID column). Other optional columns are geometry, cost, hierarchy_level, parent_node_id, node_name, node_type and active.

Link table (see Table 2) describes all links in the network. Each link has a unique numeric identifier (the LINK_ID column) and contains the identifiers of the two nodes it connects. Other optional columns are geometry, cost, bidirected, parent_link_id, active, link_level, link_name and link_type(Kothuri et al. 2010).

In this study we use spatial network with SDO_GEOMETRY type for representing points and lines.

Oracle Spatial Network Data Model is composed of a data model to store networks inside the database as a set of network tables, SQL functions to define and maintain networks (SDO_NET), network analysis functions in Java and network analysis functions in PL/SQL (SDO_NET_MEM) which is a "wrapper" over Java API that executes inside database.

Table 1 Columns of Node table in network model

NODE_ID	230
NODE_NAME	NODE-230
GEOMETRY	MDSYS.SDO_GEOMETRY(3001,NULL,MDSYS.SDO_POINT_TYPE (42.2019449799705,100.382921548946,-3.7),NULL,NULL)
ACTIVE	Y

Table 2 Columns of link table in network model

LINK_ID	15
START_NODE_ID	452
END_NODE_ID	455
LINK_NAME	Link-452-455-Corridor
GEOMETRY	MDSYS.SDO_GEOMETRY(3002,NULL,NULL,MDSYS.SDO_ ELEM_INFO_ARRAY(1,2,1),MDSYS.SDO_ORDINATE_ARRAY (115.306027729301,85.9775129777152,1.8,115.306027729301, 82.9483382781573,1.8))
LINK_LENGTH	3,029174699557899
ACTIVE	Y
LINK_TYPE	Corridor

There are two ways to define data structures for a network. One is to create network automatically by calling CREATE_SDO_NETWORK procedure defined in SDO_NET package in Oracle Spatial. This procedure creates all the tables and populates the metadata. This procedure is not atomic. If it fails to complete it may cause a half created network. The automatic network creation method gives very little control over the actual structuring of the tables and gives no control at all over their physical storage (table spaces, space management, partitioning and so on). But this procedure is easy to use and makes sure the table structures are consistent with metadata.

The other and more flexible way is creating tables manually. Creating tables is not enough to define a network in Oracle Spatial. The actual naming of the tables that constitute a network and their structure should be defined in a metadata table called USER_SDO_NETWORK_METADATA as shown below by an insert statement ensuring that the table structures are consistent with metadata.

```
INSERT INTO USER_SDO_NETWORK_METADATA
 (NETWORK,NETWORK_CATEGORY,GEOMETRY_TYPE,
NO_OF_HIERARCHY_LEVELS,NO_OF_PARTITIONS,LINK_DIRECTION,
NODE_TABLE_NAME,NODE_GEOM_COLUMN,NODE_COST_COLUMN,
LINK_TABLE_NAME,LINK_GEOM_COLUMN,LINK_COST_COLUMN,
PATH_TABLE_NAME,PATH_GEOM_COLUMN,PATH_LINK_
TABLE_NAME,
NETWORK_TYPE)
 VALUES('CORPORATION_PUTRAJAYA','SPATIAL','SDO_
GEOMETRY','1',
```

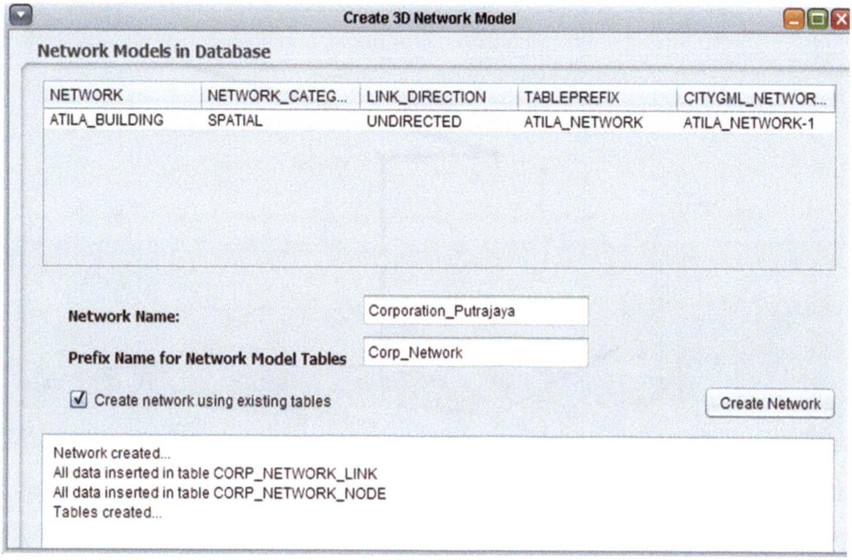

Fig. 4 Automating network creation

'1','UNDIRECTED', 'CORP_NETWORK_NODE','LOCATION',NULL,
'CORP_NETWORK_LINK','GEOMETRY', 'LINK_LENGTH',
'CORP_NETWORK_PATH','GEOMETRY',
'CORP_NETWORK_PATH_LINK','Corp_Network')

Our implementation automates network definition in Oracle Spatial database using manual network creation method presented in this section. As soon as the 3D model of a building in LOD2 and its linear network model in LOD0 opened from CityGML format, the network model creation menu of the implementation gets active to be used (Fig. 4). Network creation tool reads CityGML data, creates tables to define network, inserts proper data into tables and defines network.

4 Performing Network Analysis

The implementation performs network analysis based on Java API provided by Network Data Model of Oracle Spatial. Many kind of network analysis such as shortest path, travelling salesman, given number of nearest neighbors, all possible shortest paths between given nodes, all nodes within given distance, finding reaching nodes to a given node, finding all possible paths between two nodes and finding shortest paths to a node from all other nodes in the network can be performed as well as under some kind of constraints like avoided nodes, links and so on.

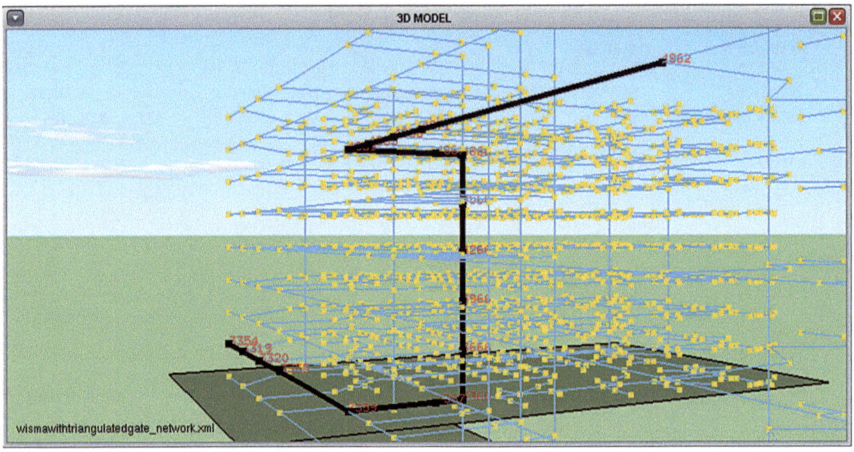

Fig. 5 Shortest path between two nodes without any constraint

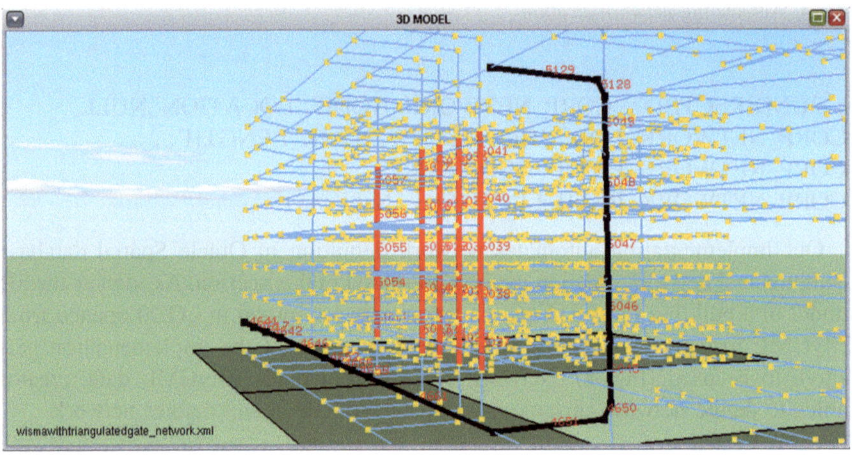

Fig. 6 Updated shortest path when the elevators are all out of order in a part of building

In this section some examples for network analysis will be presented. Figure 5 shows a shortest path analysis without any constraint and Fig. 6 shows how the shortest path is updated after links associated with elevators are avoided showed by red lines which means elevator is not in use any more in that part of building.

Figure 7 shows all nodes within 40 m distance to node 3354 and Fig. 8 shows the updated result of the same analysis after events occurred in all associated links used to go upstairs.

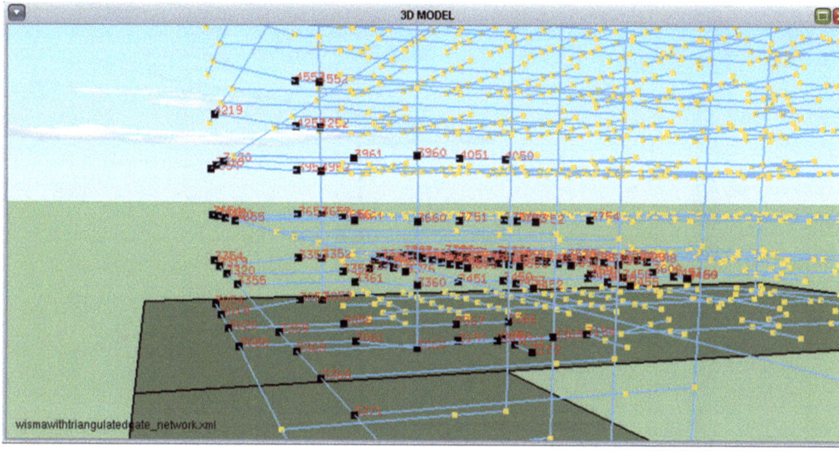

Fig. 7 All nodes within 40 m distance to node 3354

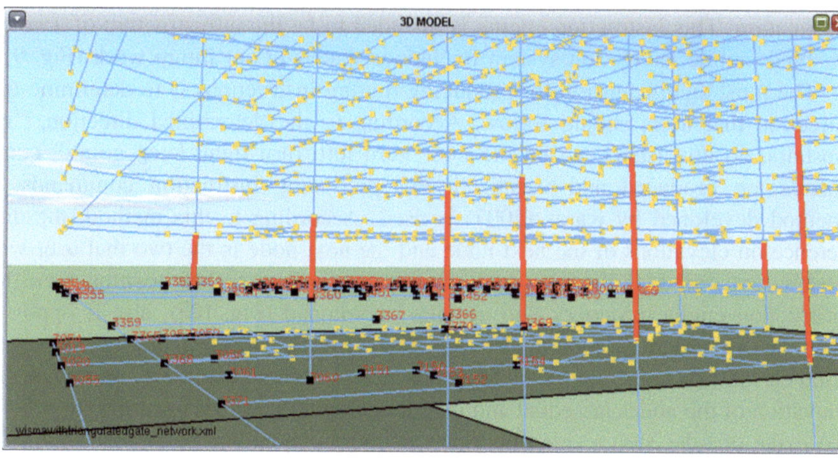

Fig. 8 All nodes within 40 m to node 3354 after events occurred that avoid going upstairs totally

5 Routing Instruction Engine

One of the most important components of an ideal evacuation system is an instruction engine which should produce real time instructions for the users to assist them accurately till they arrive destination. Our implementation has such an instruction engine which is integrated into simulation module to produce voice commands and visual instructions for assisting users dynamically on the way to the

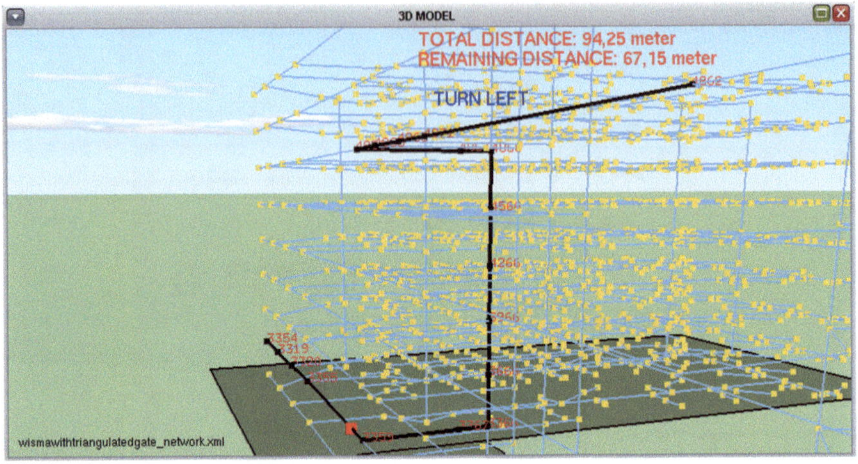

Fig. 9 Simulation process of instruction engine

destination. This instruction engine is intended to be the infrastructure of a voice enabled mobile navigation system for indoor spaces in our future work (Fig. 9).

The most significant job for producing routing instructions is to determine the direction that users should follow. According to the determined direction, "go upstairs, go downstairs, go on the floor, turn left, turn right, keep going" commands are produced and vocalized. For producing instruction commands, a method developed by Karas (2007) is used. According to this method, the difference on elevations of the next node and the next node in the two that user will visit is compared. If the next node in the two is in a higher position then the instruction engine produces "go upstairs" command (Fig. 10a), else if opposite then "go downstairs" command is produced (Fig. 10b), else if the user has just used elevator or stairs, command to be produced is "go on the floor", otherwise the elevation of the nodes are equal and instruction engine decides to turn right or left. To make this decision a side direction calculation should be performed.

Side direction calculation determines if a point is on the right side or on the left side of a line segment. If we suppose the end points of a line segment as A and B, then if a point C is on the right side of this line segment then result is positive(+), otherwise negative(−). Assuming point A is the node user stands, point B is the next node to A and the point C is the next node to B, if we calculate the length and sign of perpendicular distance of point C to line segment AB then the final direction can be determined. If the sign of the result is positive(+) then the command should be "turn right" on the point B (Fig. 10c), otherwise the command is "turn left" (Fig. 10d). If the calculated distance is zero the instruction engine produces "keep going" command (Fig. 10e).

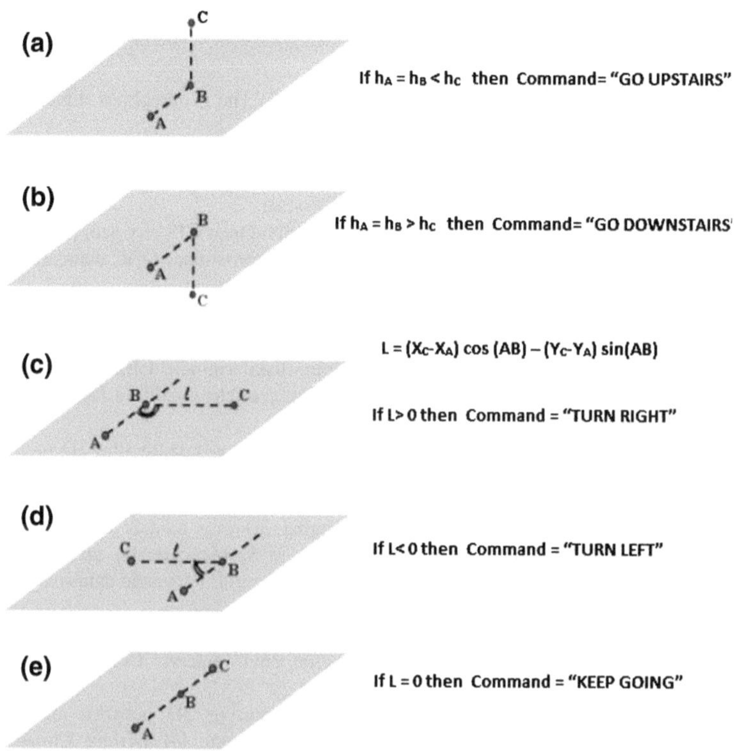

Fig. 10 Determining routing instructions

6 Conclusions

This paper presented a Java based 3D-GIS implementation which can visualize 3D building and network models from CityGML format and automate 3D network definition in Oracle Spatial's Network Data Model. We showed some samples of performing 3D network analysis with visualized results supporting both graph based and geometric constraints applied. We also elaborated a method for producing voice commands and visual instructions for assisting people dynamically on the way to the destination which is intended to be the routing engine infrastructure of our intelligent evacuation system work in progress. Our experiments successfully showed that our 3D-GIS implementation could be improved to design an ideal navigation system for evacuation purpose.

References

Cutter S, Richardson DB, Wilbanks TJ (eds) (2003) The geographical dimensions of terrorism. Routledge, New York and London, pp 75–117

Gartner G, Huang H, Schmidt M, Li Y (2009) Smart environment for ubiquitous indoor navigation. In: Proceedings of the 2009 international conference on new trends in information and service science, Beijing, China, pp 176–180

Gröger G, Kolbe TH, Czerwinski A, Nagel C (2008) Open GIS city geography markup language (CityGML) encoding standard, version 1.0.0, international OGC standard. Open geospatial consortium

Jun C, Kim H, Kim G (2009) Developing an indoor evacuation simulator using a hybrid 3D model. In: Lee J, Zlatanova S (eds) 3D geo-information science. Springer, Berlin, pp 173–178

Karas IR (2007) PhD thesis, Objelerin Topolojik İlişkilerinin 3B CBS ve Ağ Analizi Kapsamında Değerlendirilmesi. YTÜ FBE Jeodezi ve Fotogrametri Anabilim Dalı Uzaktan Algılama ve CBS, pp 99–101

Karas IR, Batuk F, Akay AE, Baz I (2006) Automatically extracting 3D models and network analysis for indoors. In: Abdul-Rahman A, Zlatanova S, Coors V (eds) Innovation in 3D-Geo Information System. Springer, Berlin, pp 395–404

Kolbe TH (2008) Representing and exchanging 3D city models with CityGML. In: Lee J, Zlatanova S (eds) 3D Geo-Information Science. Springer, Berlin, pp 15–31

Kothuri R, Godfrind A, Beinat E (2010) Pro oracle spatial for oracle database 11 g. Apress, New York

Kwan MP, Lee J (2005) Emergency response after 9/11: the potential of real-time 3D GIS for quick emergency response in micro-spatial environments. Comput, Environ Urban Syst 29:93–113

Musliman IA and Rahman AA (2008) Implementing 3D network analysis in 3D GIS. International archives of ISPRS, vol 37, part B, comm. 4/4, Beijing, China

Musliman IA, Rahman AA and Coors V (2006) 3D navigation for 3D-GIS—initial requirements. Innovations in 3D geo information systems, Springer, New York p 125–134

Pu S, Zlatanova S (2005) Evacuation route calculation of inner buildings. In: van Oosterom PJM, Zlatanova S, Fendel EM (eds) Geo-information for disaster management. Springer, Heidelberg, pp 1143–1161

Zlatanova S, van Oosterom P and Verbree E (2004) 3D technology for improving disaster management: geo-DBMS and positioning. In: Proceedings of the 20th ISPRS congress, Istanbul, Turkey

A Three Step Procedure to Enrich Augmented Reality Games with CityGML 3D Semantic Modeling

Alborz Zamyadi, Jacynthe Pouliot and Yvan Bédard

Abstract 3D representations have been recognized as an essential component of Augmented Reality (AR) oriented applications. However, not many examples of AR-oriented applications employ structured 3D data models despite the existence of standard 3D information models like CityGML. One of the reasons for this shortcoming can be explained by lack of a semantic-based modeling method for enriching AR-oriented data models with 3D features. Therefore, a three step procedure is proposed to address this limitation as (1) back-ward engineering of an AR-oriented application to its current data model, (2) enriching the current data model with 3D features, (3) and mapping the enriched model to a standard 3D information model. A notable contribution of this work is that the procedure of data modeling has been subject to the UModelAR meta-model which has brought a complementary standpoint to the employment of 3D geospatial modeling in AR environments. Furthermore, the 3D enriched data model has been mapped to CityGML information model with CityGML Application Domain Extension (ADE) concept. To demonstrate the feasibility of this approach, an operating mobile AR-oriented game has been used for the case study.

Keywords Augmented reality · 3D model · Semantic · CityGML · Application domain extension (ADE)

A. Zamyadi (✉) · J. Pouliot · Y. Bédard
Département des Sciences Géomatiques, Université Laval, 1055, avenue du Séminaire, Québec City, QC G1V 0A6, Canada
e-mail: alborz.zamyadi.1@ulaval.ca

J. Pouliot
e-mail: jacynthe.pouliot@scg.ulaval.ca

Y. Bédard
e-mail: yvan.bedard@scg.ulaval.ca

J. Pouliot et al. (eds), *Progress and New Trends in 3D Geoinformation Sciences*, Lecture Notes in Geoinformation and Cartography, DOI: 10.1007/978-3-642-29793-9_15,

1 Introduction

Augmented Reality (AR) has been defined as a live scene of reality in form of video streaming being supplemented by adding numeric representations such as graphics, annotations, images, sounds, videos, and other comments (Azuma 1997; Milgram et al. 1994). For instance, one can see restaurant names, menus, and working hours at the corresponding locations using an AR-oriented tourist guide while walking on a street in an urban area. The AR tourist guide application augments the live video capture on a user's smart phone with certain information which are not accessible to the user on the street unless checking each restaurant door by door. Indeed, the displayed information (i.e. restaurant names and menus) are virtual with respect to the user's observable reality and may appear as markers (2D or 3D graphics), annotations, or even voice guides.

There are various types of AR-oriented applications like tourist guides such as Layar,[1] games such as ARhrrrr (Oda and Feiner 2010; Yang and Maurer 2010), urban design and monitoring such as VIDENTE (Schall et al. 2009). Höllerer and Feiner (2004) have indicated the principal elements of AR-oriented 3D capabilities as visualization, registration, and interaction. Visualization considers rendering 3D graphics improving the visual quality of augmentation with rich visualization features. 3D registration is the ability to comprehend 3D position and orientation of real world objects and augmentation features respectively to resolve occluding objects considering observers' standpoint (e.g. identifying the hidden parts of an augmenting 3D graphic which are covered by the foremost buildings). 3D interaction considers the ability to interact with objects explicitly like retrieving the information for specific doors or windows of a building. These elements could be considered as the base for examining and comparing the AR-oriented 3D perspectives.

For instance, several 3D AR-oriented research initiatives have become subject to 3D visualization (Schall et al. 2007a; De Souza e Silva 2009; Lee and Park 2005; Liestol 2009; Zlatanova and Van Den Heuvel 2001). Some others AR prototypes are focusing on 3D data acquisition (Thomas et al. 2010) or 3D registration (Höllerer and Feiner 2004; Schall et al. 2007a). Finally few AR research initiatives were interested in data exchange (Badard 2006; Reitmayr and Schmalstieg 2003), or advanced behaviour-aware interactions (Harrap and Daniel 2009).

A review of various AR-oriented implementations shows that they have been mainly about 3D visualization using computer aided design (CAD) formats (Lee and Park 2005), 3D graphics like 3ds and obj (Liestol 2009), and 3D textual markup formats like KML and X3D (Thomas et al. 2010; Wojciechowski et al. 2004). However, as suggested by Zlatanova and Van Den Heuvel (2001), the use of semantics and explicit accessibility to 3D features like geometry and appearance is a requirement if the accomplishments of full 3D capabilities of AR perspectives

[1] http://layar.pbworks.com/w/page/7783211/3D-objects-in-a-layer

are considered. Stadler et al. (2009) have indicated that the mentioned 3D sources from the actual implementations are weak in semantics and explicit definition of geometries. In response, the use of geospatial database systems (Schall et al. 2007b) and standard semantic information models like CityGML and Building Information Model (BIM) (Woodward et al. 2010; Hasse and Koch 2010) have been suggested. But our investigation has shown that the relative experiments are mainly technology driven and have not been conceptualized for global diffusion to other developers.

In this context of involving a semantic model with AR applications, one of the well-known open standards for presenting a structure 3D semantic model, City-GML, leaded by the Open Geospatial Consortium (OGC), is certainly of interest. CityGML (Kolbe et al. 2005; OGC 08-007rl 2008) addresses the geometric and semantic representation of model objects explicitly as well as the appearance information having in mind the requirements for 3D AR-oriented capabilities. CityGML has been successfully examined by several types of applications such as Geographic Information System (GIS), web mapping, and navigation operating in various domains such as architecture, urban planning, transportation, and tourism (Dollner et al. 2006; Kolbe et al. 2008; Vanclooster et al. 2009). CityGML has been also a preferable choice for upgrading existing geospatial- applications from 2D to 3D since CityGML information model is extendible in different ways.

A recent study by Hasse and Koch (2010) has employed an extended adaptation of CityGML for Electronic Nautical Charts (ENC) alongside an AR engine. The output of their work has been a CityGML ADE upgrading the pre-known and well defined 2D thematic model (i.e. ENC) with 3D features. Finally, an AR-oriented interface has been developed employing the ENC ADE for displaying 3D representation of ENC objects and the corresponding electrical information. The actual tendency in employing 3D representations and CityGML ADE seems sufficient for visualization aspect of AR and domain specific thematic information. However, their work does not consider various application oriented needs (e.g. geospatial game events, constrains and behaviours) and the procedure to equip them with 3D representations.

Indeed, this paper presents a procedure for enriching an AR application (a game in our case) with 3D capabilities conforming to AR and 3D geospatial data standards. Three steps are proposed as a modus operandi. At first, the current data model of a selected AR game is figured out based on the back-ward engineering of the game scenario. Second, the data model is extended with the essential AR-oriented 3D features. An important distinction of this step is that the procedure of data modeling has been subject to the UmodelAR meta-model which has brought a complementary standpoint to the adoption of 3D geospatial modeling by AR environments (Zamyadi et al. 2011). Third, the extended model (now supporting AR-oriented 3D capabilities) is mapped on the CityGML information model.

This work is part of the GeoEduc3D project funded by the Canadian Network of Excellence—GEOIDE (GEOmatics for Informed Decisions).[2] GeoEduc3D is a major academic project carried out by several Canadian universities aiming at designing learning-oriented tools that foster enriched and augmented player experiences in real geographies. Energy Wars Game, one of the outcomes of the GeoEduc3D Project[3], is an award winning educational game prototype (Dumont et al. 2011) deploying a mobile AR interface in known urban geographies like a university campus. The current version of Energy Wars Mobile runs a game scenario (known as Energy Wars mini scenario) and employs 2D geospatial maps, interacts with 2D coordinates values and allows 3D graphics rendering. Consequently, the current AR interface of the game does not support full 3D capabilities. However, the final features of the Energy Wars project like interactions with various parts of buildings such as windows and walls, managing occlusions in the AR-oriented game area, and interactive computer generated game personas would benefit from comprehending the semantics and 3D position and orientation of real world objects and the augmenting features respectively. Thus, the main objective of the work presented in this paper is to propose an approach for enriching the Energy Wars mini scenario with 3D features and CityGML semantic modeling.

The paper is organized as follows. Section 2 presents the three step procedure proposed to enrich the Energy Wars mini scenario. Section 3 offers a discussion about the experiment, the advantages, and the limits of the proposed approach. Section 4 concludes this paper and gives the future possibilities ahead.

2 The Three Step Procedure for Enriching an AR Application with 3D Semantic Modeling

2.1 Back-ward Engineering of Energy Wars Game

As indicated, the Energy Wars Mobile prototype is a learning oriented game which is played in known urban areas employing an AR interface (Dumont et al. 2011). The goal of this game is to help teenagers to learn about some of the issues of energy consumption and Geomatics. At this time, a simple version of this game has been implemented and tested in Laval University known as Energy Wars mini scenario. This game is played on a field (i.e. the university campus) where certain buildings have been provided with energy consumption information. Each team, consisting of three gamers with different roles and role specific smart phone based mobile interfaces, must find the buildings with low energy efficiency and earn enough money to improve their energy consumption coming back to their whereabouts. Money can be earned by answering to relevant questions about

[2] http://www.geoide.ulaval.ca/

[3] http://geoeduc3d.scg.ulaval.ca

Fig. 1 Screen shot of energy wars mini scenario AR interface with augmented game hint cubes (image provided by Energy Wars mini scenario development team in Laval University)

energy saving or finding bonuses tagged at different locations on the field. Each location represents a game hint which might carry one or several questions, a money bonus, or a money loss. These hints can be observed on the screen of the smart phone as graphical cubes augmenting the reality and gamers can find out their content by arriving at their location. The augmented hint cubes can be observed only by specific gamers (according to the given roles) inside the visibility range of the gamer. Figure 1 shows a screen shot from the Energy Wars mini scenario AR-oriented interface.

The current version of the game uses 2D dimensional geospatial representations for locating buildings (i.e. footprint of the building with 2D coordinate values) and game hints (i.e. points with 2D coordinate values). Furthermore, the 3D coordinates for each vertex of every graphical cube is stored for every game hint. These coordinates are used by the rendering engine to create the corresponding 3D graphics. Since the augmenting cubes are supposed to hover above the ground surface, the base elevation of each cube (the Z coordinate of the lowest vertices) has been derived by adding a constant value to the average elevation of the ground surface of the Laval University campus. The use of the average elevation of ground surface has been sufficient for the Laval University campus since the target area on which the game has been tested is flat.

The diagram in Fig. 2 shows the current Energy Wars mini scenario data model in UML formalism. Since we have not been involved in the implementation of the Energy Wars mini scenario, this data model is the result of back-ward engineering and has been created based on the discussions with the developers of the game. In the current data model, the building whereabouts is spatially described by the *Footprint* class which is a part of the building composed of one or several connected line segments. The building energy properties are addressed by the corresponding attributes of the *Building* class indicating the actual energy consumption (the Energy Consumption attribute), the classification of energy consumption (the Energy Efficiency Level attribute and the Energy Efficiency Level domain value), and the required amount of money for improving the building energy consumption (the Needed Money attribute). Furthermore, the *Game Hint* class represents the Energy Wars mini scenario game hints. The *Hint Type* attribute defines the type of the game hint and has three possible values indicated in the Game Hint Type

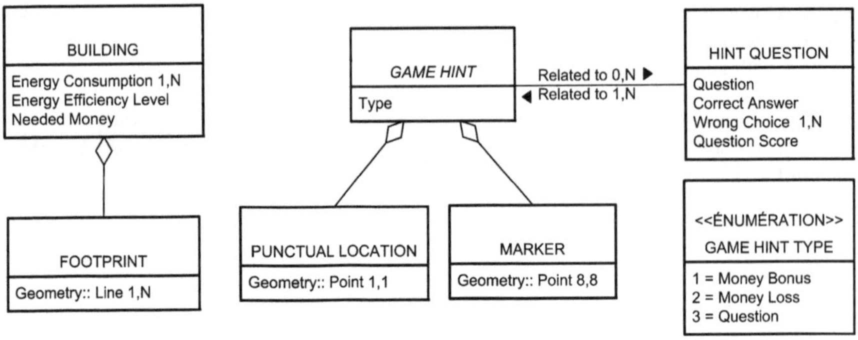

Fig. 2 The current data model of Energy Wars Mini scenario

domain value. The *Game Hint* class has two components representing the game hint location (the *Punctual Position* class) and the coordinates of the augmenting cube vertices (The *Marker* class). Finally, the Hint Question class enables storing the corresponding question of each game hint (*Question* attribute), the question right answer (*Correct Answer* attribute), the wrong choices (*Wrong Choice* attribute), and the amount of the money that can be earned choosing the right answer (*Question Score* attribute).

Based on the current data model, the Energy Wars mini scenario only supports interactions at certain locations with 2D coordinate values and does not comprehend the 3D position and orientation of real world objects and the augmenting features respectively. However, the final features of the Energy Wars project aiming at interactions with building parts or computer controlled game persona with graphical augmentation capable of hiding behind real world objects would benefit from them.

2.2 The Basis for Enriching AR-Oriented Games with 3D Features

The next step in the proposal method aims at enriching the current data model with essential 3D features comforting to AR. Azuma (1997) has defined three conditions that should be satisfied within an AR environment as:

- combining real and virtual;
- real time interactions at certain coordinates or with certain objects;
- being registered in 3D (i.e. comprehending 3D position and orientation of the real world objects and the augmenting features).

Accordingly, Höllerer and Feiner (2004) have classified the principal roles of 3D models in AR environments as providing detailed information for visual rendering, 3D registration of augmenting features, and resolving occluding objects with (e.g. identifying the hidden part of a game hint which is covered by the foremost building).

In order to design 3D data models comforting to the aforementioned, Zamyadi et al. (2011) have suggested meta-modeling as the prerequisite for modeling AR environments. Meta-model is the structured abstraction of model concepts that defines the specifications of model components prior to developing new models or upgrading the existing ones (Caplat 2008). The principal idea has been derived from the additional features and considerations that exist in AR environment comparing to the ordinary reality. There are different thematic points of view for modeling the reality. However, in AR, complementary meanings and relations are superimposed to the real themes requiring an additional outlook by which 3D models could adopt AR concepts.

The output of this proposal, named as UModelAR, is a meta-model which has brought a complementary standpoint to designing 3D semantic models with respect to AR (Zamyadi et al. 2011). UModelAR follows a top down order for representing the relation between virtuality and the associated reality in AR environments trying to bring the perspectives of AR and 3D model conceptualizers, developers, and users closer to each other. UModelAR meta-model has considered the various aspects of data modeling for AR-oriented 3D capabilities introducing the real and virtual semantics of features, localization of the AR environment and its conforming geometries. Furthermore, UModelAR defines a mechanism to create the application specific hierarchies within which the events and action occur. Such hierarchy is complementary to the common classifications of real world themes creating various semantic spaces within the AR environment. UModelAR helps defining the type and order of 3D data model components considering specific scenario and common AR needs. UModelAR consists of five packages as (1) Augmented Reality Interface (2) World and Associated Representations (3) Acquisition of Physical Reality (4) Model Localization (5) Model Description.

The following section will present the Energy Wars mini scenario data model which has been enriched with AR-oriented 3D features mapped on CityGML.

2.3 *Mapping the Enriched Data Model of the Game to CityGML*

As mentioned, the Energy Wars mini scenario is a good example of an AR application which operates in a known urban area. Hence, it has the potential to be played in various urban locations all around the world (i.e. in any university campus or city neighbourhood providing 3D geospatial urban representations). On the other hand, the final features of the Energy Wars project would benefit of having 3D geospatial supporting data layer. Therefore, the current data model has been upgraded with the essential AR-oriented 3D features according to the

UModelAR approach on modeling AR environments and by exploiting a global 3D geospatial information model (i.e. OGC CityGML standard).

CityGML addresses the thematic properties of city objects alongside their geometric properties. The base class of all thematic objects is *_CityObject* (an abstract class) from which they inherit the basic properties of the CityGML core module such as Appearance feature and External References to corresponding objects in external datasets. The actual thematic module of CityGML is composed of the most relevant urban fields including Digital Terrain Models, Buildings, Vegetation, Water Bodies, Transportation Facilities, and City furniture. CityGML can represent objects in five consecutive Levels of Detail (LOD) where objects become more detailed as LOD increases. In other words, the same object may be represented by one or several LoDs enabling different degrees of resolution (OGC 08–007rl, 2008; Kolbe et al. 2005).

The UML class diagram in Fig. 3 shows the Energy Wars mini scenario data model enriched with 3D features from CityGML information model based on the UModelAR approach on modeling AR environments. The aim of this paper is not to explain this diagram in detail. But in brief, here are some explanations.

- *White Thin-Edge* classes are from the current data model. For instance, The Footprint class corresponds to building outset extent in the Energy Wars mini scenario around which the associated events occur when the users arrive at its whereabouts from each building side.
- *Gray* classes represent the features which have been supplemented according to UModelAR (i.e. 11 classes). The new *Game Element* class organizes the Energy War application theme (i.e. game). This means that game features inherit from *Game Element*. For instance:
  - The *Game Hint* class is added under *Game Element* with two localized features (the *Punctual Location* and *Marker* classes) and number of descriptive properties.
  - The *Hint Question* class from the current data model is extended by two further classes. The new *External Question* class is used for referring to questions from external URLs.
  - The building energy consumption properties are defined by the *_Building Energy Measure*.
  - The new *Augmentation Appearance* class is an AR-oriented feature for defining the appearance information of augmenting graphics.
- *White Bold-Edge* classes are CityGML features indicating the corresponding CityGML ADE extension. Building in Energy Wars is equivalent to the CityGML *Building* class that is derived from the CityGML *_Abstract Building* class. Therefore, the building energy consumption properties would be added to CityGML *_Abstract Building* from the *_Building Energy Measure* (which is a *Gray* class) using the mechanism of extension by super classing. CityGML *External Code Lists* is used for defining the domain values. The CityGML *Appearance* from the CityGML core module has been used for defining appearance information of

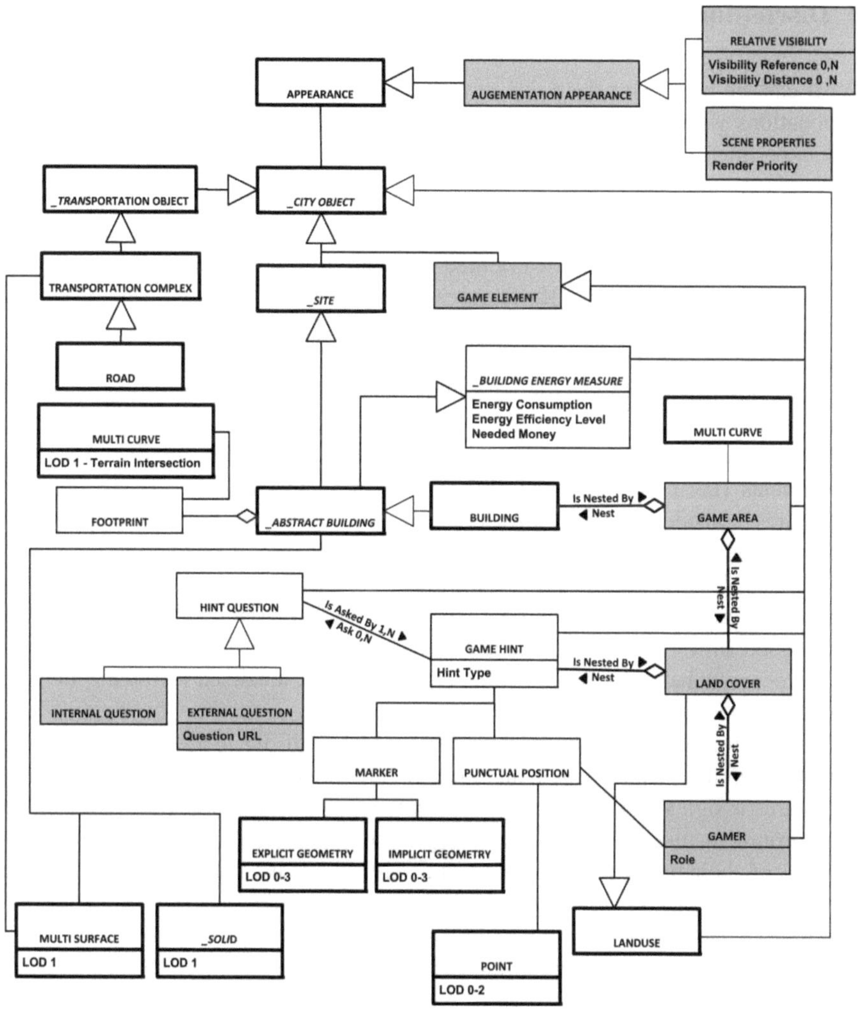

Fig. 3 Energy wars mini scenario data model enriched with 3D features according to UModelAR approach and mapped to CityGML 3D information model—*White Thin-Edge* classes are from the current data model (Step 1). *Gray* classes have been added according to UModelAR (Step 2). *White Bold-Edge* classes are used to map the enriched model to CityGML (Step 3)

the AR environment. The CityGML *_CityObject class* is associated with the CityGML *Appearance* class. *Augmentation Appearance* (which is a *Gray* class) inherits from CityGML *Appearance* and is exclusively associated with the features which augment the real scene.

3 Discussions

What can we now learn from this three-step procedure for enriching AR-oriented applications with a structured 3D data model mapped to CityGML data model? Our experiment allows us to state that:

- The backward engineering aims at creating the corresponding data model behind the AR application and depends on the original work. A developer needs to be fully aware of the target AR-oriented application design if having the intention to enrich it with further 3D features. Since UModelAR has been developed using UML conforming to the common modeling standards, it is strongly suggested to use UML formalism for creating the current data model. This would facilitate UModelAR reasoning and finding the correspondences in CityGML.
- In agreement with the first condition of Azuma (1997) for creating AR environments (i.e. combining real and virtual), it is necessary to classify the AR features considering their correspondence to the reality before mapping AR environments to 3D information models like CityGML. Considering CityGML for example, every localized feature of the AR environment which corresponds to a real urban object has an equivalent CityGML thematic object or would be extended in a CityGML thematic module. On the other hand, it is required to create a new theme from *_CityObject* for adding the localized features which are not derived from a real urban object.
- In order to enrich an AR-oriented data model with 3D features, the localized feature of the AR environment must be identified. However, since localized features assign various descriptive features like attributes and appearance to the environment, their spatio-semantic representation should conform to the notability of their measures (i.e. shape and size). It means that a localized feature should have an explicit geometric representation such as 0D, 1D, 2D or 3D primitive/aggregation, if the scale, precession, and accuracy of its location, shape and size are considered in spatio-semantic computations like occlusion or interactions. Otherwise, it would be represented by symbols. For instance, in the new data model the *Marker* class is associated with 0D to 3D CityGML *Explicit Geometry* for permitting multiple interactions with various marker parts or orientation. It is also associated with CityGML *Implicit Geometry* for visualization-only cases. The use of implicit geometry permits using external 3D graphics like VRML or X3D from any source. This may be preferable for localized features with duplicating shape like a cube. However, it should be remembered that each source of an implicit 3D geometry (e.g. VRML or X3D) has a local spatial reference which would be oriented to the CityGML dataset global spatial reference by an anchor point and a transformation matrix. For a mobile scenario with several visual augmentations, the on-the-fly metrical computations could be a drawback.
- In order to register the interactions and behaviours on AR environment features, an application theme should be created considering the localization hierarchy of

its components partitioning the AR environment. It has been noticed that 3D localized features create the various spaces where only specific game (application) events occur (i.e. UModelAR Space features). Each UModelAR Space feature, except the top most space (i.e. the global extent of the application) may partition one or many superior UModelAR Space features. The lowest level of an application theme hierarchy is the UModelAR Object feature and is not partitioned any more. It means that no other localized feature is part of an UModelAR Object feature or no application event is localized inside it. The UModelAR application theme hierarchical relationships among UModelAR Spaces/Objects are shown with aggregations in the class diagram and will be created by GML *XLinks* for implementation.

- The extent of the game, with respect to the real urban area where the game is played, is an important UModelAR Space Feature for deciding the spatial extent and the application theme features and relationships. For example, in the Energy Wars mini scenario users are only part of the game while they are within a particular extent. This particular area corresponds to the top most UModelAR Space feature and is represented by the new *Game Area* class. Furthermore, users can not enter buildings and there is no specific event that occurs inside buildings. Therefore, *Building* class corresponds to an UModelAR Object feature and is directly related to the *Game Area* class with an aggregation. Likewise, *Game Hint* is an UModelAR Object feature. The *Game Hint* must be part of a superior UModelAR Space feature having in mind that game hints cannot situate on buildings. This UModelAR Space feature is represented by the new *Land Cover* class in the new data model.

- If the interactions (e.g. game events) are immersive in a way that user position becomes an event or constrain handler, the position of user (e.g. gamer) should be explicitly defined as a localized feature. For instance in the Energy Wars mini scenario, the visual augmentation of game hints is determined according to various user positions who may have different roles. Considering the importance of the hierarchical partitions where game event can occur, *Land Cover* is the only partition of the *Game Area* that users can traverse. Therefore, the *Gamer* class corresponds to UModelAR Object feature whose superior UModelAR Space feature is *Land Cover*. Since *Gamer* is only part of *Land Cover* partition it can only interact with other parts of the *Game Area* that are part of *Land Cover* (like *Game Hint*) or are adjacent to it. Indeed, the definition of UModelAR Space and Object features would help to indicate the topological relationships of the AR environment as well.
- An AR-oriented data model may include number of localized features who are not part of the application theme but would be useful to improve the sensitivity of the computer based system to the surrounding reality. Such features are preferred to be explicitly presented avoiding symbols. In this case study, *Street* is not part of the game scenario but could improve the awareness of the system by its presence, like by warning the users about the possible dangers as they enter the street area.

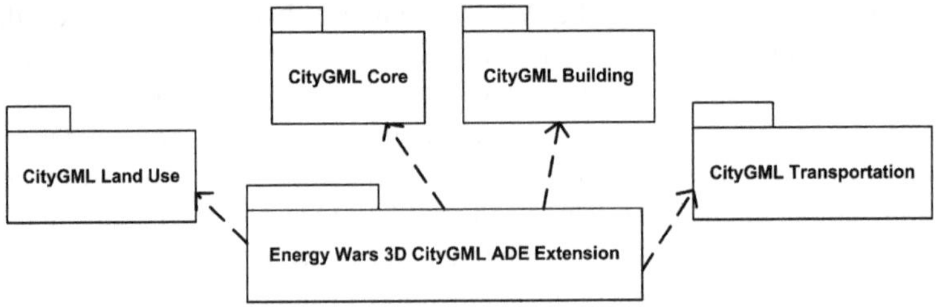

Fig. 4 The dependency model of energy wars mini scenario 3D extensions to CityGML

- The geometric representation of various localized features of an AR environment may require different resolution with respect to AR-oriented application needs and the target 3D standard (e.g. CityGML).
  - The *Punctual Location* class of the new data model is mapped to CityGML by LOD-0 to LOD-2 *Point* with the horizontal and vertical accuracy varies between 2 and 5 m. However, the LOD-0 3D point accuracy (around 5 m) would be sufficient for a mobile scenario.
  - In the current data model the augmentation component of the *Game Hint* (the *Marker* class) is limited to a box like cube. In the new data model the *Marker* class may be associated with either CityGML *Explicit Geometry* or CityGML *Implicit Geometry*, both from LOD-0 to LOD-3. An LOD-3 Marker permits various parts with different interactive or appearance capabilities (e.g. buttons on different sides of the marker) while an LOD-1 marker is limited to a box like shape.
  - The most relevant CityGML concept to building footprint is the CityGML *Terrain Intersection Curve* (*TIC*) denoting the exact position where the building touches the earth surface. Therefore, the *Footprint* class is by TIC concept. However, with the current version of CityGML it is essential to have an LOD-1 TIC since LOD 0 is not permitted.
- The package diagram in Fig. 4 shows the Energy Wars mini scenario CityGML ADE package diagram as the sketched arrows mark the dependencies to CityGML modules. Dependencies occur when classes from a package are related to or derived from classes of another package. The direction of the arrow indicates the direction of dependency. The Energy Wars mini scenario CityGML ADE has dependencies to the CityGML Core, Building, Land Use, and Transportation modules. According to this resolution, The Energy Wars mini scenario game can be implemented in every urban area around the world which has been presented by CityGML datasets improving the portability, interoperability, and reusability of existing resources. Such game (application) can be played in any of such places without any change. For instance, the gamer may connect to Delft CityGML model as soon as arriving there and play the game as he/she has been playing in Quebec City.

4 Conclusions and Future Work

In this paper we proposed a three step procedure to enrich the Energy Wars mini scenario with a structured 3D data model based on CityGML. At first, the back-ward engineering step permitted us to know the current data model behind the game. The current data model has shown the missing presence of 3D representations. Therefore, the current data model has been subject to the UModelAR meta-model analysis during the second step for being enriched with 3D features. In the third step, the enriched data model has been mapped to CityGML information model.

For the given case study, the presented steps became feasible in a two week effort which is promoting for the test on other cases. Indeed, it must be remembered that we have been working on the AR-oriented 3D perspectives. If a developer intends to follow our three-step procedure, practical experience with data model conceptualization, 3D representations, and AR domain would be an advantage. Besides, a very important base of the three-step procedure is the UModelAR meta-model conducting the AR-oriented 3D feature modeling. Therefore, the relevant documentation is required to be studied. Accordingly, this procedure has been encouraging since the outcome of UModelAR analysis facilitated identifying the required 3D features and mapping to CityGML information model. However, the developers should be noticed that UModelAR is being used to ensure that the developed data model conform to AR-oriented 3D requirements and does not rate the developed model.

CityGML has shown enough completeness and flexibility for supporting these features either by its original thematic modules and objects or the ADE concept. However, the developers may have a difficulty to choose between CityGML ADE and CityGML Generic Objects and Attributes for cases in the third step (e.g. *Game Hint* and *_Hint Question* classes which are member of the known thematic classifications). In order to resolve such difficulty, it should be remembered that the outcome of the second step of the procedure is an AR-oriented 3D data model indicating the game specific hierarchy of 3D features being reused in every instance of the game. Accordingly, CityGML ADE looks as the appropriate choice for mapping the 3D data model from the second step on CityGML for two reasons. First, ADE permits structured additions to CityGML which are portable and reusable. Second, ADE extensions are formally specified and can be validated against CityGML and respective schemas.

The next step ahead is the implementation of the enriched data model for Energy Wars mini scenario. The 3D model of the game area in the target 3D standard dataset format (e.g. CityGML) is the basic requirement to implement the output of the three-step procedure. The additional features (under *Game Element* class in Fig. 3) would be implemented conforming to the target 3D standard (e.g. the complementary XML). In order to examine the functionality of the output model, it is recommended to create the additional features dataset (the *Game Element* namespace) and run the application in various locations with separate CityGML datasets. After validation phases, one of the outcome could be a new

CityGML-based application schema for AR game, that is an important market for promoting the use of such semantic model. Indeed, CityGML is becoming more popular worldwide and such perspective would be a benefit to both 3D modeling and AR game community. At the same time being in contact with OGC would help us disseminate our results.

Finally, we are also interested to verify the achieved portability, and reusability of the proposed approach in testing its implementation for other case studies. All we need is an AR-oriented application functioning in urban areas (demanding AR 3D capabilities), the corresponding technical documentations for the application theme and design, and the 3D model of the target area (in CityGML dataset format or convertible to that). Furthermore, it would be possible to follow the three-step procedure and run the application in various locations which provide CityGML 3D model.

Acknowledgments This study was done part of GeoEduc3D project, a research initiative funded by the Canadian Network of Excellence—GEOIDE (GEOmatics for Informed Decisions). Accordingly we wish to thank all scientific and industrial partners, the Geomatics Department and Centre de Recherche en Géomatique (CRG) at Laval University for their supports. We also thank Thomas Butzbach from the developers of the Energy War mobile game for helping us with procedure of back-ward-engineering.

References

Azuma R (1997) A survey of augmented reality. Tele-operators Virtual Environ 6:355–385

Badard T (2006) Geospatial service oriented architectures for mobile augmented reality. In: Proceedings of the 1st international workshop on mobile geospatial augmented reality, Banff, Canada, pp 73–77

Caplat G (2008) Modèles & métamodèles. Presses polytechniques et universitaires romandes, France

De Souza e Silva A (2009) Hybrid reality and location based gaming: redefining mobility and game spaces in urban environments. J Simul Gaming 40:404–424

Dollner J, Baumann K, Buchholz H (2006) Virtual 3D city models as foundation of complex urban information spaces. In: Proceedings of CORP 2006 and Geomultimedia 06, Vienna, Austria

Dumont M-A, Power MT, Barma S (2011) GéoÉduc3D: Évolution des jeux sérieux vers la mobilité et la réalité augmentée au service de l'apprentissage en science et technologie. Can J Learn Technol 37:2

Harrap R, Daniel S (2009) Mobile LIDAR mapping: building the next generation of outdoor environment model for augmented reality. In: IEEE international symposium on mixed and augmented reality (ISMAR), Let's go out workshop, Orlando, Florida, USA. Available at: https://www.icg.tugraz.at/~reitmayr/lgo/harrap_lidar.pdf

Hasse K, Koch R (2010) Extension of electronical nautical charts for 3D interactive visualization via CityGML. In: Proceedings of GeoInformatik 2010, Germany

Hollerer T, Feiner S (2004) Mobile augmented reality (Chapter nine). In: Karimi H, Hammed A (eds) Telegeoinformatics: Location-Based Computing and Services. Taylor & Francis books ltd, London

Kolbe TH, Gröger G, Plümer L (2008) CityGML—3D city models and their potential for emergency response. In: Geospatial information technology for emergency response. Taylor & Francis Group, London

Kolbe TH, Gröger G, Plümer L (2005) CityGML—Interoperable access to 3D city models. In: Van Oosterom P, Zlatanova S, Fendel E M (eds) Proceedings of the international symposium on geo information for disaster management. Springer, pp 883–899

Lee W, Park J (2005) Augmented foam: a tangible augmented reality for product design. In: Proceedings of 4th IEEE and ACM international symposium on mixed and augmented reality. Vienna, Austria, pp 106–109

Liestol G (2009) Situated simulations: a prototyped augmented reality genre for learning on the iPhone. Int J Interact mobile Technol 3:24–28

Milgram P, Takemura H, Utsumi A, Kishino F (1994) Augmented reality: a class of displays on the reality-virtuality continuum. In: Proceedings of telemanipulator and telepresence technologies, vol 2351. pp 282–292

Oda O, S Feiner (2010) Rolling and shooting: two augmented reality games. In: Proceedings of the 28th international conference on human factors in computing systems. Atlanta, USA, pp 3041–3044

OGC 08-007rl (2008) OpenGIS® city geography markup language (CityGML) encoding standard. Groger G, Kolbe TH, Czerwinski A, Nagel C (eds) Open Geospatial Consortium Inc

Reitmayr G, Schmalstieg D (2003) Data management strategies for mobile augmented reality. In: Proceedings of international workshop on software technology for augmented reality systems (STARS). Tokyo, Japan, pp 47–52

Schall G, Mendez E, Kruijff E, Veas E, Junghanns S, Reitinger B, Schmalstieg D (2009) Handeheld augmented reality for underground infrastructure visualization. Pers Ubiquit Comput 13:281–291

Schall G, Mendez E, Junghanns S, Schmalstieg D (2007a) Urban 3d models: what's underneath? Handheld augmented reality for subsurface infrastructure visualization. In: Proceedings of Ubicomp 2007. Springer

Schall G, Junghanns S, Mendes E, Schmalstieg D, Reitinger B (2007b) Handheld geospatial augmented reality using urban 3d models. In: Proceedings of the workshop on mobile spatial interaction, ACM international conference on human factors in computing systems, San Jose, USA

Stadler A, Nagel C, König G, Kolbe TH (2009) Making interoperability persistent: a 3D geo database based on CityGML. In: Jiyeong L, Zlatanova S (eds) Lecture notes in geoinformation and cartography. Proceedings of the 3rd international workshop on 3D geo-information, Seoul, Korea, pp 175–192

Thomas V, Daniel S, Pouliot J (2010) 3d modeling for mobile augmented reality in unprepared environment. In: Kolbe TH, Koing G, Nagel C (eds) Proceedings of 5th international 3D GeoInfo conference, Berlin, Germany

Vanclooster A, Maeyer PD, Fack V (2009) Implementation of indoor navigation networks using CityGML. In: Proceedings of the 4th international workshop on 3D Geo-Information, Ghent, Belgium

Wojciechowski R, Walczak K, White M, Cellary W (2004) Building virtual and augmented reality museum exhibitions. In: Proceedings of the 9th international conference on 3D web technology, Monterey, USA, pp 135–144

Woodward C, Hakkarainen M, Rainio K (2010) Mobile augmented reality for building and construction: software architecture. In: Proceedings of mobile AR Summit at the 9th IEEE international symposium on mixed and augmented reality, Seoul, Korea

Yang J, Maurer F (2010) Literature survey on combining digital tables and augmented reality for interacting with model of human body. Technical report. University of Calgary, Calgary, Canada

Zamyadi A, Pouliot J, Bédard Y (2011) Improving the interoperability of 3D models among augmented reality systems: Proposal for a meta-model. In: Proceedings of the joint ISPRS workshop on 3D city modelling and applications and the 6th 3D GeoInfo Conference. In: The China academic journals electronic magazine. Wuhan, China

Zlatanova S, Van Den Heuvel FA (2001) 3D city modeling for mobile augmented reality. In: Proceedings of CIPA 2001 symposium, Potsdam, Germany

Implementation of a National 3D Standard: Case of the Netherlands

Jantien Stoter, Jacob Beetz, Hugo Ledoux, Marcel Reuvers, Rick Klooster, Paul Janssen, Friso Penninga, Sisi Zlatanova and Linda van den Brink

Abstract This paper describes the motivation and problem statements as well as the ongoing investigations regarding the follow-up activities of the 3D Pilot NL.

J. Stoter (✉)
Kadaster, Hofstraat 110, 7311 KX Apeldoorn, The Netherlands
e-mail: jantien.stoter@kadaster.nl, j.e.stoter@tudelft.nl, j.stoter@geonovum.nl

J. Stoter · H. Ledoux · S. Zlatanova
OTB, GISt, Delft University of Technology, Jaffalaan 9, 2600AA Delft, The Netherlands
e-mail: H.Ledoux@tudelft.nl

S. Zlatanova
e-mail: S.zlatanova@tudelft.nl

J. Stoter · M. Reuvers · P. Janssen · L. van den Brink
Geonovum, Barchman Wuytierslaan 10, 3818 LH Amersfoort, The Netherlands
e-mail: m.reuvers@geonovum.nl

P. Janssen
e-mail: p.janssen@geonovum.nl

L. van den Brink
e-mail: l.vandenbrink@geonovum.nl

J. Beetz
Design Systems Group, Department of the Built Environment, Eindhoven University of Technology, 5612 AZ Eindhoven, The Netherlands
e-mail: j.beetz@tue.nl

R. Klooster
Geo-information Department, Municipality Apeldoorn, Marktplein 1, 7311 LG Apeldoorn, The Netherlands
e-mail: R.klooster@Apeldoorn.nl

F. Penninga
Geo-information Department, Municipality The Hague, Spui 70, 2511 BT Den Haag, The Netherlands
e-mail: f.penninga@denhaag.nl

J. Pouliot et al. (eds), *Progress and New Trends in 3D Geoinformation Sciences*, Lecture Notes in Geoinformation and Cartography, DOI: 10.1007/978-3-642-29793-9_16,

This pilot is a large collaboration in the Netherlands aiming at pushing 3D developments in the Netherlands. The first phase resulted in a national 3D standard, modeled as CityGML Application Domain Extension. Some insights obtained during this phase are sufficiently mature to be anchored in practice such as maintaining and further developing the 3D standard by Geon ovum and the provision of a countrywide 3D midscale base dataset which is currently under study at the Kadaster. Other results need further attention in a collaborative setting, specifically how the new 3D standard works in practice. This is currently being further explored in a second phase of the 3D Pilot in which over 100 organizations are participating. The goal of the follow-up pilot is more focused than the first pilot and aims at writing best practice documents by joint effort of the 3D Pilot community. The best practice documents are based on tools and techniques that are being developed for supporting the implementation of the 3D standard. Specific attention is being paid how to align City GML to the standard in the BIM (Building information Model) domain (IFC). Initial findings and work in progress are presented.

Keywords 3D city and landscape models · 3D standard · City GML

1 Introduction

Over the past 10 years technologies for generating, maintaining and using 3D geo-information have matured. In addition many local governments have 3D models of the city, a large number of companies are providing services for constructing 3D models, and universities and research organisations are investigating 3D technologies (3D re-construction, data management, validation and visualisation). Yet many (governmental) organisations face numerous challenges in introducing 3D applications and technologies in their day-to-day processes. Despite the practical difficulties, it is clear that 3D information is becoming increasingly important in many applications. These developments motivated a pilot in the Netherlands to advance the use of 3D in this country. The pilot was initiated by the Dutch Kadaster, Geonovum (the National Spatial Data Infrastructure executive committee in the Netherlands which develops and manages the geo-standards), the Netherlands Geodetic Commission (NCG) and the Dutch Ministry of Infrastructure and Environment.

From January 2010 until June 2011 a uniform approach for acquiring, maintaining and disseminating 3D geo-information has been explored in collaboration between over 65 stakeholders in The Netherlands (Stoter et al. 2011a). A major result of the pilot was the proof of concept for a 3D Spatial Data Infrastructure (SDI), covering issues on the acquisition, standardisation, storage and use of 3D data. The findings of the pilot were formally established in a national 3D standard realised as a CityGML Application Domain Extension. The ADE completely integrates the OGC CityGML Encoding Standard (OGC 2008, 2012) with a new version of the existing national Information Model for Geo-information

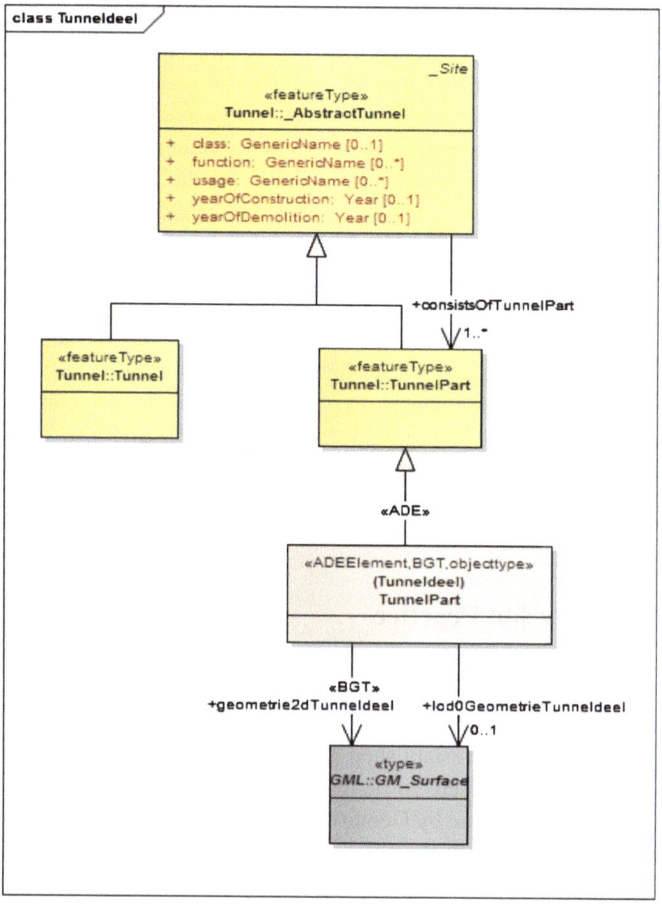

Fig. 1 TunnelPart AD element with 2D geometry

(called IMGeo; described in IMGeo 2007). IMGeo contains object definitions for large scale representations of roads, water, land use/land cover, bridges, tunnels etc. and prescribes 2D point, curve or surface geometry for all objects. As the new version of IMGeo is completely integrated with CityGML, (see Fig. 1), IMGeo version 2.0 also facilitates extensions to 2.5D representations (i.e. as height surfaces; equivalent to CityGML LOD0) and 3D (i.e. volumetric; i.e. CityGML LOD1, LOD2 and LOD3) representations of the objects according to geometric and semantic principles of CityGML.

The close integration between an existing information model for 2D geo-information and CityGML is an important step toward the practical use and re-use of 2D and 3D information. Further technical details about the ADE are reported in van den Brink (2012a, b).

Although the 3D standard is an important prerequisite for further 3D developments, wide use of 3D is still not common practice in the Netherlands. Further advances are required to assure that 3D Pilot results are implemented in actual applications. Therefore the follow-up activities have been started to make the results of the 3D Pilot further ready for practice.

These activities are described and justified in this paper. The main purpose of the paper is to describe the motivation and problem statements as well as ongoing investigations regarding the follow-up activities. On the one hand these activities elaborate on findings that are sufficiently mature to be picked in daily processes of governmental organisations (Sect. 2). On the other hand, these activities focus on further research within a similar collaborative and experimental environment as the first phase of the 3D Pilot. Section 3 describes the motivations and methodology of the second phase of the 3D Pilot and details the six activities. Initial conclusions and work in progress are finally described in Sect. 4.

It should be noted that main focus of this paper is on the construction and maintenance of 3D spatial data to support the national 3D SDI. The use of 3D data in applications was studied during the first phase of the 3D Pilot. Demonstrations of the use cases can be found at Geonovum (2012c).

2 Topics Ready for Practice

The pilot identified three main topics that are ready to put into practice in order to support further 3D developments.

Firstly, to assure that the established 3D standard NL serves as solid base for 3D innovations, the standard needs to be maintained as well as to be improved based on new insights. This is done by Geonovum and also includes studying extensions of other domain models with the notion of 3D if appropriate.

Secondly, besides the need for a national 3D standard, the pilot showed the need for a nationwide 3D base dataset. This dataset can serve as reference for (new) 3D information in the 3D virtual world and as a basis for 3D planning and management of public space, and can be further refined when a 3D project develops. Many large municipalities have 3D data sets, but these are specifically acquired for the territory of the city and in various formats and resolutions. The pilot has shown promising results for generating a 3D national topographic dataset as combination of 2D topography with high-resolution laser data, based on work of participants. Currently those results are extended to generate a national 3D topographic dataset covering the whole of the Netherlands in collaboration between University of Twente, Delft University of technology and the Kadaster (who also holds the national mapping agency).

The high-resolution data used to generate the 3D base dataset is AHN2: the National Height model of the Netherlands, obtained by airborne LiDAR systems with an average point density of 10 points per square meter. Two 2D topographic data sets are candidates for automated extension into 3D to obtain a complete 3D data coverage of the Netherlands: the large scale base data modelled according to IMGeo and TOP10NL data.

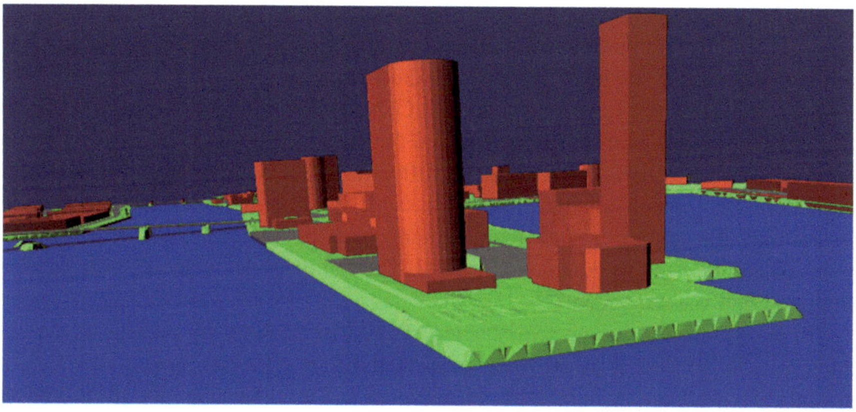

Fig. 2 First results of 3D TOP10NL (Oude Elberink 2010)

As mentioned before, the new version of the model IMGeo (focusing on scale 1:500–1:1000) has recently been established. It is expected that providers of this data—municipalities, water boards, provinces, ProRail (the manager of Dutch railway network infrastructure) and Rijkswaterstaat (Dutch Ministry for infrastructure)- will produce the data from 2015 onwards. The second candidate for the national 3D dataset (TOP10NL) is being maintained by the Netherlands' Kadaster and is available since 2005. This dataset is currently being used to generate a nationwide 3D base dataset.

The reason to focus first on the TOP10NL is not only because it is available nationwide. This less detailed scale is also better suitable for 100 % automated 3D object reconstruction since it is less demanding concerning 3D details. Consequently it was decided that 3D TOP10NL is the best option to generate and disseminate a nationwide 3D base dataset in a limited amount of time. This is currently realised by in collaboration between University of Twente and the Kadaster, see Fig. 2.

Finally, the accomplished network is considered crucial for further 3D developments in the Netherlands. Therefore the network is being maintained and supported by social media and further expanded by a continued facilitation of the 3D test bed (Stoter et al. 2011b) and through regular 3D symposia where organisations exchange ideas and experiences regarding 3D applications.

3 3D Pilot NL Phase II

In the development process of CityGML ADE IMGeo 2.0 a number of topics were identified that requires further attention before the standard can be widely implemented.

Firstly, more research is needed to understand how the national 3D standard works in practice including the consequences of this new modelling method for IMGeo when used for both 2D and 3D datasets, e.g. how to preserve the links between the different Levels of Detail (LODs) and how to upgrade 2D LOD to higher LODs. Also, knowledge is required on the ability to use 3D IMGeo data in CityGML-aware software, i.e. whether software systems are compatible with our extensions and which changes are necessary? Finally more research is needed concerning the creation and management of CityGML-IMGeo data. Which methods can be used to generate CityGML-IMGeo data? How should this data be validated and maintained?

These open issues are currently being studied in a follow-up project of the 3D Pilot. A pilot setting is again used because the first pilot has shown that fundamental 3D innovations can best be realised by an intensive collaboration of research institutes, private and public organisations. These organisations all possess unique knowledge and experiences that need to be brought together to accomplish 3D innovations. Also further agreements between many stakeholders are necessary for advances in 3D.

The goal of the follow-up pilot is more focused than the first pilot and aims at writing best practice documents by joint effort of the 3D Pilot community. The best practice documents are based on tools and techniques that are being developed for supporting the implementation of the 3D standard. Specific attention is being paid how to align CityGML to the standard in the BIM (Building information Model) domain (IFC).

In summer 2011 a new call was launched responded by over 100 organisations. These organisations, listed at Geonovum (2012b), are currently executing the six activities of the second 3D Pilot NL. The activities, including background, motivations and work in progress, will be further explained in the remainder of this section and are:

1. Generating example 3D IMGeo data for several levels of detail and classes
2. Writing example tendering documents for creating 3D information
3. Designing and implementing a 3D validator
4. Describing a generic approach for maintenance, update and dissemination of 3D IMGeo data
5. Collecting examples of 3D killer applications
6. Align CityGML and IFC/BIM

3.1 Generating Example 3D IMGeo Data

To understand how IMGeo works for the integrated 2D and 3D approach example 3D IMGeo data is being built and made available to wider audiences. The example data is also used to check whether CityGML compliant software is capable to understand the 3D IMGeo data. The example data will be specifically useful to:

Fig. 3 LOD concept applied to trees (Clement 2011)

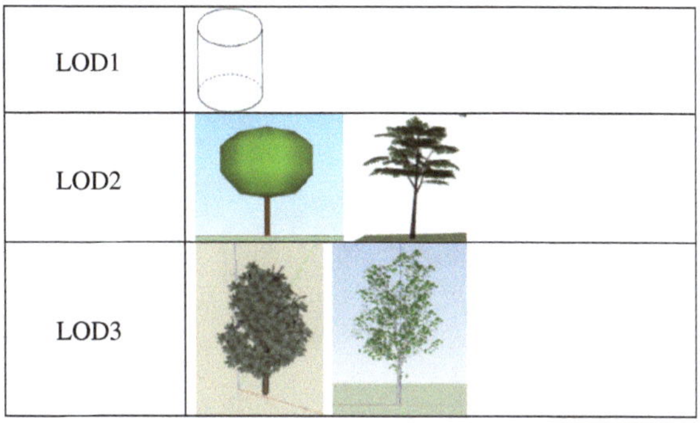

- Obtain insights into the 3D aspect of our approach including different LOD's, i.e. for buildings the LOD concept is well-defined, but how does the LOD concept applies to other object such as trees, see Fig. 3;
- Provide the possibility for third (new) parties to experiment with 3D IMGeo data;
- Provide test data for 3D validation tests (see Sect. 3.3);
- Show how the standard is interpreted when applied to real data (helpful for future data providers).

3.1.1 Test Area

In the previous 3D Pilot the test area was located in the City of Rotterdam. For this phase we selected a test area which is more familiar to an average municipality: situated in the municipality of Den Bosch (southern part of The Netherlands) containing a usual living area with common houses, a river and a bridge:

The source data that has been made available on the 3D Pilot data server (hosted by the Delft University of Technology) are:

- IMGeo compliant 2D data (see Fig. 4a, provided by the municipality of Den Bosch;
- Stereo photos (30 march, 2011), 10 cm resolution, provided by the municipality of Den Bosch;
- Point cloud (3 points per m2), DTM&DHM, date: April 2009, provided by the municipality of Den Bosch;
- High resolution laserdata (selected from a data set available for the whole country: *Actueel Hoogtebestand Nederland*, AHN), provided by Het Waterschapshuis;
- Ortho photos (provided by Cyclomedia);
- High resolution point cloud obtained from terrestrial laser scanning (by Cobra, see Fig. 4b
- Point clouds generated from aerial photographs (Imagem)
- Oblique photos (Slagboom en Peeters)

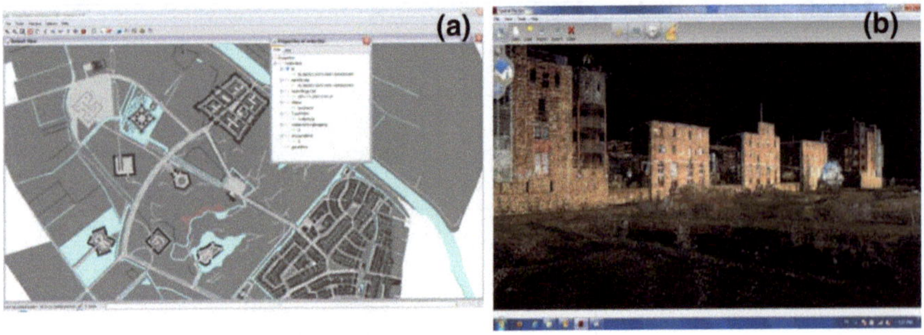

Fig. 4 Example source data available for the 3D Pilot test area. a 2D IMGeo data. b High resolution laser data, obtained by terrestrial laser scanning

To get thorough insight into the key aspects of 3D IMGeo data including how the LOD concept applies to several themes and how these data can be created accordingly, the following 3D information will be created:

- LOD3 of a selected number of buildings as combination of AHN2, stereo photo's, texture information and 2D IMGeo data;
- LOD3 of both the bridge and the lock situated in the test area (see Fig. 5a, b); This is being done by the company "Coenradie". The modelling of the bridge are visualised in Fig. 5c and d.
- LOD0 of the complete test area (as combination of AHN2 and 2D IMGeo);
- LOD1 of all buildings in test area (as combination of AHN2 and 2D IMGeo);
- LOD2 of city furniture (traffic signs);
- LOD2 of trees;
- LOD3 of a selected number of points of interest.

During the work of activity 1, decisions have to be made and tips and tricks will be formulated. These experiences gained from the technical operation of IMGeo 2.0 will be supportive for future use and creation of 3D IMGeo data. And this will be important input for the example tendering documents (Activity 2).

3.2 Tendering Documents for Creating 3D Information

Usually a municipality will outsource the 3D data acquisition for 3D IMGeo data. For most municipalities 3D is a new domain and example-tendering documents may help them to precisely specify what to ask the market. In a next step, when the data is delivered, the specifications can be used as acceptance criteria, i.e. to check the quality of the data. For private companies these documents are helpful since they both clarify and unify the demand for their services.

Apart from the experiences gained from building example data (activity 1), the tendering documents will be based on experiences of cities that have already invested in 3D city models, i.e. The Hague and Rotterdam. Both cities faced difficulties in

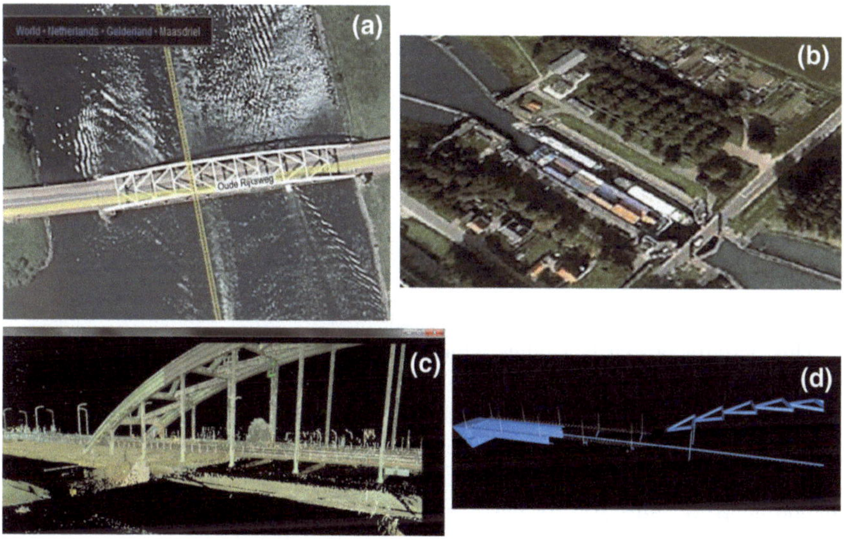

Fig. 5 Objects in test area to be modelled at CityGML LOD3. **a** Bridge. **b** Lock. **c** Point cloud of the bridge. **d** Model of the bridge (in progress)

comparing offers from different companies because the specifications appeared to be interpretable in several ways and this also caused problems in setting up acceptance criteria for the delivered product. The result is that the CityGML datasets differ between the two cities but it is not always clear whether this was intended.

Since the example tendering documents will be a joint effort of the 3D pilot community, they will be based on knowledge, interests and experiences of research institutes, private and governmental organisations and not only based on the information available at the bidder as currently practiced.

Several variants of the tendering documents are possible based on the available source data (i.e. point clouds or high resolution photographs) and the ambition level (i.e. which information at which LOD).

3.3 Design and Implementation of a 3D Validator

A validator is necessary as an independent tool to verify whether a dataset is compliant with IMGeo 2.0, or not. This also applies for the 3D extensions. When validating objects, it is necessary to validate both the semantics and the geometry. The former is according to the classes of CityGML and/or of the IMGeo extensions, and the latter is according to the international specifications (e.g. ISO19107 and GML). Geonovum has already built a validator for IMGeo datasets [the software is available as open source software, see Geonovum (Geonovum 2012a)], but it is only for two-dimensional primitives.

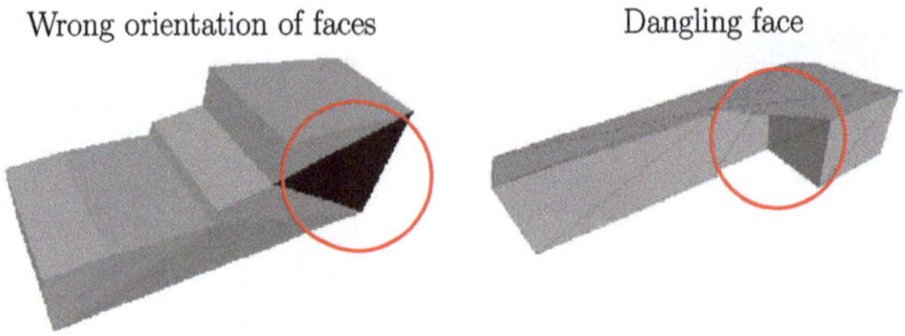

Fig. 6 Two real-world invalid buildings

This activity primarily studies which functionalities are necessary to validate the geometry of 3D solids. During the first 3D Pilot we have noticed that several real-world datasets have objects that appear to be valid when looking quickly at them, but in reality they are not. Figure 6 shows two examples. These (often small) problems prevent users from, for instance, convert their objects to other formats (including BIM and CAD formats, see Sect. 3.6) and also to analyse them (the volume of an invalid solid could be impossible to calculate).

While different definitions of a valid 3D object are used in different disciplines, we focus on the definition given in the ISO standards (ISO 2003) and implemented with GML (OGC 2007). A GML Solid: "The extent of a solid is defined by the boundary surfaces as specified in ISO 19107:ISO 2003. gml: exterior specifies the outer boundary, gml: interior the inner boundary of the solid" (OGC 2007). Without going into all the details, we can state that a solid is represented by its boundaries (surfaces), and that like its counterpart in 2D (the polygon), a solid can have 'holes' (inner shells, or cavities) that are allowed to touch each others, or the outer boundary, under certain circumstances. To be considered a valid solid, a solid must fulfil several properties. The most important are: (1) it must be simple (no self-intersection of its boundary); (2) it must be closed, or 'watertight'; (3) its interior must be connected; (4) its boundary surfaces must be properly oriented; (5) its surfaces are not allowed to overlap each other.

We are currently building an open-source 3D validator. This is because none of the surveyed GIS packages that provide functionalities for validating 3D objects was fully compliant with the definition of the ISO. Our validator is based on the work of Ledoux et al. (2009) and is ISO-compliant. It uses advanced data structures and operations to analyse the topological relationships between 3D objects. Furthermore, it will be built as an extension to the current validator for 2D (developed by Geonovum) so that the geometry of 3D objects can be taken into account while using the same website with the same workflow.

Finally, other validation issues will be investigated. We plan to investigate the validation of not only solids, but also 3D MultiSurfaces as these are often used (buildings are often modelled without the ground floor for instance).

3.4 *Maintenance, Update, Visualisation and Dissemination of 3D IMGeo*

After having invested in a good 3D model, the next question is how to maintain and update the model. Can mainstream DBMSs be used? How to update: integrated with the existing data processes, renewed creation when the 2D data changes or a mix of both? For the maintenance of the data it is a relevant question how to guarantee that 3D IMGeo data remains synchronized with the 2D data. The challenge differs if the 3D data is managed separately from 2D (how to maintain 3D data? In a 3D spatial DBMS?) or generated on the fly.

An important first step is to obtain more insight into how CityGML data encoded in CityGML files can be maintained and updated.

Therefore a challenge was organized in order to study the state-of-the-art of 3D editing in commercial software. Four neighboring CityGML data sets (courtesy of the Municipality of The Hague) were provided and the following challenges were defined:

1. Integration of CityGML files

   Create one 3D model of the four adjacent neighbourhoods by integrating the eight CityGML files. The resulting 3D model can either be stored in a database, a CityGML file or another file format (without loss of information).

2. Editing in CityGML files

   File 13_buildings.xml contains a building with id {B65F9980-76C8-4F8C-8449-243FE4FD168E}. Select this building, add another storey on top of it and save the results in another CityGML file

3. 3. Enrichment of CityGML files from other sources

   File 12_buildings.xml contains o.a. the "Binnenhof" in The Hague (houses of parlement) in CityGML format. Show how the two more detailed KMZ models of the Binnenhof can be used to enrich the CityGML files and save this enriched model as a CityGML files.

4. 4. Bonus question

   For those vendors that encounter no problems with the challenges above: pick a more complex operation and demonstrate this

In addition to these challenges it was mentioned that it was up to the vendors to decide in which environment or format the actual edits were made, as long as both input and output were in CityGML format without any loss of data.

Up till now the following companies demonstrated their capabilities: StrateGis, Toposcopie, CPA Systems, Safe and Bentley. Results and findings are presented in the remainder of this subsection.

3.4.1 StrateGis—Gebiedsontwikkelaar

StrateGis is a Dutch company, founded in 2006, focusing mainly on decision support systems for urban planning. Their system "Gebiedsontwikkelaar" (which roughly translates as "Space developer") supports interactive planning and provides insight in the costs and benefits of different versions of plans. Although originally based on Microsoft Excel, with the emphasis on financial consequences, StrateGis now also supports 3D planning. The 3D modeling module is based on Sketch Up.

Challenge Results

Importing the separate citygml files turned out to be a straightforward operation, although it took a significant amount of time (30–60 min). Since the Gebiedsontwikkelaar incorporates the SketchUp API for editing, the challenges on building edits and KMZ texturing were easy. Exporting the results back to CityGML is also possible with the export to citygml functionality of Sketch Up. So from a modelling point of view the Gebiedsontwikkelaar does not offer any additional functionality over Sketch Up, But the product enables financial analysis based on the 3D model of The Hague, although it turned out to be rather time consuming. Outputs are visualized in Fig. 7.

3.4.2 Toposcopie—Toposcopie

Toposcopie is a small Dutch company that has developed 3D modelling software based on terrestrial photogrammetry, using inexpensive regular cameras. Already in 2007 the support of CityGML was added to te existing VRML support.

Challenge results

The Toposcopie module Append is designed for this purpose. As a result, integrating the CityGML files was easy.

Toposcopie also uses SketchUp for 3D edits. Two different approaches were selected for this challenge. This first option is to separate the specific building from the CityGML file and import only this building into Sketch Up. After editing, the results are exported to CityGML format and integrated in the CityGML file. Although this approach is the fastest, it does require specific knowledge of CityGML in order to be able to separate and later integrate the specific building. The second option does not require specific CityGML knowledge, as it converts the entire file into Sketch Up. After editing it converts the entire file back into CityGML format. Although easier, it is obviously more time consuming.

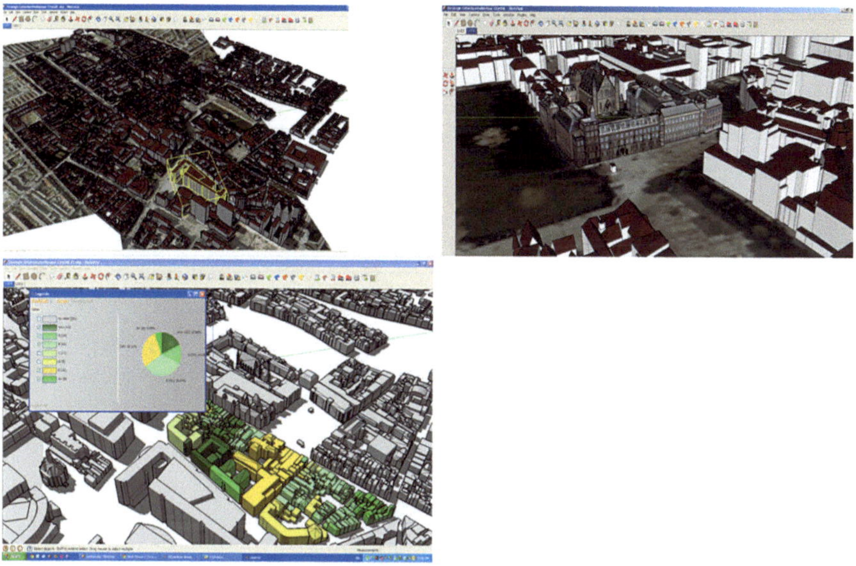

Fig. 7 Screenshots of the strateGIS solutions

This time the KMZ is directly imported into SketchUp and than exported together with the 3D model exported to CityGML. In order to position the KMZ-model Toposcopie uses its module Convert And Translate KML.

Although the conversions between CityGML and Sketch Up include the ID's it has to be checked whether the other attributes are also preserved during these conversions. Outputs are visualized in Fig. 8.

3.4.3 CPA Systems—SupportGIS

CPA systems is a German company that focuses on OGC compliant geo DBMS's, 3D city models and municipal applications. With Support GIS CPA offers a database solution, independent of any specific GIS software, DBMS manufacturer and operating system (see Fig. 9). Their solution is based on ISO and OGC standards in order to achieve interoperability. Data models can be incorporated through XSD schema's, geo-information is im- and exported in GML, the JDBC kernel is used to create OGC compliant web services. Support GIS consists of a database, an editor and a viewer. Support of 3D data is accomplished through the Google Sketch Up API. In cooperation with GeoRes, one of CPA's partners, exports to Google Earth and Bing are created.

Challenge results

Merging the files was done by loading all CityGML files into the database. This turned out to be easy, since there were no gaps or overlaps between the separate

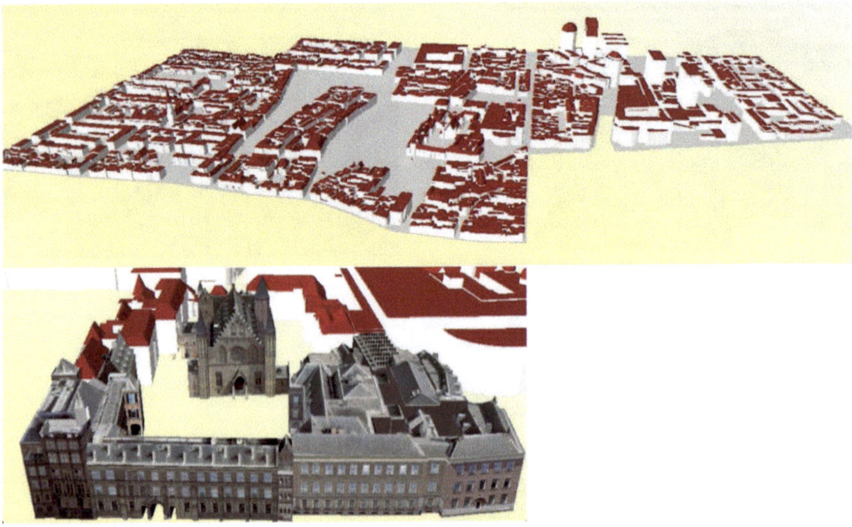

Fig. 8 Screenshots of the toposcopie solutions

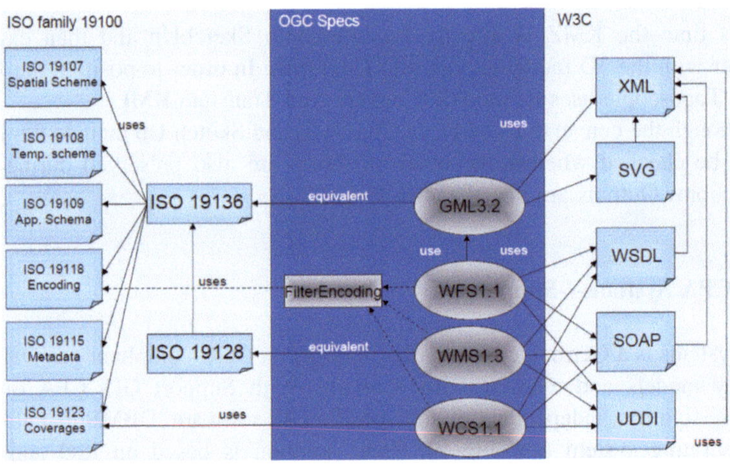

Fig. 9 Schema of the CPA approach

cityGML files and all ID's were unique. Editing is also enabled in a separate Editor. Since the building consists of multiple building parts, it was decided to select all building parts of this building and raise their height with a standard function of the editor. Integrating the KMZ model was performed again trough the Sketch Up API, followed by an export to CityGML format.

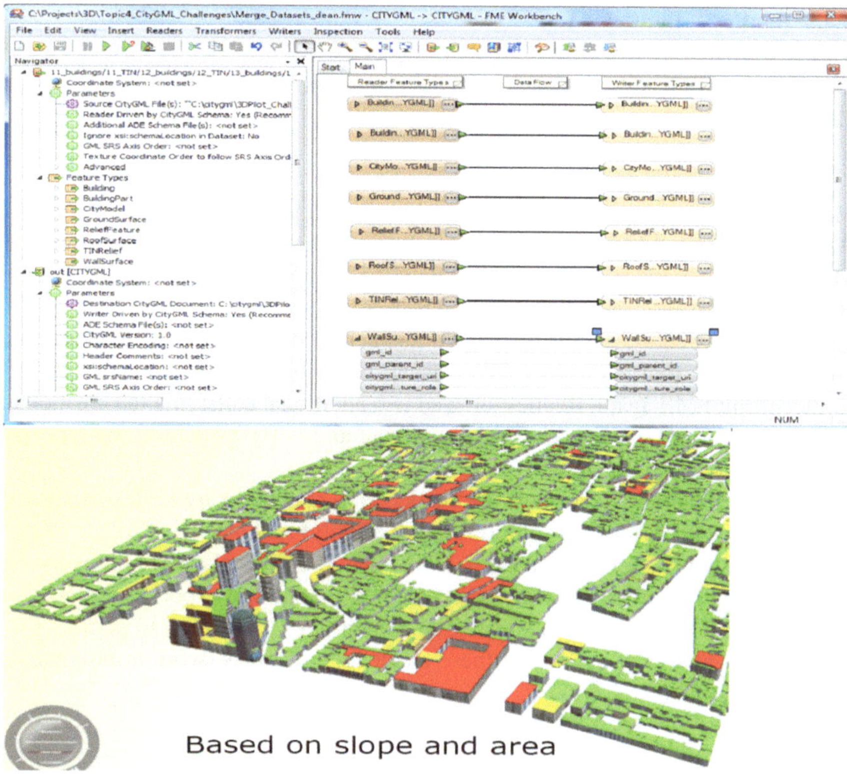

Fig. 10 Screenshots of the safe software approach

3.4.4 Safe—FME

Safe offers with FME a solution for data transformation issues, supporting over 275 different data formats. Transformation issues include both transforming between formats and coordinate systems and transforming data models and schemas.

Challenge results

Merging the separate files is done by a simple workbench, with multiple readers and writers. As FME is not a intended as an editing environment, the editing challenge was not presented.

The integration of a KMZ model was also performed with a simple workbench, although it turned out that the FME Data inspector did not show texture. In the CityGML file itself however, the textures were present. As an additional challenge a filter was presented to identify high risks in case of huge snow loads, based on roof area and slope. Screenshots by Safe Software are visualized in Fig. 10.

3.4.5 Bentley—Bentley Map

Bentley (a GIS/CAD vendor) used their module Bentley Map to face this challenge.

Challenge results

Merging the files was accomplished by importing all files into Bentley Map. Since Bentley Map uses FME to do so, the results were the same as the results of Safe. Modifying structures is well supported with the drawing functionality from Micro station. An edit was demonstrated in which a surface was extruded first, then a center line was added and as a last step this center line was extruded in order to create a saddle roof. Converting the results back to CityGML format was performed using FME again. Bentley showed two additional edits: first the creation of cross sections of 3D models to simplify interpretation of 3D situation and second a solar exposure analysis.

The preliminary conclusion of the challenge to maintain City GML data is first that the five vendors showed solutions that (partially) rely on either Google Sketch Up (or the Google Sketch Up API) or FME. In addition database solutions for 3D data are rare. Therefore the availability of good import and export functionalities for CityGML (and the ADEs) is essential, which gave motivation to plan a "City GML relay" as follow up step of these challenge-outcomes (work in progress).

3.5 Collecting Examples of 3D Killer Applications

Although a 3D application are common practice for many professionals, 3D is new and considered as "complex" and "expensive" to others. To show the need for 3D to policy makers and new comers in the field, this activity is collecting examples of 3D applications that are already practised by the 3D pilot participants and make them available in an easy accessible format (flyer, PPT, Website). Specific attention is paid to the integration-role of 3D information, i.e. as base information for many applications, see Fig. 11.

3.6 Align CityGML and IFC/BIM

In both GIS and BIM domains it is acknowledged that the integration of both types of data is beneficial. BIM data is commonly modelled in the IFC standard and 3D GIS data can be encoded in CityGML. The two standards model similar object types. Therefore it is relevant to see how these two standards map, integrate and interact with each other.

BIM (i.e. design) data can feed GIS data and GIS can serve as reference for BIM data. However, integration should acknowledge the differences between both

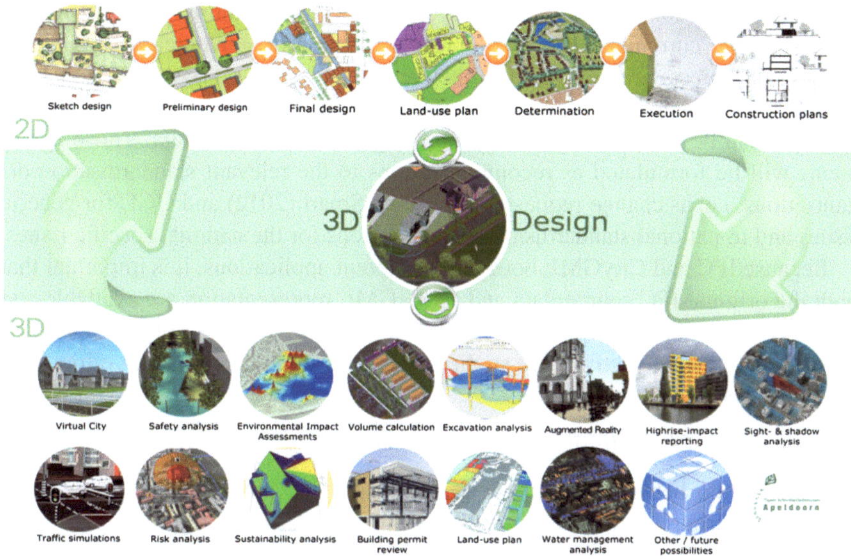

Fig. 11 Screenshot of the 3D pilot website with uses cases (the *circles* represent the different uses cases)

types of data. To start with, the object description of BIM and GIS (e.g. CityGML LOD4) differs significantly. In addition GIS is characterised by coverage of large areas (e.g. a complete city) and lower precision, while BIM is characterised by its local and very detailed approach, the limited number of construction models usually available in a city and high precision necessary for reliable construction calculations. Also the modelling approaches of CityGML and IFC differ significantly, i.e. IFC models much more classes and allow also non-hierarchal relationships, where CityGML contains a limited number of classes structured via hierarchical relationships. Another core issue for bidirectional transformations are additional geometry types that are handled in the building industry and can be captured in IFC instances (Nagel 2007). Among them are Boundary Edge Representations (BRep) and Constructive Solid Geometry (CSG), which are frequently used as implicit capturing formats while CityGML is limited to explicit polygonal representations. While polygonal representations can be derived from these geometry types in a straightforward manner (thus transforming IFC to CityGML), it is impossible to generate e.g. efficient CSGs from triangulated surface representations.

Several studies have shown that a conversion between IFC and CityGML is possible, see (Isikdag and Zlatanova 2009; Berlo and De Laat 2010; Bormann 2010; El-Mekawy et al. 2010). However, because of the different modelling approach of both information models, there is not one optimal or uniform conversion.

Therefore, based on experiments and a study on best practices, this activity is working on making agreements how to best realise the alignment between the two standards; e.g. agreements on a unique mapping between IFC and CityGML to make sure that a conversion always happens in the same meaningful way. This will also avoid the currently common situation that the rich semantics of IFC is lost because all objects are converted in the GenericObjectClass. Also it may help to model according to specific rules in IFC to make sure that specific CityGML concepts can be derived (e.g. LOD2 Buildings) from the IFC data. Those agreements will be formulated as recommendations to the relevant standardisation organisations, i.e. as change requests to Building Smart (2012) and OGC for generic issues and to national standardisation organisations for the national specific issues.

Because IFC and CityGML both serve different applications, it is important that both the original IFC source data and a CityGML representation are available and that CityGML objects explicitly refer to their interrelated IFC objects and vice versa. In this specific activity we studied how this can be done by joint effort from IFC and CityGML experts.

Referring from CityGML objects to IFC objects

The integration between the source IFC data and the 3D CityGML data can be maintained through a link between the two representations.

In CityGML one can refer to an external object via "external references". This reference maintains the link to the external objects from which the CityGML data was derived. One CityGML object can contain more of such external references.

The external reference is a URI (either URL or URN). Every object that is a subclass of IfcRoot (all semantic classes) has an UUID identity that is compressed into a unique ID within the specific dataset, for example 3QbcAsYsg7Hvx$4VHzijdF. This ID can be converted into a 128-bit UUID via a publicly available method and can be used in the CityGML external reference.

For example linking a CityGML Building to a IfcBuilding can be done via an URN based on the decompressed GUID of the IfcBuilding:

urn:uuid:[UUID]
or based on the compressed ID:
urn:ifc-guid:3QbcAsYsg7Hvx$4VHzij

It is still not clear if the IFC GUIDs should be used in the reference or the uncompressed UUID. Both seems possible because a compressed ID can be converted into an uncompressed ID and vice versa.

Another option is to use an external Reference that contains a http URL. The advantage is that it is both identification and a reference to the location where more information can be found about the object. In contrast, a URN is only an identification; to find more information about the object an extra step is required to resolve the URN to a location on the internet. An example is the BIM Server (www.bimserver.org) where every IFC object has a URL. This could simply be used as CityGML external Reference for every object that was derived from an Ifc object.

This next example shows a CityGML XML fragment with in **bold** an external reference:

```
<core:cityObjectMember>
  <bldg:Building gml:id = "Build0815">
    <core:externalReference>
      <core:externalObject > .
   <core:uri > urn:uuid:550e8400-e29b-41d4-a716-446655440000 </core:uri>
   </core:externalObject>
  </core:externalReference>
   <bldg:function
codeSpace = "http://www.sig3d.org/codelists/standard/building/2.0/_Abstract
Building_function.xml">1000</bldg:function>
   <bldg:measuredHeight uom = "#m" > 8.0 </bldg:measuredHeight>
   <bldg:storeysAboveGround > 2</bldg:storeysAboveGround>
   <bldg:storeyHeightsAboveGround    uom = "#m" > 2.5    2.5 </bldg:storey
HeightsAboveGround>
   <bldg:lod2Solid > … </bldg:lod2Solid>
  </bldg:Building>
</core:cityObjectMember>
```

A similar approach of referring to external objects is available in IFC and therefore this solution can establish an integration on the semantic level. It should be noted that this reference mechanism does not solve the problem of mapping the boundary-presentations of CityGML to the component-assemblage representations of IFC. Instead, the external references make it possible to use IFC as a kind of additional LOD5 representation of CityGML objects. This is a simpler approach, than modeling IFC as Application Domain Extension (Berlo and De Laat 2010).

4 Initial Conclusions and Work in Progress

This paper presents the follow-up of the 3D Pilot NL, which is a large collaboration in the Netherlands aiming at pushing 3D developments in the Netherlands. The first phase resulted in a national 3D standard. Some results and insights obtained during the first phase are sufficiently mature to be anchored in practice such as maintaining and further developing the 3D standard by Geonovum and the provision of a countrywide 3D midscale base dataset which is currently under study at the Kadaster (collaboration with University of Twente). Other results of the first 3D Pilot NL phase need further attention, specifically how the new 3D standard works in practice. This is currently being further explored in a second phase of the 3D Pilot in which 100 organizations are participating.

The main conclusion of running the 3D Pilot is the change of vision concerning 3D in the Netherlands. At the start of the 3D Pilot (March 2010) many saw that 3D had potentials, but did not know how to deal with 3D. In the course of the pilot the ambitions for 3D have become much more focused, also supported by the national 3D standard. These ambitions are further developed now the second phase of the pilot is running. Several aspects appear to be crucial for the adoption of the 3D standard. Firstly, the engagement of many stakeholders is important to gain the necessary support. Secondly, aligning to the ongoing 2D efforts makes that 3D applications become in reach for governmental organizations. In addition, collaborating is important because the issues of 3D are complex and sharing knowledge between different 3D experts is therefore important to realize innovations. Finally, it has been important that some national organizations took the responsibility to facilitate the process. Although the pilot is a joint effort and 'owned' by the community, national organizations have to initiate and facilitate such a network organization and they are important for anchoring the results.

Currently the work on the six activities of the 3D Pilot NL II is running in parallel, supported by discussions within the 3D Pilot NL LinkedIn group (over 500 members) and frequent meetings. 3D test data have been prepared for the test area and several participants are currently working on generating different LODs and different themes for 3D IMGeo data. The 3D validator is being developed, a contest for maintaining and updating 3D CityGML data has been launched and killer applications for 3D are being collected. In addition the information models IFC and CityGML are studied for possible integration, and the possible mappings, alignments and conversions are discussed in dedicated working sessions. Also the integration of 3D IMGeo with the subsoil (i.e. geology and cables & pipelines) is being studied (see also the work of Becker et al. 2010; Hijazi et al. 2010; Zobl and Marschallinger 2008).

The 3D Pilot will finish in summer 2012. Among the end results are: examples of 3D IMGeo data, a 3D validator, best practice documents of how to acquire, maintain, update and disseminate 3D IMGeo data, demonstrators that show the potentials of 3D, and recommendations for further developing CityGML compatible with 3D standards in other domains and with the established 2D information models. The results will be presented to the wider (professional) public in a special issue of a Dutch professional magazine on geo-information and a national 3D symposium.

From our national pilot we have observed that 3D is increasingly vital for managing and planning our densely built environment. Therefore 3D information will be more and more important for governmental organisations. To move forward in the highly complex domain of 3D information, we consider agreements and collaborations essential. In addition a national consensus on a generic 3D approach supported by a 3D standard diminishes the risks of investment for individual organisations. This is accomplished in the 3D Pilot NL.

Acknowledgments We would like to acknowledge the contributions of all 3D Pilot participants which is very important for the achievements in our country. This research is supported by the Dutch Technology Foundation STW, which is part of the Netherlands Organisation for Scientific Research (NWO) and partly funded by the Ministry of Economic Affairs, Agriculture and Innovation (project code: 11300).

References

Becker T, Nagel C, Kolbe TH (2010) Integrated 3D modelling of multi-utility networks and their interdependencies for critical infrastructure analysis, 5th international 3D geoinfo conference, Berlin, Nov 2010

Berlo L, de Laat R (2010) Integration of BIM and GIS: the development of the cityGML geoBIM extension, 5th international 3D geoinfo conference, Berlin, Nov 2010

Bormann A (2010) From GIS to BIM and back again—a spatial query language for 3D building models and 3D city models, 5th international 3D geoinfo conference, BerlinV, Nov 2010

BuildingSmart (2012) [online] available from. www.buildingsmart.nl/

Clement J (2011) Presented at the second meeting of 3D pilot phase II [online] available from. http://www.geonovum.nl/sites/default/files/3D/december.zip

El-Mekawy M.S, Östman A, Shahzad K (2010) Towards interoperating cityGML and IFC building models: a unified model based approach, 5th international 3D geoinfo conference, Berlin, Nov 2010

Geonovum (2012).[online]. http://validatie-dataspecificaties.geostandaarden.nl/genericvalidator/content/standard/19

Geonovum (2012b) [online] available from. http://www.geonovum.nl/dossiers/3d-pilot/deelnemersvervolg

Geonovum (2012c) Demonstrations of the 3D pilot uses cases [online available]. www.geonovum.nl/dossiers/3dpilot/bibliotheek/presentaties#films (YouTube)

Hijazi I, Ehlers M, Zlatanova S, Becker T, van Berlo L (2010) Initial investigations for modelling interior Utilities within 3D geo context: transforming ifc-interior utility to cityGML/UtilityNetworkADE, 5th international 3D geoinfo conference, Berlin, Nov 2010

IMGeo (2007). Informatiemodel Geografie, oktober 2007[online]. www.geonovum.nl/sites/default/files/IMGeo_rapport_definitief_versie_1.0.pdf

Isikdag U, Zlatanova S (2009) Towards defining a framework for automatic generation of buildings in CityGML using building information models. In: Lee J, Zlatanova S (eds) 3D geo-information sciences. Springer, Berlin, pp 79–96

IS0 (2003). ISO 19107:2003 Geographic information—Spatial schema, [online]. Available from www.iso.org/iso/catalogue_detail.htm?csnumber=26012

Ledoux H, Verbree E, Si H (2009) Geomatric validation of GML solids with the constrained Delaunay tetrahedralization. In: De Mayeer, Neutens, De Rijck (eds) Proceedings of the 4th international workshop on 3D-Geo- Information, Ghent, pp 143–148

Nagel C. (2007). Ableitung verschiedener Detallierungsstufen von IFC Gebäudemodellen (Derivation of different levels of detail from IFC building models). Master Thesis. Karlsruhe University of Applied Sciences, Germany

OGC (2007) OpenGIS® Geography Markup Language (GML) Encoding Standard. Version 3.2.1, doc # OGC 07-036. http://portal.opengeospatial.org/files/?artifact_id=20509

OGC (2008) OpenGIS® City Geography Markup Language (CityGML) Encoding Standard, version 1.0.0, document # 08-007r1. http://portal.opengeospatial.org/files/?artifact_id=28802

OGC (2012) OpenGIS® City Geography Markup Language (CityGML) Encoding Standard, version 2.0. www.opengeospatial.org/standards/citygml

Oude Elberink SJ (2010) Acquisition of 3D topography: automated 3D road and building reconstruction using airborne laser scanner data and topographic maps. Enschede, University of Twente, Faculty of geo-information science and earth observation ITC dissertation 167, p 171. ISBN: 978-90-6164-288-6

Stoter J, Vosselman G, Goos J, Zlatanova S, Verbree E, Klooster R, Reuvers M (2011) Towards a national 3D spatial data infrastructure: case of the Netherlands. PFG Photogrammetrie, Fernerkundung, Geoinformation, (6):405–420

Stoter J, Tijssen T, Verbree E, Zlatanova S (2011b) Het 3D-testbed van de 3D-Pilot: ceci n'est pas CityGML (in Dutch). In: Geo-Info, Volume 8, 5, 2011, pp 19–22

van den Brink L, Stoter JE, Zlatanova S (2012a) Establishing a national standard for 3D topographic data compliant to CityGML, in: International Journal of Geographical Information Science, in press.URL: http://www.tandfonline.com/doi/abs/10.1080/13658816.2012.667105
van den Brink L, Stoter JE, Zlatanova S (2012b) Modeling an application domain extension of CityGML in UML. In: Proceedings 3D geoinfo symposium, Quebec, Canada, 16–17 May
Zobl F, Marschallinger R (2008) Subsurface geobuilding information modelling GeoBIM. GEOinformatics 8(11):40–43, December 2008

Open Building Models: Towards a Platform for Crowdsourcing Virtual 3D Cities

Matthias Uden and Alexander Zipf

Abstract Within the last years, Volunteered Geographic Information (VGI) has developed rapidly and influenced the world of GIscience significantly. Most prominently, the Open Street Map (OSM) project maps our world in a detail never seen before in user-generated maps. Particularly within urban areas, the focus recently shifts from only streets towards buildings and other objects of the environment such as parks or street furniture. However, this innovation is mostly restricted to 2D so far. In order to come closer to the Digital Earth, it needs to be discussed, how the 3D aspect can be integrated into such VGI-projects. This article has two objectives that are closely related: firstly, the current situation of 3D-VGI is reviewed and crucial issues for future development are pointed out. This leads to the concept of defining a free and open web repository for architectural 3D building models. Therefore secondly, the concept of such a new web platform called Open Building Models is presented. This is an important effort towards 3D-VGI. The models can be linked to OSM objects and displayed by a dedicated 3D viewer. This can extend the possibilities to crowd source 3D city models in the future.

Keywords Volunteered geographic information · Open street map · 3D city models · 3D spatial data infrastructures · 3D building modelling

M. Uden (✉) · A. Zipf
Department of Geography, University of Heidelberg, Berliner Straße 48, 69120, Heidelberg, Germany
e-mail: uden@uni-heidelberg.de

A. Zipf
e-mail: zipf@uni-heidelberg.de

J. Pouliot et al. (eds), *Progress and New Trends in 3D Geoinformation Sciences*, Lecture Notes in Geoinformation and Cartography, DOI: 10.1007/978-3-642-29793-9_17,

1 Introduction

It's now been over 4 years that Goodchild (2007) has introduced the term "Volunteered Geographic Information" (VGI) describing the recent revolution of collaboratively created spatial information on the Web 2.0. The increasing availability of smart phones as devices for creating, using and sharing geoinformation spatially enables our society more and more (cf. Williamson et al. 2011). This leads to an ever growing amount of various VGI-related activities which also have an impact on research in geographic information science (e.g. WikiMapia 2012; Cloudmade 2012; Panoramio 2012).

The most popular and successful VGI project is probably Open Street Map[1] (OSM). Recent investigations on its completeness and quality have shown that in particular urban areas in Central Europe have already been mapped with an impressive level of detail (cf. Haklay 2010; Neis et al. 2012). In those areas, OSM is meanwhile well ahead of only mapping the street network. For a continuous improvement of OSM it is crucial to enable the mapping of even more detailed, 3D spatial information.

Another phenomenon which is rapidly gaining relevance over the last decade not only in the geodomain but also in the general public is 3D city models. This has been an important research field in GIscience since a couple of years now (cf. Förstner 1999; Früh and Zakhor 2004; Kolbe 2009). Furthermore, numerous efforts exist from both companies (e.g. *Google* 2012) as well as public administrations (e.g. Stadtmessungsamt 2012). The era of mapping the Digital Earth (Gore 1998) in 3D has long begun. Based on the recent developments in VGI it now becomes possible to investigate the potential of applying crowd sourcing also to 3D geodata.

The dominant parts of 3D city models are buildings. In early 2012, the total number of mapped building footprints in OSM exceeded 50 million.[2] It even surpassed the number of mapped streets. This shows the gradual shift in the purpose of this project, which is no longer limited to only the streets but also includes buildings and other components of urban environments. However, the footprint is only a very rough representation of a building and much more information lies in its detailed 3D structure. The idea of letting the crowd assemble comprehensive 3D city models is very promising.

This paper firstly examines the current state and future directions of user-generated 3D spatial information. Secondly, the concept of Open Building Models, one possible way of how to advance 3D-VGI, is introduced thereafter. We investigate the following questions: How can the potential of VGI be exploited for generating 3D city models beyond what has been reached so far? What are the main scientific and practical questions and problems in this leap forward from 2D

[1] http://www.openstreetmap.org

[2] extracted from our internal, regularly updated OSM database.

to 3D with respect to crowd sourcing? What are the means to enable voluntary users to contribute rich 3D information?

The remainder of this paper is structured as follows: the review part starts in Sect. 2 with related work on user-generated spatial 3D content. Subsequently, the current situation of 3D-VGI is critically reflected and crucial issues for the future development are pointed out. In Sect. 3, the concept and a first prototype of Open Building Models (OBM) is described. Furthermore, a discussion of advantages and drawbacks of the OBM approach as well as the implemented prototype is given. The last chapter summarises important aspects of this article, pointing out the main insights and limitations as well as potential future work.

2 The 3D Aspect in VGI

2.1 Related Work

Letting the crowd generate spatial 3D information is still in its early stages, however, the idea is not entirely new. This section contains related work which deals with this topic, approaching it from different directions.

One of the most prominent examples is *Google*'s 3D *Warehouse.*[3] This shared repository contains user-generated 3D models of both geo-referenced real-world objects such as churches or stadiums and non-geo-referenced prototypical objects like trees, light posts or interior objects like furniture. The former also appear in *Google Earth.* In order to voluntarily contribute, users have to have a certain level of 3D modelling skill. The main focus of this repository does not lie on assembling 3D city models as the non-geo-referenced objects seem to be more important in related work. They are for example used to improve methods of automatic object recognition in the field of laser scan classification (Lai and Fox 2009) or robotic vision (Klank et al. 2009). Also, the 3D warehouse models are being integrated in several commercial systems like design tools (Render Lights 2012) or simulation software (Simio 2012).

Google also developed the *Building Maker,*[4] which provides a model kit to create buildings, deriving the 3D geometry from a set of oblique (and proprietary) birds-eye images of the same object from different perspectives. In contrast to the 3D *Warehouse*, this tool specifically aims at geo-referenced 3D building models only. It is intended for people who do not have knowledge in 3D modelling, but still want to contribute. Willmes et al. (2010) and Yiakoumettis et al. (2010) have demonstrated, how this tool can be used to create 3D buildings rather quickly, even of an entire university campus. Drawbacks are the potentially inaccurate modelling due to image errors or obstructions, the current little availability of the required

[3] http://sketchup.google.com/3dwh

[4] http://sketchup.google.com/3dwh/buildingmaker.html

oblique aerial imagery as well as limited usage of the result due to restrictions of *Google*'s proprietary data.

Even though both introduced methods are based on collaboratively collected data, it is *Google* who stands behind it and claims usage and distribution rights for the contributed contents. Hence, this is far away from being open source or open data. However, there are also numerous free-to-use 3D object repositories on the internet, for example *Open Scenery X,*[5] *Archive* 3D[6] or *Shapeways.*[7] These projects emerged from entirely different communities with interest in e.g. flight simulators or 3D printing. The contents usually lack the connection to the real-world but can nonetheless also be useful to enrich real 3D city model visualisation.

More and more ideas for collaboratively mapping the third dimension are also being discussed recently in the Open Street Map community. Several approaches exist which basically all try to utilise the crowd sourced data for deriving 3D city models from it.

The OSM2World[8] software takes into account various 3D-related information that is available, such as the building height or the basic roof shape. It offers different export possibilities like obj-files, direct output via an OpenGL binding for Java (JOGL) or a .pov-file for the *persistence of vision*[9] ray tracer. It also experiments with the integration of terrain data from a Digital Elevation Model (DEM), however, this is currently still under development. Also, an easy-accessible web or desktop interface or standardisation of output formats is missing.

Another example which is intended to support the user in generating 3D data for OSM is the Kendzi 3D plug-in (Kendzi 2011) for the widely used Java Open Street Map-Editor (JOSM).[10] It directly converts several 3D-related information into a 3D model during an edit session on the screen, allowing the user to immediately see the result of their annotation. This makes model creation in the OSM context much easier. It is, however, still under development and many improvements in usability and the supported 3D-related OSM annotations have to be carried out in the future.

The most advanced work in the context of creating 3D city models from VGI data is the OSM-3D project (cf. Over et al. 2010; OSM-3D 2012). It combines the extrusion of the building footprints into the 3D with a detailed integrated terrain model derived from SRTM[11] height data. It provides the 3D data in a standardised manner through a Web 3D Service[12] (W3DS), which is currently a discussion paper at the Open Geospatial Consortium (OGC). At present, the OSM-3D W3DS

[5] http://www.opensceneryx.com

[6] http://www.archive3d.net

[7] http://www.shapeways.com

[8] http://osm2world.org

[9] http://www.povray.org

[10] http://josm.openstreetmap.de

[11] http://srtm.csi.cgiar.org

[12] http://www.w3ds.org

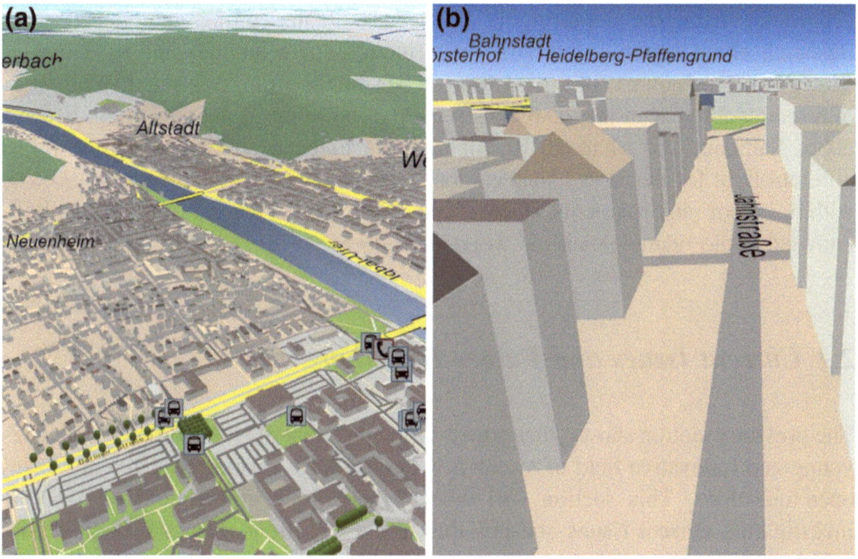

Fig. 1 **a** OSM-3D overview of Central Heidelberg in X Navigator. **b** Detailed view of a street with its buildings

supports different terrain generalisation levels and provides tiled 3D scenes, based on the requested point of view, in VRML, X3D, COLLADA or KML format. There has also been developed a tailored client software called XNavigator,[13] which automatically requests the data from the W3DS server and assembles complex 3D landscapes worldwide. This client also allows the integration of other OGC Web Services such as a Web Feature Service (WFS), the Open GIS Location Services (Open LS, Schilling et al. 2009) or the Sensor Observation Service (SOS, Mayer and Zipf 2009). Thus, for example POIs or 3D routes can be included. The interoperability with different data sources (e.g. also CityGML), web services and targets has recently been examined within the OGC 3D Portrayal Interoperability Experiment (3DPIE, OGC 2011). The wide applicability of the W3DS and X Navigator with heterogeneous data could be demonstrated. Figure 1 exemplarily shows two scenes of Central Heidelberg rendered in X Navigator.

All these approaches are mainly different available tools in this highly innovative field of 3D volunteered geographic information and only little research on it exists. Some related work shows, however, that the preconditions for voluntary 3D data capturing are already mostly fulfilled. With low-cost sensors and cameras which are often available in today's smart phones, the acquisition of 3D data becomes possible for a wider audience. This allows for example the reconstruction of 3D objects based on 2D images taken by low-cost sensors (Rocchini et al. 2001; Pomaska 2009; Wang 2011). According to the typical processing workflow of any

[13] http://XNavigator.sourceforge.net

geographic data, the capturing is followed by data editing and visualisation. Research about suitable data structures, modelling techniques and visualisation strategies in the context of crowd sourced 3D data is strongly needed. While a lot of research about accurate 3D reconstruction of buildings is available (e.g. Brenner 2005; Sampath and Shan 2010), all this has to be examined in an altogether different light for the crowd sourced approach. Apart from some work on 3D-VGI within indoor environments (Goetz and Zipf 2012), there is, to the author's knowledge, no related scientific work in this area.

2.2 Current Issues and Future Directions

The previous section shows, that crowd sourced spatial 3D modelling is still a very young and innovative field of research with only little existing research and many open questions. This section will take a closer look at the field of 3D-VGI, investigating current issues and possible future directions more thoroughly. This will be done with a focus on Open Street Map, since it is currently the most elaborated platform for crowd sourced geoinformation. Apart from the above mentioned approaches, there are also various other ideas and concepts in the OSM community. Numerous people are working on this topic and lively discussions on how to advance it are taking place (cf. OSM Wiki 2011a).

The step from 2D maps to 3D models is not a small one and there exist a lot of issues that have to be tackled. In particular, three main aspects can be pointed out which currently prevent a faster development into the 3D within OSM:

- The missing support of 3D geoinformation in the simple OSM data model
- The lack of a mature and disseminated 3D viewer for OSM
- The lack of appropriate mechanisms that allow users to contribute various kinds of 3D environmental information with different levels of detail

2.2.1 No Suited 3D Data Model

The main issue regarding the first point is that the OSM data model is deliberately kept rather simple in order to attract as many users as possible. It tries to model the world with only nodes, ways and relations. Apart from this geometry, every kind of additional information for a certain feature has to be expressed by so called *tags* in the form of key-value pairs (cf. Haklay and Weber 2008). Whilst these tags can in general be freely defined by any user, there exists an agreement on common tag names and values, whose usage is recommended (OSM Wiki 2011b). With this ontology it is possible to semantically annotate many features of our environment. For instance, building footprints can be modelled with closed way geometries which are annotated with the tag *building* = *yes*. That is, 3D geometry is not inherently supported in the data model, but can so far only be expressed with appropriate tags. Simple building properties like its height or number of floors can be modelled this way. However, for more complex geometrical features such as detailed roof structures, dormers or balconies, this approach is severely limited. Nevertheless, there are efforts trying to express 3D phenomena with this simple data model. Simple 3D-related tags are partly already being used in OSM (cf. Goetz and Zipf 2012 in press). Also, proposals for appropriate tagging schemas for more complex 3D buildings are currently being discussed in the community (cf. Strassenburg-Kleciak 2011). Most of the proposed tags are nonetheless only rarely used so far since no consensus on tag-based 3D modelling has been reached yet. One could argue that the basic OSM data model should be extended to better allow for 3D modelling. However, such a major change is currently not very likely to be accepted by the community, since there are also people disapproving of the idea to bring 3D into OSM. It is preferable to find out how far the approach to build 3D information on top of the simple OSM model brings us closer to the goal of supporting crowd sourcing 3D, and at which point a different approach with a dedicated data model is needed.

2.2.2 No Widespread 3D Viewer

Another reason for the little usage of 3D-related attributes is the absence of a widespread and established 3D viewer that visualises the 3D data. The appearance of one's specially created cartographic object on a worldwide available online map is one of the main motivations why people are committed to the OSM project in the first place (cf. Coleman et al. 2009). An essential point in order to push forward 3D mapping is therefore the development and dissemination of such a 3D viewer which displays the objects modelled by the users. If already existing approaches like OSM-3D or OSM2World are improved and promoted, there will probably be a boost in the interest of 3D mapping. Without it, it is currently not surprising that only few people are willing to contribute 3D information, even though it is already possible to a certain extent.

2.2.3 Not Enough Support for Voluntarily Contributing 3D Information

Finally, there are currently too little possibilities for the crowd to contribute 3D information to OSM apart from the aforementioned tag-based modelling. This aspect of missing capturing and editing mechanisms will be reviewed more thoroughly in the following.

The range of possible 3D-related information in our environment that could be mapped by the users is very wide. It starts with simple building characteristics such as the number of floors, the facade material or a rough roof structure. This information is easy to obtain by any mapper, neither 3D modelling skills nor specialised equipment are required, and it is already supported in OSM. However, it should also be possible

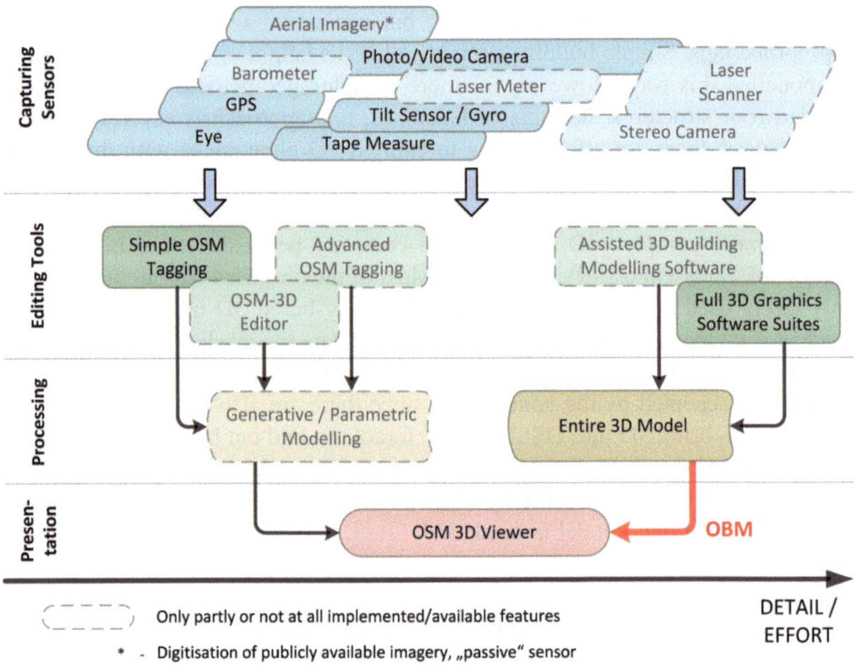

Fig. 2 Methods for capturing and processing 3D volunteered geographic information for Open Street Map on different levels of detail

to map more details about a 3D object, up to entire architectural models, and link them to the OSM world. The above mentioned example of the *Google* 3D *Warehouse* has shown that there are people who have the skills and interest to do this. Accordingly, there should be a wide range of tools and concepts which support the motivated mapper in contributing various 3D information in different levels of detail.

On the top of the diagram in Fig. 2, different sensors are shown with which spatial 3D information can be obtained by voluntary mappers. Whilst simple building properties do not require sensors with high accuracy but are simply measureable with the eye, more complex models can only be created by means of various sensors such as laser meters, terrestrial and/or aerial imagery, GPS or even terrestrial laser scanning. Many of these sensors are now-a-days included in modern smart phones, making them a multi-sensor-system which is pretty well-suited for crowd sourced 3D data capturing. In the future, further sensors like barometers, stereo cameras (e.g. *Microsoft Kinect*, cf. Elgan 2011) and maybe also laser meters and little laser scanners will possibly be included into smart phones, making them even more all-round tools for 3D-VGI.

Once the data is captured, it needs to be edited. Between simple OSM-tagging and the creation of entire 3D models with 3D graphics modelling software, there are further (planned or existing) tools inbetween, which support mappers with different skills and ambitions in terms of the level of detail. For instance, an editor like the aforementioned Kendzi 3D JOSM-plugin (Kendzi 2011), which is specialised for advanced OSM 3D-building modelling, makes it easier for the users to assemble parametric 3D models without caring about the rather complicated and cumbersome tagging itself. Such an editor could also avoid incorrect modelling and ensure topological consistency in complex 3D objects. Besides the manual creation of entire 3D models with 3D modelling software, there is also the possibility of reconstructing 3D buildings from terrestrial photographs. In research, many photogrammetric and digital image processing approaches exist (e.g. Debevec 1996; Müller et al. 2007). This reconstruction could traditionally only be accomplished with complex and expensive photogrammetrical software systems such as *ERDAS LPS*.[14] However, as already mentioned in Sect. 2.1, there recently also emerged free-to-use "assisted" 3D modelling software like *Autodesk 123D Catch*[15] or *My3DScanner*[16] which offer a low-cost alternative for creating 3D models from photographs. This could evolve to an important feature for crowd sourced 3D modelling.

There are basically two types of models which emerge from the editing layer: On the one hand, we get parametric building models, which are based on a geo-referenced footprint plus various tags and can be generated dynamically by the 3D viewer. On the other hand, there are finished models, created by dedicated (or assisted) modelling software, which have to be placed properly in the 3D viewer. Attributes describing their geo-reference and scale are then required.

The currently available 3D viewers do only partly include the desired functionality and flexibility shown in the diagram. Generative modelling based on OSM tags is currently limited to some simple 3D-related attributes. The possibility to integrate *entire* 3D models is in general already available in the OSM-3D viewer X Navigator (cf. Fig. 3). However, this process includes many difficult steps which have to be carried out manually. Also, no real link to the OSM database is being established, nor is the model stored in a publicly available online repository. Hence, other viewers cannot display this model.

Many parts of the diagram in Fig. 2 are visionary and only partly implemented, if at all. There are a lot of things to be done in order to assist the realisation of VGI in the 3D. In this article, we start with a concept for linking entire 3D models to the OSM database. As Over et al. (2010) mentioned, the development of a free 3D repository with architectural building models would be a major step forward.

3 The Concept of Open Building Models (OBM)

In this chapter, a first prototype of Open Building Models is introduced. Its objective is to create a web-based platform for uploading and sharing entire 3D building models. Many complex buildings like churches or other landmarks cannot

---

[14] http://www.erdas.com/products/LPS/LPS/Details.aspx

[15] http://www.123dapp.com/catch

[16] http://www.my3dscanner.com

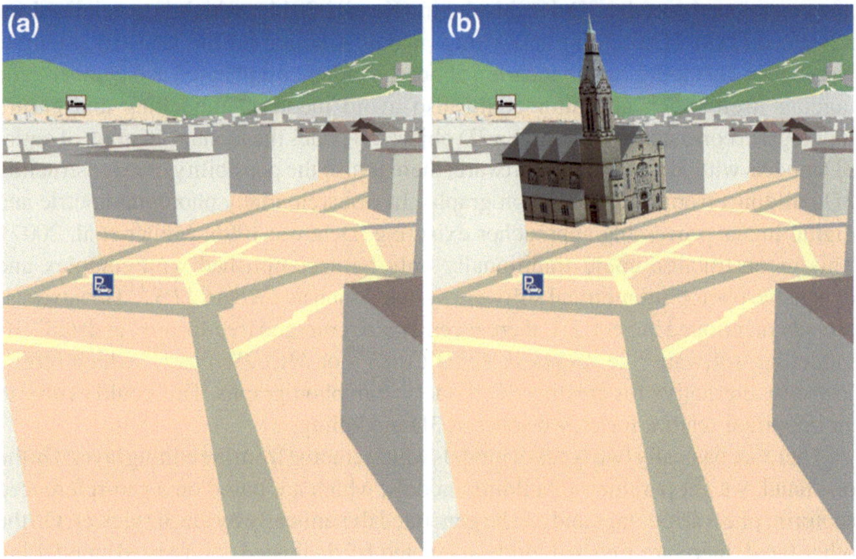

Fig. 3 An OSM-3D scene shown in X Navigator without (**a**) and with (**b**) a manually integrated architectural 3D model of a church

be modelled in detail with a tag-based, parametric approach. Instead, 3D models from the OBM repository should be linked to OSM, so they can be rendered by OSM 3D viewers subsequently. Thus, crowd sourced 3D city models can be greatly improved.

The processing of the OSM data and set up of a model repository in the first prototype comprises several steps, which are briefly described in the following. The user should be able to interactively choose the building of interest from a map. Therefore, the ground plans first have to be derived from the OSM data separately and overlaid as vector layer. This is achieved by first importing the OSM data into a PostgreSQL/PostGIS database with the *Osmosis* tool and converting the closed *ways* which are tagged as buildings to polygon geometries. Native SQL along with several PostGIS functions is used for this. More complex footprints that contain inner holes and therefore consist of more than one closed way can also be converted (cf. Goetz and Zipf 2012; Goetz et al. 2012). This conversion step is part of the pre-processing illustrated at the top of Fig. 4. For the provision of the processed building polygons to the web-client, we set up a GeoServer.[17] The web-client requests the geometries via an OGC-compliant Web Feature Service.

Figure 5 shows a screenshot of the current OBM web-client. The main component is an Open Layers[18] map, which shows the building footprints as a vector

[17] http://geoserver.org

[18] http://openlayers.org

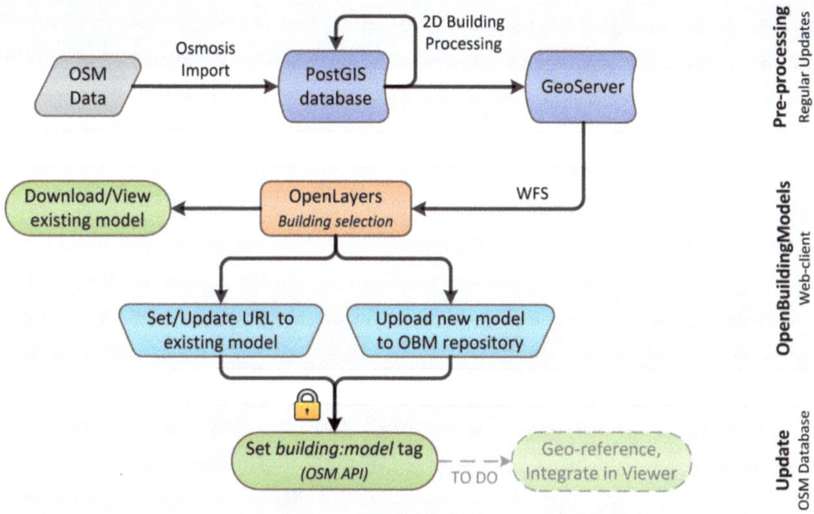

Fig. 4 Workflow of the Open Building Models prototype

overlay. When a building is selected, the client requests general information about the OSM way with the corresponding ID over the OSM API[19] via a HTTP GET request. The API delivers a small XML file which contains for example the user who created that geometry and all associated tags (key-value pairs). The XML is parsed by a PHP script and the information is displayed to the user. Since this operation only reads contents from the database, no OSM authentication is necessary.

The connection of a separately created 3D model with the selected OSM building is being achieved in the current prototype by simply setting the tag *building:model* = *URL*. When a new building model is uploaded, this tag is set to the new URL automatically. Alternatively, the user can set or update the URL manually, if a model for the selected building already exists on some other publicly available server. For these operations, an OSM user account is required, since new information is written to the database. If the model-tag is already set for a given building, it is also possible to directly download the model, e.g. in order to edit it or convert it to a different format.

4 Discussion

After the technical components of the OBM prototype have been explained, the general approach as well as the client is reflected critically in the following.

[19] http://api.openstreetmap.org

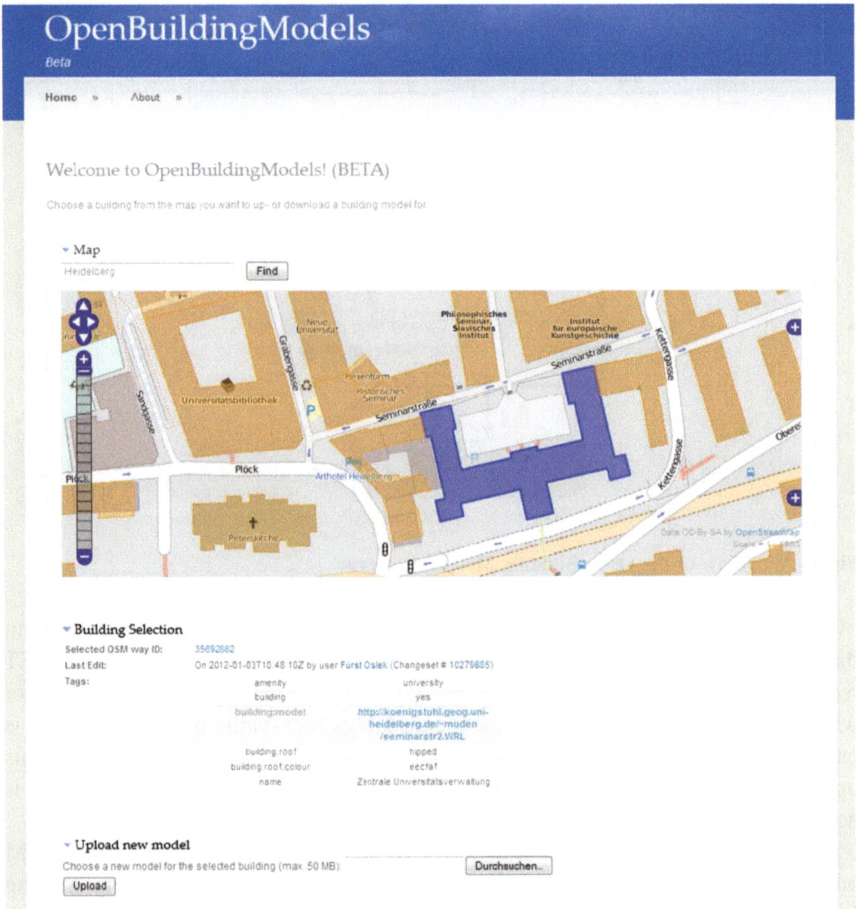

Fig. 5 Screenshot of the OBM web-client. The OSM properties of a building are shown on selection and a 3D model can be up- or downloaded to our repository

Linking 3D building models with the OSM database was in general already possible before by manually setting the *building: model*-tag and storing the model on some publicly available server. However, the OBM prototype makes this process much easier and integrates building selection and uploading in an interactive easy-to-use client. Once the models appear in a viewer like OSM-3D, OBM can potentially attract valuable users from the 3D modelling domain with the necessary skills to build complex models. It is expected, that predominantly complex landmark buildings will be uploaded to the OBM repository first. This is beneficial, since these kinds of buildings are particularly important for possible applications like pedestrian navigation. The improved visualisation in the 3D viewer with enhanced landmark models will lead to better representations of the cities. This can be seen in Fig. 3b, where the integration of a recognisable, realistic 3D model of the church is a massive improvement.

While the 3D visualisation can be greatly enhanced with this approach, it has also some drawbacks. As good as architectural 3D objects might look, they often lack topology and semantics, unless they are created in a dedicated format like CityGML. Apart from an improved visualisation, models without semantic annotation will have only little added value for applications that require semantic, standardised 3D city models. Similar to existing repositories, users will certainly create 3D models of varying complexity. Some might include in their model each and every detail of the outer structure or also model the interior. Others might create 3D models with a lower level of detail. On top of the entire 3D models, parametric models based on 3D-OSM-tagging will also exist with varying detail. Hence, one could generally argue that offering various tools and approaches for contributing 3D information could lead to a rag rug with a low level of standardisation and that an agreement on only one defined method to map 3D buildings is preferable. However, such heterogeneous modelling is a general phenomenon inherent in VGI. The current 2D OSM also consists of many differently mapped features and the consensus on common mapping techniques is rather low. And this is not necessarily a drawback, because only this keeps the barrier for beginners as low as possible. The introduction of mandatory mapping standards for quality assurance would repel most people from participating and contradict the general idea of the openness of VGI.

The described client is only a first prototype and there are lots of open issues that have to be addressed in the future. Figure 4 indicates the most important next step: the models have to be geo-referenced in order to place them correctly in the real-world. This could be accomplished by a semi-automatic approach which tries to reference the object by fitting it into the given OSM footprint first and asks the user for further manual refinement. In order to ensure a correct alignment between 3D objects and OSM footprints in the first place, it would be helpful to initially offer a download of the existing building outline as a basis for the modelling of a new building. OBM should support various 3D formats in the future in order to guarantee a high level of flexibility. At the same time, a 3D viewer has to support all these formats. This will lead to issues about format conversion and interoperability. Currently, textures are not supported by OBM. Since textures are an essential part of high-detail building models, this should be made possible in the future. Another important aspect is the usability of the web-client. It is desirable to provide not only the upload of 3D models for buildings whose footprint is already part of OSM, but also for such that have not been mapped at all. The user could properly place the 3D model on the map and the corresponding 2D footprint could be derived and added to the OSM database automatically. Furthermore, an important improvement would certainly be to support models with different levels of detail for the same object. This is currently not possible due to the bijective approach that only uses the *building: model*-tag. Also, performance will play a crucial role once a couple of detailed models will have been added. Ensuring smooth visualisations and efficient storage of the models will be a challenging task due to the enormous data load.

5 Conclusion and Future Work

In this paper, the current situation of 3D volunteered geographical information was discussed and a new concept of how to push it forward was introduced with Open Building Models.

In the first part, a review about related work in this context was given, showing that besides some available tools, only little research exists in this innovative field. Subsequently, the current situation was critically discussed and main issues regarding further progress in this area were pointed out. In the second part of this article, a basic prototype of Open Building Models was introduced. This approach aims to build up a free repository of 3D building models which can be linked to the OSM database. Advantages and shortcomings of this approach were discussed thereafter. OBM has the potential to attract users with 3D modelling skills to the geodomain and the Open Street Map project in the future. This is an important step for the progression of 3D-VGI, because it is impossible to model each and every building based on the rather low-level data schema of OSM. However, there are still many challenges to be tackled. Most importantly, these include a correct geo-referencing of the models. Also, there are many different 3D formats in which the models can be created and therefore questions regarding the conversion of formats and interoperability are to be answered as well. Furthermore, issues arise about how complex 3D models can be effectively edited by multiple users, since this is not as straightforward as it is for standard 2D map features. The possibility to upload multiple models for one object to the repository with different levels of detail must be considered. Finally, performance as well as effective storage and compression of complex models will be further issues.

A VGI-based approach to 3D city modelling is very promising and can potentially lead to an added-value in this field of application. User-generated approaches have already proven their potential to be capable of capturing high-quality spatial information. The interest in 3D mapping is rising in the OSM community. Since OSM is an established and successful platform, it is currently most qualified for 3D-VGI, although there are some shortcomings like the rather unsuitable data model. One key question will be how much 3D is capable and sensible in such a project and when new approaches and platforms are needed. Open Building Models is one of several possible means to enable voluntary users to contribute rich 3D information. It is a first effort to push forward the 3D mapping and apart from it, many other possible issues could be tackled in the future. For instance, the tag-based modelling and therefore the parametric building model creation could be advanced. Dedicated OSM-3D editors have to be developed in order to make it as easy as possible for inexperienced mappers to contribute 3D information. There should be a wide range of mechanisms available to allow crowd sourced 3D mapping on different scales. Also, an extension of the OBM repository to not only buildings but also other objects of our environment like street furniture, prototypical landscape objects or the like is conceivable and would lead to more detailed and usable crowd sourced 3D city models.

Acknowledgments The authors would like to thank all proofreaders who helped to improve this article and particularly all contributors to the OSM-3D project. This research has been partially funded by the Klaus-Tschira Foundation (KTS) Heidelberg.

References

Brenner C (2005) Building reconstruction from images and laser scanning. Int J Appl Earth Obs Geoinf 6(3–4):187–198

Cloudmade (2012) Application gallery. http://cloudmade.com/application-gallery. Accessed 09/01/2012

Coleman DJ, Georgiadou Y, Labonte J (2009) Volunteered geographic information: the nature and motivation of produsers. Int J Spat Data Infrastruct Res 4:332–358

Debevec PE (1996) Modeling and rendering architecture from photographs. Dissertation, University of California, Berkeley, CA

Elgan M (2011) Kinect: microsoft's accidental success story. http://www.computerworld.com/s/article/9217737/Kinect_Microsoft_s_accidental_success_story. Accessed 11/01/2012

Förstner W (1999) 3D-city models: automatic and semiautomatic acquisition methods. In: Fritsch D, Spiller R (eds) Photogrammetric Week '99. Wichmann, Heidelberg

Früh C, Zakhor A (2004) An automated method for large-scale, ground-based city model acquisition. Int J Comput Vision 60(1):5–24. doi:10.1023/B:VISI.0000027787.82851.b6

Goetz M, Lauer J, Auer M (2012) An algorithm based methodology for the creation of a regularly updated global online map derived from volunteered geographic information. Paper presented at the 4th international conference on advanced geographic information systems, applications and services geoprocessing, Valencia, Spain, 31/01/2012

Goetz M, Zipf A (2012) Towards defining a framework for the automatic derivation of 3D cityGML models from volunteered geographic information. Int J 3D Inf Model (IJ3DIM) 1(2):496–507

Goetz M, Zipf A (2012) The evolution of geocrowd sourcing: bringing volunteered geographic information to the 3D. In: Sui D, Elwood S, Goodchild M (eds) Volunteered geographic information, public participation, and crowdsourced production of geographic knowledge. Springer, Berlin (in press)

Goodchild M (2007) Citizens as sensors: the world of volunteered geography. GeoJournal 69:211–221

Google (2012) Earth. http://www.google.com/earth. Accessed 09/01/2012

Gore A (1998) The digital earth: understanding our planet in the 21st century. Aust surveyor 43(2):89–91

Haklay M (2010) How good is volunteered geographical information? a comparative study of open street map and ordnance survey datasets. Environ Plan B Plan Des 37(4):682–703. doi:10.1068/b35097

Haklay M, Weber P (2008) Open street map: user-generated street maps. IEEE Pervasive Comput 7(4):12–18. doi:10.1109/MPRV.2008.80

Kendzi (2011) 3D plug-in for josm. http://wiki.openstreetmap.org/wiki/Kendzi3d. Accessed 13/12/2011

Klank U, Zeeshan M, Beetz M (2009) 3D model selection from an internet database for robotic vision. In: Proceedings of the IEEE international conference on robotics and automation (ICRA 2009), Kobe, Japan

Kolbe TH (2009) Representing and exchanging 3D city models with cityGML. In: Lee J, Zlatanova S (eds) Lecture notes in geoinformation and cartography: 3D geoinformation sciences, Springer, Heidelberg, pp 15–31. doi:10.1007/978-3-540-87395-2_2

Lai K, Fox D (2009) 3D laser scan classification using web data and domain adaptation. In: Proceedings of robotics: science and systems, Seattle, USA

Mayer C, Zipf A (2009) Integration and visualization of dynamic sensor data into 3D spatial data infrastructures in a standardized way. Paper presented at the geoviz 2009, contribution of geovisualization to the concept of the digital city, Workshop, Hamburg, Germany, 05/03/2009

Müller P, Zeng G, Wonka P, Van Gool L (2007) Image-based procedural modelling of facades. ACM Trans Graph (TOG) 26(3):85–93

Neis P, Zielstra D, Zipf A (2012) The street network evolution of crowdsourced maps: open street map in Germany 2007–2011. Future Internet 4:1–21. doi:10.3390/fi4010001

OGC (2011) Open geospatial consortium 3D portayal interoperability experiment (3dpie). http://www.opengeospatial.org/projects/initiatives/3dpie. Accessed 13/12/2011

OSM-3D (2012) The open street map 3D project. http://www.osm-3d.org/home.en.htm. Accessed 27/03/2012

OSM Wiki (2011a) 3D development. http://wiki.openstreetmap.org/wiki/3D_Development. Accessed 16/12/2011

OSM Wiki (2011b) Map feature list. http://wiki.openstreetmap.org/wiki/Map_Features. Accessed 14/11/2011

Over M, Schilling A, Neubauer S, Zipf A (2010) Generating web-based 3D city models from open street map: the current situation in Germany. Comput Environ Urban Syst 34(6): 496–507. doi:10.1016/j.compenvurbsys.2010.05.001

Panoramio (2012) Map. http://www.panoramio.com/map. Accessed 09/01/2012

Pomaska G (2009) Utilization of photosynth point clouds for 3D object reconstruction. In: Proceedings of the 22nd CIPA symposium, Kyoto, Japan

RenderLights (2012) Viewer for google 3D warehouse. http://www.renderlights.com/?p=103745. Accessed 10/12/2012

Rocchini C, Cignoni P, Montani C, Pingi P, Scopigno R (2001) A low cost 3D scanner based on structured light. Comput Graph Forum 20(3):299–308. doi:10.1111/1467-8659.00522

Sampath A, Shan J (2010) Segmentation and reconstruction of polyhedral building roofs from aerial lidar point clouds. IEEE Trans Geosci Remote Sens 48(3):1554–1567. doi:10.1109/TGRS.2009.2030180

Schilling A, Over M, Neubauer S, Neis P, Walenciak G, Zipf A (2009) Interoperable location based services for 3D cities on the web using user generated content from open street map. Paper presented at the 27th urban data management symposium (UDMS 2009), Ljubljana, Slovenia, 24—26/06/2009

Simio (2012) Enhancing 3D animation with google warehouse. http://www.simio.com/resources/videos/Enhancing-3D-Animation-with-Google-Warehouse.htm. Accessed 10/12/2012

Stadtmessungsamt (2012) Stuttgart 3D-stadtmodell. http://www.stuttgart.de/item/show/21491. Accessed 09/01/2012

Strassenburg-Kleciak M (2011) Roof table proposal. http://wiki.openstreetmap.org/wiki/DE:Roof_table. Accessed 14/11/2011

Wang Y-F (2011) A comparison study of five 3D modelling systems based on the sfm principles. Technical Report 2011-01. Visualize Inc., Goleta, USA

WikiMapia (2012) http://wikimapia.org. Accessed 09/01/2012

Williamson I, Rajabifard A, Wallace J, Bennett R (2011) Spatially enabled society. Paper presented at the FIG working week 2011, Marrakech, Morocco, 19/05/2011

Willmes C, Baaser U, Volland K, Bareth G (2010) Internet based distribution and visualization of a 3D model of the university of cologne campus. Paper presented at the 3rd ISDE digital earth summit, Nessebar, Bulgaria, 14/06/2010

Yiakoumettis CP, Bardis G, Miaoulis G, Plemenos D, Ghazanfarpour D (2010) Virtual globe based collaborative 3D city modelling. Intell Comput Graph 321:165–184

MIX
Papier aus verantwortungsvollen Quellen
Paper from responsible sources
FSC® C105338

If you have any concerns about our products,
you can contact us on
ProductSafety@springernature.com

In case Publisher is established outside the EU,
the EU authorized representative is:
Springer Nature Customer Service Center GmbH
Europaplatz 3, 69115 Heidelberg, Germany

Printed by Libri Plureos GmbH
in Hamburg, Germany